AF873108

WILLIAM SCOTT TAGGART

COTTON SPINNING

VOLUME III

Elibron Classics
www.elibron.com

Elibron Classics series.

© 2005 Adamant Media Corporation.

ISBN 1-4021-4181-5 (paperback)
ISBN 1-4212-8367-0 (hardcover)

This Elibron Classics Replica Edition is an unabridged facsimile of the edition published in 1916 by Macmillan and Co., Ltd., London.

Elibron and Elibron Classics are trademarks of Adamant Media Corporation. All rights reserved.

This book is an accurate reproduction of the original. Any marks, names, colophons, imprints, logos or other symbols or identifiers that appear on or in this book, except for those of Adamant Media Corporation and BookSurge, LLC, are used only for historical reference and accuracy and are not meant to designate origin or imply any sponsorship by or license from any third party.

COTTON SPINNING

MACMILLAN AND CO., Limited
LONDON · BOMBAY · CALCUTTA
MELBOURNE

THE MACMILLAN COMPANY
NEW YORK · BOSTON · CHICAGO
DALLAS · SAN FRANCISCO

THE MACMILLAN CO. OF CANADA, Ltd.
TORONTO

COTTON SPINNING

BY

WILLIAM SCOTT TAGGART, M.I.MECH.E.

AUTHOR OF
'COTTON MACHINERY SKETCHES' AND 'COTTON SPINNING CALCULATIONS'
EXAMINER IN COTTON SPINNING TO THE CITY AND GUILDS OF LONDON INSTITUTE
EXAMINER IN TEXTILE MATHEMATICS, MECHANICS, DRAWING AND ENGINEERING
TO THE UNION OF LANCASHIRE AND CHESHIRE INSTITUTES

VOLUME III

WITH ILLUSTRATIONS

FOURTH EDITION

MACMILLAN AND CO., LIMITED
ST. MARTIN'S STREET, LONDON
1916

COPYRIGHT

First Edition 1898. *Second Edition* 1902
Reprinted 1907. *Third Edition* 1911
Fourth Edition 1916

PREFACE TO THIRD EDITION

SEVERAL corrections have been made in the body of the book, and some important additions made which will be found in the Appendix. These additions include practically a full description of a self-acting mule that is used extensively for the productions of fine numbers. Interesting details of another type of mule are also added. A very complete set of gearing illustrations of the chief types of self-acting mules, together with full calculations of each, will be found in *Cotton Spinning Calculations* recently published.

W. S. T.

BOLTON, 1910.

PREFACE

In the two previous volumes the preparing processes in cotton spinning have been fully treated. In this volume the subject of spinning and the preparation of yarns is treated in an equally exhaustive manner, with, I trust, an avoidance of some defects that existed in the earlier books.

In a work of this kind, which covers so much ground and deals with many features upon which other men have written, it is perhaps unnecessary to suggest that originality is a difficult matter to obtain in one's treatment of the subject. In my efforts to do so I may have, occasionally but unconsciously, adopted similar methods of other writers; when such has been pointed out to me I have tried, and I hope successfully, to prevent this being an offence, and I sincerely trust that readers and writers alike will find in my efforts a desire to simply present the study of cotton spinning in an interesting and instructive manner, so that it may prove of value to those whose well-being is dependent on the success of that part of the industry it represents.

The various parts of the subject have been treated in

such a manner that the young student may with great benefit to himself use them as a text-book, whilst the older reader will undoubtedly find in them much to interest him and develop a desire for a fuller knowledge and a more perfect grasp of the principles underlying many of the processes and much of the mechanism of cotton-spinning machinery.

Completeness is impossible, and defects must exist in such a work as this; but publishers, printer, and writer have done all they could to render it of more than ordinary value to those interested, and suggestions, corrections, and advice to make the books more serviceable will be fully appreciated.

I beg to thank several machine firms who have generously helped me by supplying me with sketches of parts of their machines, which it would otherwise have been difficult for me to obtain, and the *Textile Mercury* is specially deserving of recognition for the excellent reproductions of my drawings.

WM. SCOTT TAGGART.

BOLTON, 1898.

NOTE TO 2ND EDITION.—Owing to the size of the book, it has not been considered advisable to include all the smaller improvements recently made to the machinery dealt with in this volume. Additions and corrections, however, have been made so as to bring the book up to date and render it useful to the practical man and to the student.

CONTENTS

CHAPTER VIII

CHAPTER IX

ILLUSTRATIONS.

ILLUSTRATIONS IN VOLUME I.

ILLUSTRATIONS IN VOLUME II.

CHAPTER I

THEORY OF SPINNING

In the course of the two preceding volumes, it has been considered necessary several times to examine into the principles underlying the various operations of the different machines, and the effect these machines have upon the cotton passing through them. The remarks already made may now be extended into a more detailed examination; and as they will serve the purpose of an introduction to a description of the Self-acting Mule, on which the final operation of spinning—that of making the cotton into yarn—is performed, the conclusions arrived at will materially assist in making some of the operations of that machine more readily understood.

The ideal state of the cotton, to obtain which every effort has been made, may be summed up as follows:—Absolute cleanliness; equality in length of fibre; perfect parallelisation of the fibres; a disposition of the fibres among themselves such as to ensure the strongest result; uniformity throughout the length in diameter, weight, and strength; and a round solid yarn as the ultimate result of the whole series of operations.

Sufficient has been said already to show that the first three conditions have been almost satisfactorily attained, especially when the cotton has been combed; otherwise it cannot be assumed that the parallelisation of the fibres is so nearly perfect as is generally supposed. It will, therefore, be necessary to confine the examination to the last three ideal conditions of a perfect yarn; and to see whether they are possible of attainment, and how far present machinery is capable of achieving them.

Uniformity of the Yarn.—In the first place the question of the uniformity of yarn will be considered. Uniformity applied to yarn generally means that the yarn is uniform in diameter and in weight; the uniformity of strength is not as a rule the occasion of so much observation as the other two, provided a very long length doubled many times answers to a satisfactory test for breaking weights (this point will be dealt with later on).

Regularity of Diameter.—There are several ways of testing yarn as to its uniformity of diameter, the common one being to wrap a certain number of lengths side by side on a black slip of wood or cardboard. The contrast of colour thus afforded gives a very good idea of inequalities, and a judgment based on such an examination is generally considered sufficient for practical purposes. It is, however, at best only a crude method, and when what are considered good results are passed under the microscope with low power, the great difference between adjacent diameters is instantly recognised. The fault of the "sight" method of judging yarn is due to our inability to see small differences in small diameters. We give an example:—

Suppose a 20's yarn is $\frac{1}{100}$ inch in diameter, this we can see is a very small dimension; if a thinner one is taken, say $\frac{1}{110}$ inch diameter, and a thicker one, say $\frac{1}{90}$ inch diameter, the difference between the three yarns is so little that it would require an unusually good eyesight to detect it. When, however, such a yarn is passed under a microscope and magnified say 100 times, the $\frac{1}{100}$ inch would become of such a size as to show clearly any difference of other diameters when compared therewith; if $\frac{1}{100}$ of an inch became enlarged to one inch, the two other dimensions would become $\frac{11}{10}$ and $\frac{9}{10}$ of an inch respectively—the difference in each case being $\frac{1}{10}$ inch. Such a difference is really enormous and represents a large percentage of variation, and yet it is one that would not readily be noticed by the ordinary testing method, simply because of the eye's failure to judge of such small differences. If a finer yarn is taken, say 60's, its diameter of say $\frac{1}{200}$ inch would render it even more difficult to discern variations of diameters unless they were unusually large. We thus see that the usual method of judging yarns is by no means perfect; it evidently satisfies ordinary requirements of trade, but we ought not to ignore the fact that very unequal yarn is still made in spite of all that has been done to perfect the machinery for making it. Combed yarns among the higher numbers display almost as great an inequality of diameters as the low numbers do, mainly because of the fact mentioned above; but combed 60's compared with ordinary 60's is much superior, although, as already noted, a very large percentage of variation exists.

Another way of rendering very apparent the variation that exists in the diameter of yarns is to double together two rovings of the same hanks, and the same cotton, but one of them dyed, the other white, or of a contrasting colour. Each roving by itself will probably show very little variation; but when doubled, the mere fact of twisting will bring out every thin and thick place in a remarkable manner. The writer had recently a striking object-lesson on this point, while in a spinning mill on the Continent. The specialty of the mill in question is coloured and mixed yarns made from doubled rovings. A cop formed of double roving, one white and one black, at the mule, while generally even in appearance at the first glance, was in reality one whole length of irregularities. These were made apparent by the distinct character of the twists, which could easily be seen, owing to the contrast in colour of the two rovings; the twists lay very close together in places, drawing the yarns tightly together and making a thin hard place; at others they were correspondingly separated, and at these spots a thick fuzzy place was formed. Such irregularities existed and followed each other in varied lengths from $\frac{1}{4}$ to $1\frac{1}{2}$ inches throughout the cop. At first the suggestion was made that the dyed roving was perhaps the chief offender; but when two dyed rovings were used, similar results followed, and an examination of the rovings only showed that they were good average results of "good middling" cotton obtained after passing through modern preparing machinery. The same two rovings put through a ring frame gave a cop that was scarcely distinguishable from that of the mule so far as the marked character of the variations was to be seen. A strange thing about it was that when double rovings of white, or two of the same colour were used, the yarn was remarkably good and even in appearance; but no sooner were the twists made apparent by a contrast of colour than the unreliability of one's judgment by sight was immediately emphasised.

Regularity of Length and Weight.—In close connection with the uniformity of diameter is that of length. Owing to the universal use of the wrap reel and scales, any variation in this direction is quickly noted, and the judgment has little if anything to do with the decision. But even with the wrap reel it is only average results that are dealt in; long lengths are always taken, varying from 120 to 840 yards, and the weights of the same lengths from different cops are compared. This rough method, however, fails to show whether the yarn is uniform, for if fifty cops can be taken from different parts of the same mule, wide variations will be noted in their weighings. Such variations, however, will be intensified if a number of wrappings be taken from the same cop and carefully compared. Differences like the one just suggested are of a distinct practical character, and being very well known are always allowed for; but if the examination be continued by splitting up say 840 yards into pieces of 10 yards each, or even less, and weighing them, the same average result for the whole length will be given, but the individual weighing will vary to an extent that is astonishing. It is a difficult matter to say how it happens that this state of things exists; it is probably due to errors in the previous machines, principally in the card and scutcher, and the reason for its non-detection at these machines is the too great reliance that is placed on average weighing in the bulk, and the fact that a slight variation under such conditions is not considered of importance for practical purposes.

To show what is meant, let it be supposed a scutcher makes laps that vary only within ¼ lb. in a lap of 32 lb.; this would be a very good result indeed, and if it represented the actual variation of the laps, there would be an unusual degree of uniformity in the yarn. But when we consider that a difference of ½ lb. in a 32 lb. lap causes a variation of a single hank at 60's, it will be readily understood that uniformity of scutcher laps in the bulk is not a good foundation on which to base anticipations of uniform yarn. By taking periodically very short lengths of the lap, and weighing them, a much better idea of the variation would be arrived at, and means could then be taken to ensure more uniform results. Practical tests in this direction of weighing short lengths of what seemed to be a good lap, have shown variations of as much as 25 per cent. It is therefore not surprising to find that yarn is not uniform; to a large extent variations will always exist, but much could be done to remedy them if a correct judgment were formed by individual observation instead of depending so much on large average results.

Although irregularity of diameter is such a noticeable feature, it by no means follows that it corresponds to the variations in weight, except in the case of sliver and rovings; in yarns the twist put in has an all-powerful influence in affecting the diameter. There is no doubt from even a casual observation that variations exist, but they are so distinctly brought to view by means of the twist put in the yarn, that a little consideration of this feature will not be out of place.

Twist and Weft.—The object of twisting has already been explained. From the fact that the twist can be put in the yarn in two directions, the terms "twist and weft way" are general. The term weft, however, is not applied so much to the direction of the twist as to its condition. It implies less twist and a softer yarn, and as a rule weft yarn is made from cotton that gives a soft and more pliable effect. Twist is as a rule formed by turning the spindle in the same direction as that in which the hands of a clock turn, and it gives to the yarn a spiral twist, corresponding to that seen on a right-handed screw. Weft has its twist put in generally in the opposite direction. It does not always follow, however, that the direction of the twist gives the yarn its character of twist and weft.

Effect of Twist.—The tendency of the twists to fly to

the thin places in the yarn is a well-observed fact, and several suggestions have been made as to its cause, the chief one being the greater difficulty of twisting a thick place than a thin one. Whether the thick place be caused through a larger number of fibres existing at the place, or through the fibres being coarser, it is highly probable that the above reason is the correct one; and if so, it resolves itself into a purely mechanical fact that the twists should fly to the thinnest places of the yarn.

The following illustrations will serve to make this point clear, and every reader can readily convince himself of its

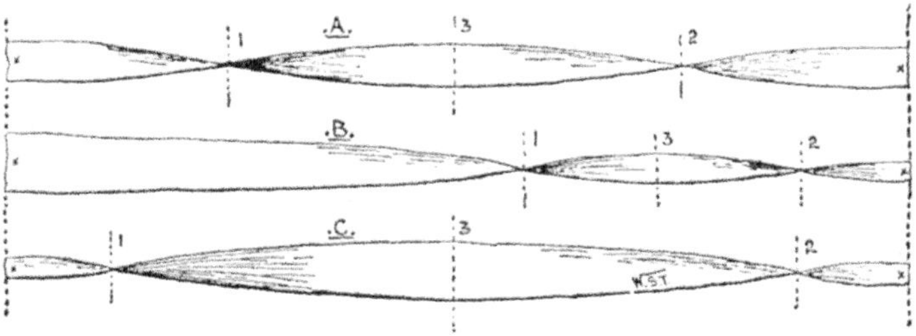

FIG. 1.

truth. Take three lengths of narrow tape (or even slips of paper), cut to the shapes shown in Fig. 1 at A, B, and C. A is a uniform narrow slip, and it has been twisted one complete turn: a perfectly uniform twist is the result, because the resistance to twisting is the same throughout the strip. If a second slip be taken, wider at one end than at the other, as at B, the complete turn does not give a uniform result, the wide end of the slip being more difficult to turn, and as a consequence the twist is confined to the narrow end. By making the slip wide in the middle and thin at the ends, as at C, we have a similar effect; but in this case the thick portion, while it has evidently turned and transferred the twists from one thin end to the other, has failed to be twisted itself, the narrower ends only receiving the twists. This is very conclusive evidence of the effect of a thick or thin place in the yarn, and, as can be seen, it is one of a purely mechanical nature. If a similar problem were presented in regard to wire, or a shaft, its solution would be found at once, and definitely, on the above lines; and the fact that yarn is not so homogeneous as iron does not interfere very materially with the reasoning; it only prevents a definite conclusion as to the amount of the result being arrived at.

As will be seen a little later, the peculiar action of the mule—and it is one of its chief advantages—has a beneficial effect in modifying the extreme result of twist; nevertheless, it is always considerable, and the only remedy is to be sought in more uniform results in the preparing processes.

Strength of Yarn.—The strength of yarn depends upon two principal factors, namely—the kind of cotton, and the arrangement of the fibres among themselves. The strongest-fibred cotton does not make the strongest yarn: firstly, because it is shorter, and therefore not capable of being bound into as strong a yarn as the longer but weaker fibres; and secondly, because its greater diameter does not allow of as many fibres in the cross section of the yarn as is the case when finer fibres are used; the percentage of extra fibres in such a case is greater than the percentage of weakness in the individual fibre: consequently, if, say, 30's be spun out of Indian and Sea Island cottons, the weaker Sea Island fibre would make the stronger yarn—for the two reasons given above.

Arrangement of the Fibres in the Yarn.—The disposition of the fibres in the yarn is rather an important matter, and it is quite obvious that—other things being equal—the strongest yarn is that which has its fibres arranged to the best advantage in respect to one another. The actual arrangement of fibres in yarn is of course practically unknown, but we may reasonably argue from some of the known facts, and conjecture. For instance, cotton that retains 15 to 20 per cent of its shorter fibres is clearly bound to produce weaker yarn than if those fibres were removed by the combing process; and in the same way it is reasonable to suppose that the haphazard arrangement of the fibres taken from the doffer must yield poorer results as to strength than the ordered condition of the fibres after passing through the comber. Both sets of fibres, however, are modified as to their arrangement in the subsequent processes, and it is most probable that the former is greatly improved, whilst the latter loses somewhat of its advantages. Nothing definite is known, however, and this opinion is only expressed after a careful examination of the drawing process, as seen in such machines as are used for jute and flax, where the operation —owing to the long length of fibre—is easily seen, and its action readily followed.

In order to demonstrate what might be considered an ideal state in the disposition of the fibres, it will be necessary to make use of the diagrammatic method, similar to that used by Mr. Nasmyth in his book on *Cotton Spinning*. The reasoning and conclusions arrived at, however, are different, the similarity being simply in the diagrams. Fig. 2 shows several possible arrangements of the fibres; but it must be thoroughly understood that none of them are probable, the actual conditions most likely partaking of a combination of all of them. At A an arrangement is shown which gives a perfectly uniform thickness of yarn; but it is absolutely without strength, for the obvious reason that the fibres are simply end-on-end, and are not bound together in any way. It may be taken as representing one extreme in any combination that may take place, and the probability of its happening to a certain degree, if only a small one, introduces a possible cause of the well-known weakness of yarn compared with the strength of the individual fibres. At B the

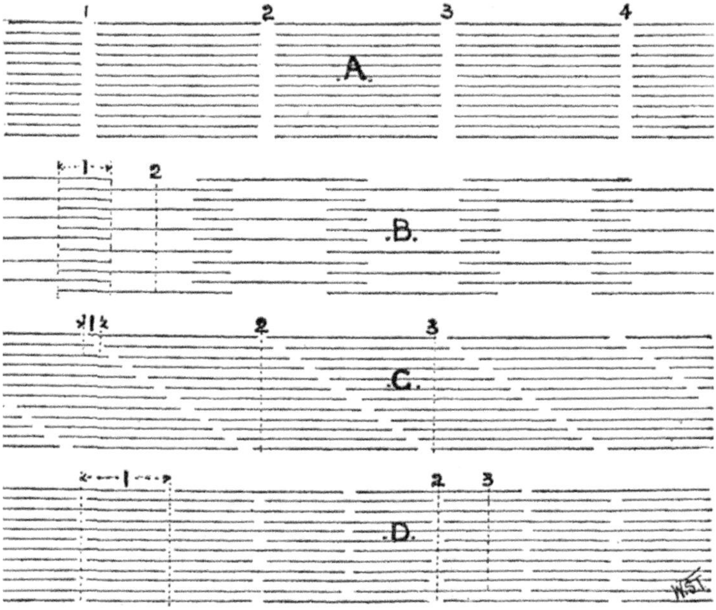

FIG. 2.

fibres are shown with a short overlap; when twisted together a certain strength would be obtained, but it would clearly be only of a slight character, and it is highly probable that its weakness would lie in the slipping of the fibres over one another owing to the insufficient lap. A more serious evil, however, is seen in the unevenness of the yarn that would be made; at 1 the thickness is that of twelve fibres, while at 2 only six fibres are twisted. This arrangement most certainly exists in yarn, and is the cause of unevenness. The well-known action of the comber arranges the fibres on this plan, but of course with a much greater overlap. At C is shown a modified form of B, in which the fibres produce a uniform thread, and equally as strong. It is an unknown point what proportion of the length of the fibres ought to be twisted in order that the weight to cause rupture should just equal that necessary to produce slippage. If one-eleventh of the length be sufficient to resist slippage when a number of fibres (say twelve) are twisted together, the arrangement shown at C would be the strongest possible one. At 2 a section is given of the weakest place, and yet it is only 9 per cent less than the theoretical value of all the fibres; at 3 the full value is obtained, but since the strength of the yarn is that of its weakest spot, rupture would take place probably at 2. At D the arrangement is one in which the greatest possible adhesion is given to the mass of fibres in the yarn when twisted, the possibility of slippage being reduced to a minimum, and from this point of view it has an advantage over C; it is also uniform, but a glance at the diagram will show that the weakest spot of such a combination of fibres contains only six fibres (see 2) and is therefore 50 per cent weaker than the strongest place (as at 3), which has twelve fibres in cross section. Next to A, D is the poorest combination that can be given to the fibres, on the assumption, of course, that we are treating of fibres all of which are of equal length, and that half the length of the fibre is requisite for twist in order to equal the breaking weight. It is the opinion of the writer that a much smaller proportion of length is sufficient, and of course the smaller it is the stronger is the yarn; the object of attainment seems to be to lay the fibres in such a way as to break as much as possible the joint caused by the ends coming together. Such an assumption as that mentioned above is, however, far from being correct in practice. In the best combed cotton a large percentage of variation exists, and this means that the overlapping of the fibres follows no strict law; moreover, when we know that a very large draft is given to the sliver after passing the comber, before it is made into yarn, it is clearly impossible to suppose that the apparent regularity with which the comber does its work results in the fibres of the yarn being arranged as at C or D. If a single end of combed sliver, with its fibres arranged as at C (which is quite possible), be made into yarn, the nearest approach to the aggregate strength of its component fibres will be obtained, but when several slivers are doubled, the overlappings of the fibres in the different slivers do not correspond, and a condition is produced which prevents dependence on the original arrangement. We may, however, conclude that a stronger yarn will be made, both from the greater uniformity in the length of the fibres as well as from their better disposition, which is a source of strength when twisted.

The above remarks will have prepared the reader for the conclusion that the strength of yarn is a very variable factor; that the disposition of the fibres follows no fixed arrangement; that it is impossible to arrange them in a manner to obtain more than a relatively small percentage of the strength of the individual fibres; and that the probable arrangement of fibres is a mixture of those shown in Fig. 2.

Rotundity of Yarn.—In considering the question of the rotundity of the yarn after it has been twisted, it ought to be remembered that it is not simply one of a number of objects sought for in the making of good yarn: it really represents the sum and substance of all of them combined. Granted that ideal conditions in material and processes existed, perfectly round yarn would be the natural result; but in the absence of ideal conditions, round, or rather sectionally round, yarn is still possible.

The roving as it passes between the rollers is compressed into a thin flat ribbon of fibres, and on issuing from them is immediately twisted into a strand in which all the fibres are more or less bound together. Considering the number of twists given to the yarn it is natural to expect a cylindrical form as a result. The only thing that interferes with this conclusion is the homogeneity of the fibres as a whole, and it is upon this feature that the question depends. Roundness is the result of twisting. If the yarn were homogeneous throughout its length it would have a circular appearance in a sectional elevation, but this rotundity would not necessarily be perfectly cylindrical, because, as we have already pointed out, the sliver from which the yarn is spun is unequal, therefore there would exist different diameters at various points. In spite of this a sectional view would give a circle. It is quite obvious that thick places, whether containing more fibres in cross section, or the same number of fibres each of a greater diameter, can be made round, just as readily as in the case of a small number of fibres. The fact that yarn is far from being round must be sought for on the assumption that any given section of it is not homogeneous, which assumption can be easily verified by the microscope. Suppose that in the thin ribbon of fibres which issues from the front rollers there are two or three fibres slightly thicker than the rest, the presence of those fibres will cause that particular part to offer a greater resistance to twisting than that of the weaker and thinner fibres, and as a consequence an irregular shape will be produced. Now it is fully well known that thick and thin fibres exist throughout the best of cotton. In some classes this is more so than others, and it is the fact that a few of these thicker or even unusually thin fibres can be found in the cross section of any yarn that causes the irregular shape it is found to possess. In regard to the round form of section of yarn and of fibres, it is as well to observe that it may have two distinct meanings. It may mean that the cross section itself is round or that the general view from the cross section is round. These are two very widely different things.

Elasticity.—Elasticity is all-important in the characteristics of yarn, and this to a greater or less extent exists in all textile fibres. It may be defined as a property which enables a substance to be distorted to a certain extent and yet to return to its original condition without having suffered injury. If all the fibres in Fig. 3 were packed closely together, there would be very little elasticity, because the fibres have no room in which to yield; the yarn cannot lengthen unless the diameter becomes smaller at the same time, so that if the smallest diameter is obtained by close packing, the yarn ceases to have elasticity in the sense understood in cotton spinning. The drawing shows that the fibres are not arranged in any close order, and, as a consequence, if the yarn is stretched slightly, the diameter is reduced; the fibres come together, and in doing so cause a lengthening to take place. A yielding of this kind naturally relieves the yarn of any shock that may come upon it and thus prevents rupture. At the same time the fibres themselves, in the aggregate, possess sufficient elasticity to cause them to spring back into their original position when the pressure is removed from the yarn.

Whilst recognising that elasticity and strength are not convertible terms, it must be understood that they are entirely dependent upon each other. The maximum strength of any given yarn depends upon a certain degree of elasticity, and this in its turn depends upon the character and number of the twists put into the yarn. Confining our attention to mule yarn it will be seen that a less number of twists than what is considered normal will

FIG. 3.

increase the liability to lengthen when pressure is applied, but such a reduction in twist will weaken the yarn, and, therefore, a considerably less pressure will cause rupture or slippage. Consequently nothing is gained by this procedure in the way of strength. It happens, however, that strength is not the all-important factor in some classes of yarn. A yielding thread is often desired to be used in material or for purposes where it is not subjected to forces that will cause rupture, so that we find large quantities manufactured to serve such special conditions.

On the other hand, an unusual degree of hardness in the yarn is sometimes desired, and in such a case elasticity is sacrificed, and extra twists put in the yarn. It must be borne in mind, though, that extra twist means additional strains on the fibres, and these naturally are a source of weakness; but since circumstances demand hard twisted yarn, it is necessary to make it.

In further consideration of the subject it will be noted that between the two cases mentioned above it is possible to obtain a yarn with a maximum strength combined with such a degree of elasticity as to satisfy the best conditions of the two factors. Exactly at the moment when rupture takes place the yarn should cease to stretch, and, simultaneously with this, slippage of the fibres over each other ought to begin. Under these circumstances a standard yarn would be produced. It need scarcely be remarked that our present knowledge absolutely prevents such a high degree of excellence in the making of yarn, partly from the fact that the cotton fibre is an ever-varying element, and also that little, if anything, has been done in the way of investigation into the best means of obtaining a basis upon which to work.

The following table, taken from *The Textile Mercury*, will give some idea of the elasticity of yarn:—

For Nos.	20 to 30	. .	4·5 to 5	per cent
„ „	30 to 40	. .	4·0 to 4·5	„ „
„ „	40 to 60	. .	3·8 to 4·0	„ „
„ „	60 to 80	. .	3·5 to 3·8	„ „
„ „	80 to 120	. .	3·0 to 3·5	„ „
„ „	120 to 140	. .	2·5 to 3·0	„ „
„ „	140 to 170	. .	2·0 to 2·5	„ „

In measuring the diameter of yarn it is often overlooked that a maximum diameter and minimum diameter may exist at the same part of any given section, and yet if this were used for a basis upon which to obtain an average diameter, absurd results would follow. To use the example of a twist drill, it is palpably incorrect to estimate its diameter from the average of its least and greatest diameters. Paradoxical as it may seem, its average diameter is certainly its greatest diameter. This comes about because the greatest diameter is uniform. In yarn the greatest diameter is not uniform, consequently the average diameter in such a case must be obtained from a large number of measurements of the larger diameters obtained from sections in which the least dimensions at those points can also be observed.

Rule for the Diameter of Yarn.—$\frac{0{\cdot}0278}{\sqrt{\text{Nos.}}}$ = dia. in inches. This rule is, of course, based on finding the volume of a certain weight and length of yarn and then calculating the diameter.

The Principle of the Spinning Action in the Mule.—In the mule, as in all spinning machines, the characteristic action is that employed for putting the twist into the roving. "Spinning" is the general name applied to this action when the amount of twist is in excess of that required to strengthen the roving so as to enable it to be taken from one process to another: in other words, spinning transforms the loose fibrous roving into the finished yarn. There are several important methods of performing this operation, which will receive attention subsequently. The one now to be dealt with is that applicable to the mule.

It is an exceedingly simple operation in itself, but, as will be seen later, the mechanism necessary to perform it automatically, and the actions associated therewith, are of a very complicated character. It will therefore be advisable to explain first the principle underlying the action of twisting, and afterwards to deal with the various features connected with and dependent upon it.

In effect, the twists are put into mule yarn by first winding it upon a thin steel spindle, and then drawing it off from the end. This results in giving one twist for

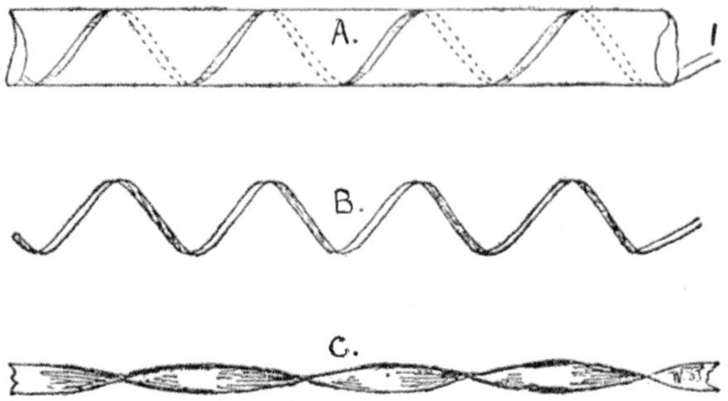

FIG. 4.

every turn the yarn has been wound round the spindle. The accompanying sketch fully explains the action. At A, Fig. 4, a spindle is shown with yarn wound round it a number of times. If the end at 1 be drawn off, the portion previously on the spindle will appear as at C, the

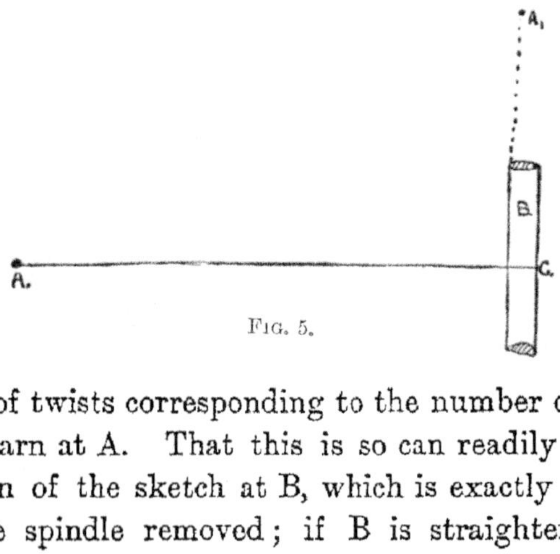

FIG. 5.

number of twists corresponding to the number of the turns of the yarn at A. That this is so can readily be seen by inspection of the sketch at B, which is exactly like A, but with the spindle removed; if B is straightened it will appear twisted as at C. This example serves to demonstrate the effect of drawing yarn from the end of a spindle after it has been wound thereon. In the mule this action is taken advantage of, but in an improved and modified form. In the first place a method is adopted of winding the yarn on the spindle and unwinding it in such a manner as to obtain a continuous action for a long length

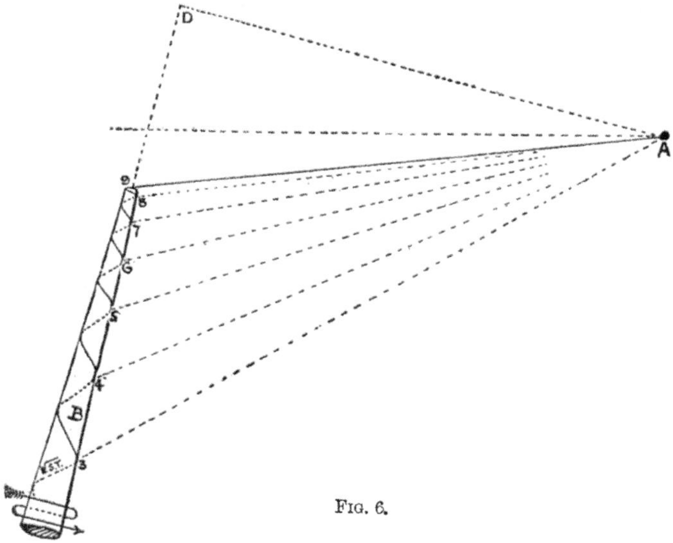

Fig. 6.

of yarn. To a casual observer it appears as a single operation, but in reality it is composed of two distinct actions. This is absolutely necessary if the yarn is to be twisted; it must first be wound on the spindle and afterwards drawn off, as was shown in Fig. 4. On reference to Fig. 5, a spindle B is placed at right angles to the source A from which the yarn is delivered; if B is revolved, it is clear that the yarn will be wound on at C only, and when it is desired to twist it by drawing it off from the end of B, A must be removed to A_1. This, of course, is impracticable, but the same effect is obtained by placing the point of delivery A above the point at which the yarn passes to the spindle, as in Fig. 6; in this position the yarn is not wound on at right angles to B, but by virtue of its inclination to the spindle its tendency is to assume a position at right angles; in doing this it naturally rises up the spindle in a series of spiral turns, each turn bringing it more into the desired position, which would be at D if the spindle were sufficiently long.

It is in connection with this feature that the characteristic of mule-spinning is seen. If the end of the spindle B is arranged to be below the point D, there will be no interference with the tendency of the yarn to rise to that point as the spindle revolves, and consequently when the end of the spindle, at 9, is reached, the yarn continues its upward course, and naturally slips off the end and instantly drops to 8. This is equal to having one turn off the spindle-point, and that turn of course puts one twist in the yarn between the spindle and A. As the spindle continues its revolution another turn is wound on from 8 to 9, by virtue of the tendency to reach D, and another slippage over the spindle-point takes place. This goes on until the desired number of twists have been put in, after which another operation comes into action. (An interesting experiment to illustrate this explanation can be made by winding a thin narrow tape on the spindle and noticing the effect as it winds itself up the spindle and slips over at the end.)

It has just been stated that the yarn must not be allowed to pass to the spindle during the twisting process at right angles to the axis. To prevent this, the nip of the front roller at A is placed above the spindle-point, and still further to improve matters, as well as to prevent the vertical distance between the points being unduly large, the spindle itself is inclined.

So far it has been assumed for the purpose of explanation that the twists are put into a fixed length between A and 9 on B, but this is only partially true. The spindles during the twisting operation are caused to move slowly away from the front rollers, which at the same time revolve and deliver almost sufficient roving to compensate for this movement. As the spinning continues while the spindles move from P to P^1 (Fig. 8), the full length of the yarn between the points has the twists comparatively well distributed. As an aid to this distribution of the twists, the vibratory motion given to the yarn as it slips over the spindle-point is rather important; the shaking which it receives in this way causes the twists to assume a perfectly natural position in the yarn, instead of being instantly fixed at the point where the twist was given. A further and highly characteristic feature is also to be observed as the movement of the spindles takes place. The slight excess of the traversing movement of the spindle over the amount of roving given out by the front rollers causes a little stretching to take place in the yarn; the tension to which it is in this way subjected causes the thicker and softer portions to be drawn out, and, as already explained, this tends to equalise the twist, which would otherwise leave the thicker parts with a less proportion of twists than the thinner portions receive.

Inclination of Spindle.—From Fig. 8 the influence of the inclination of the spindle can also be observed. If the spindle were vertical, as in Fig. 7, its inclination with the yarn near the rollers at A would probably be enough for spinning easily; but when it reaches its extreme out ward position, at B, the angle has been considerably reduced—to almost a right angle—and the slippage of the yarn over the point would not be so easily performed—especially considering the vibratory motion of the yarn which might readily cause the two to be momentarily at right angles, in which case spinning would cease and the ends would break. By inclining the spindles, suitable conditions exist throughout the traverse of the spindles, and although the angle

FIG. 7.

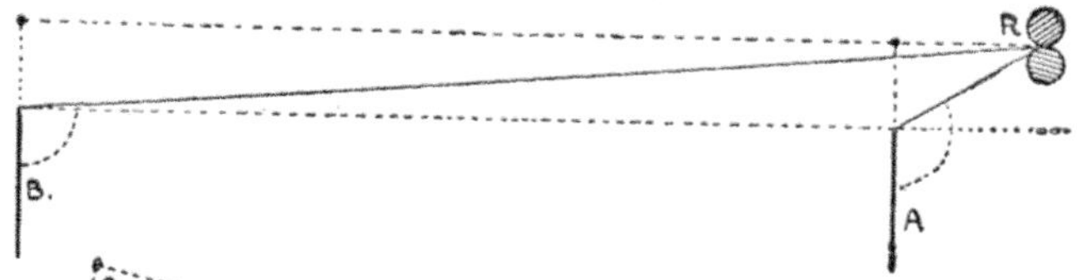

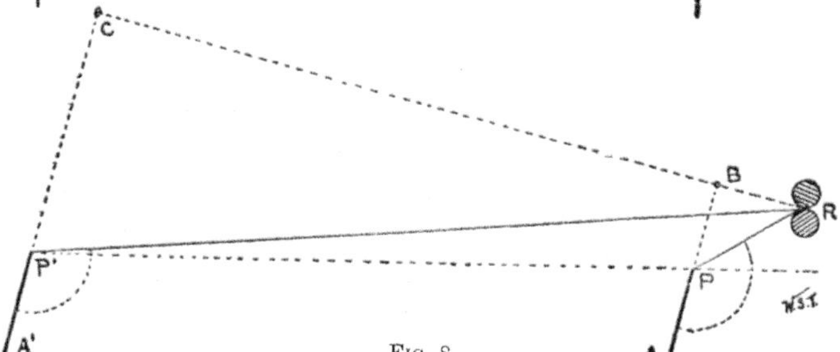

FIG. 8

is reduced it is still considerable, and the positions of the points at right angles to the spindle in the extreme positions at B and C are so high above the end of the spindle that there is no danger of non-slippage of the yarn.

The Taper of the Spindle.—The taper of the spindle, as shown in Fig. 6, is due partly to the fact that this form enables the cop to be readily withdrawn, but primarily because as fine a point as is consistent with rigidity is necessary in order to get the best result in the slippage of the yarn over the end. If the end be thick, slippage would be bound to take place; but it will be seen that the one turn unwrapped from a large diameter would cause a slackness that would be inconvenient in several ways: the slackened yarn might run into snarls, or disturb the turns that are on the spindle just below the point, and thus introduce variations that would destroy the value of the result; excessive vibration might also be easily caused; a quarter of an inch diameter of spindle gives three quarters of an inch of yarn in one turn, and this being set free at the rate of 5000 to 10,000 times a minute is not likely to prove beneficial, consequently the point is made much thinner than the body, and for very fine work it is frequently only a little over one-sixteenth of an inch in diameter.

CHAPTER II

MECHANISM AND WORKING OF THE MULE

General Arrangement.—Before giving a description of the mechanism of the self-acting mule, it will be advisable to briefly point out the disposition of the various parts

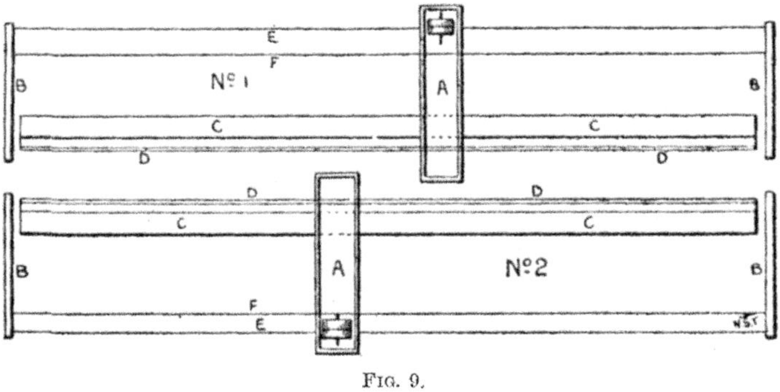

FIG. 9.

which go to make up the complete machine. For this purpose a sketch plan is given in Fig. 9, which is somewhat similar to the diagrammatic representation usually shown on mill plans. Its main features consist of the headstock A, which contains practically the whole of the mechanism, and from which point the machine is driven; extending for some distance on either side of the headstock is a strong wooden structure C, called the carriage, which carries the spindles, faller rods D, etc. The creel E and rollers F are arranged parallel to the carriage and extend in a similar manner on each side of the headstock A. The ends of the machine are terminated by a frame B firmly bolted to the floor. The accompanying illustration, Fig. 10, will now enable a general description of the mule to be given. It represents a section through the essential parts of the machine, and from it an outline of its action can be obtained.

The bobbins A are taken from the last passage of fly-frames and placed in the creel at the back of the mule; from here the rovings are guided over wires and passed through three lines of rollers which are arranged to give it a suitable draft. From the front rollers it is now led on to the spindles, and after receiving the requisite amount of twist it is wound on in the form of a cop.

The headstock is a strong framework consisting of two frames similar to that shown in Fig. 11. The two portions are firmly connected by cross pieces, and within the rectangular structure thus formed the mechanism is placed. This mechanism is of a very complicated character, and in the descriptions of the various actions that take place during a cycle of operations repetition will be unavoidable and in many cases necessary. This is rendered more so by the fact that most of the actions are directly connected, or depend upon each other for their performance and in several instances are working simultaneously.

When the carriage commences the twisting operation it is brought as close to the rollers as possible, the spindles occupying the position shown at L; this distance is usually from 3 to 5 inches. As already explained, the twisting continues by causing the spindles to revolve at a rapid rate, and at the same time moving them gradually away in

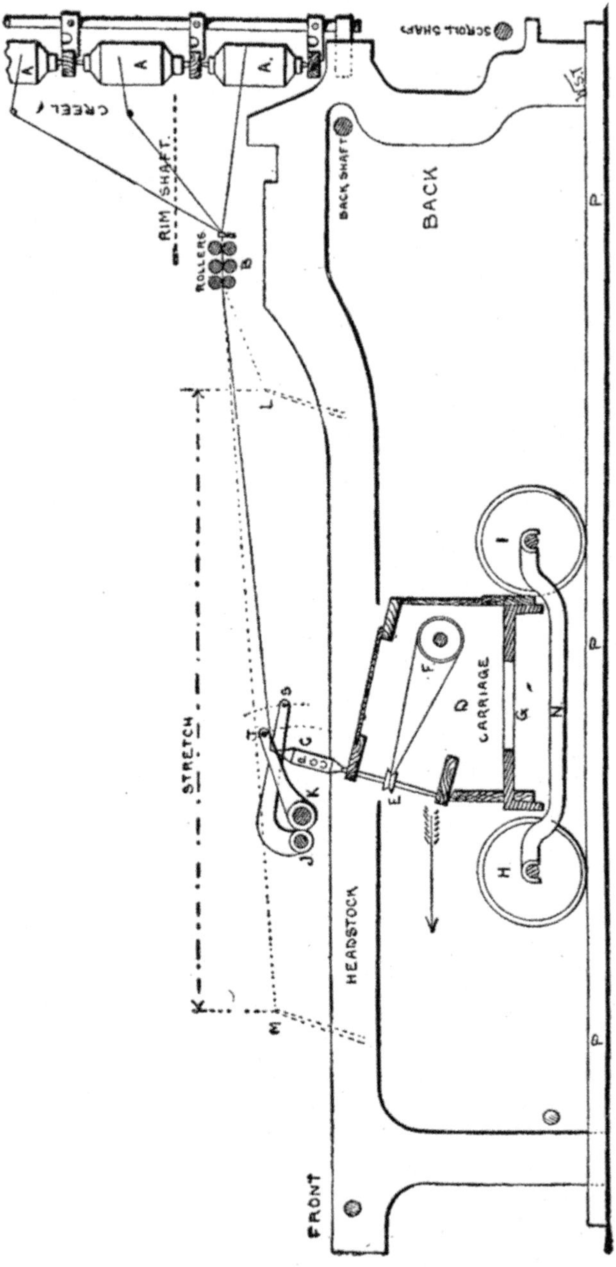
CREEL
RIM SHAFT.
ROLLERS
BACK SHAFT
SCROLL SHAFT
BACK
STRETCH
CARRIAGE
HEADSTOCK
FRONT

FIG. 10.

the direction of the arrow, until they arrive at M; when this position is reached the spindles cease twisting, an

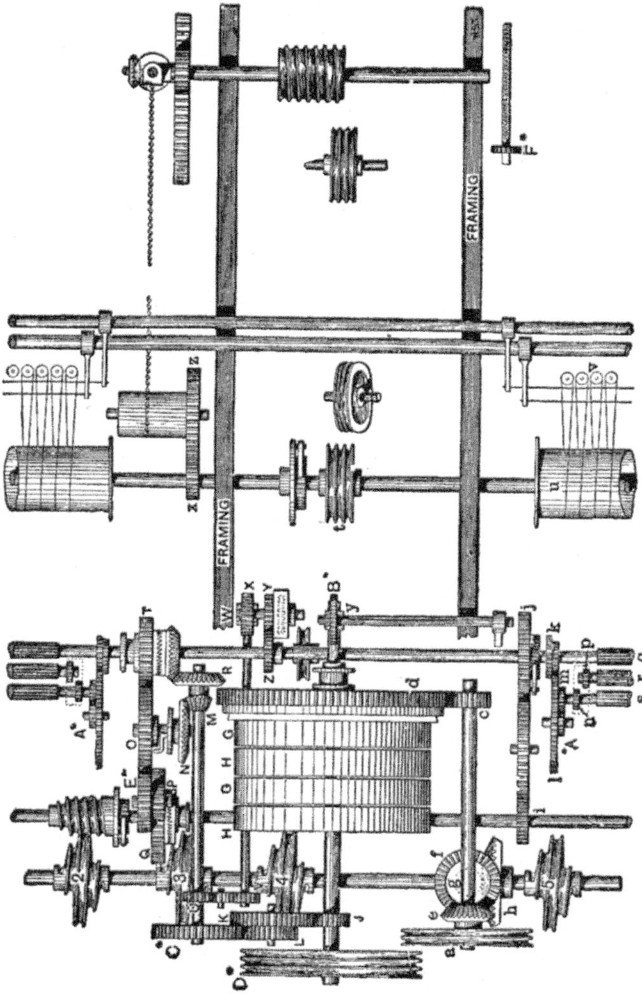

GEARING OF S.A. MULE

Fig. II.

action called "backing off" comes into play, and immediately following this the carriage begins its return journey to the rollers; whilst this is being performed the yarn which was twisted during the outward run is wound on the spindle.

The distance traversed by the carriage from L to M is termed the "stretch"; its length varies for different purposes, ranging from 48 up to as high as 68 inches, the most usual length, however, being about 64 inches. A "draw" is generally understood to mean one complete action, *i.e.* from the commencement of spinning when the carriage is at L to its return to the same position after the "outward run" and the "run in." If a mule works, say, four draws in one minute, it means that the carriage has started from L and returned to it four times in the course of one minute; in other words, the machine has gone through the whole of its actions four times in sixty seconds.

The carriage is mounted on a series of bearings N supported by bowls I and H; they are placed at suitable intervals along the length of the carriage and run on iron rails P. The spindles are driven by the tin cylinder F carried by the carriage. This arrangement, however, only drives the spindles whilst twisting; when winding, or building the cop takes place, they receive a special motion. The cop is formed through the medium of the wire T carried by a lever, centred on the copping faller K, and during the building, tension is maintained in the yarn by the wire S carried by a similar lever, but which is connected to the shaft J and called the counter-faller. All the above-mentioned features are carried by the carriage and will be dealt with subsequently in detail, and fully illustrated.

As will be observed from Fig. 9, mules are worked in pairs, arranged so that they can be attended to by one set of workers. The spindles of each mule approach each other in their outward run to within such a distance as will permit of freedom of movement for the workers, who in the course of their duties pass to and fro along the passage between the faller rods D of each mule.

In order to convey an idea to the reader of the mechanism of the headstock, or at least the general features of it, in plan view, an illustration is given in Fig. 11. The principal driving of the machine takes place through the pulleys H, G, driven from a counter shaft above; from the same counter shaft is driven the "drawing-up" pulley *a* by means of a band. The spindles *v* are driven by the rim pulley D through *t*, and the tin cylinder *u*; the carriage is actuated by means of strong bands through the scrolls 2, 3, 4, and 5. The gearing for the driving of the rollers can be readily traced from the wheel J on the rim shaft. The turning of the spindles for "winding" during the inward run of the carriage is produced by means of the quadrant, a chain from which passes over the winding drum and transfers the motion through the wheels *z* and *x* to the tin cylinder and on to the spindles. A reference to this drawing in connection with the further descriptions that will be given, will be of great assistance in explaining much that might otherwise appear vague.

The Creel.—Although the arrangement of the creel is not of much importance as a detail of a machine, yet it ought to be noticed, especially in connection with the mule. There are obvious advantages to be gained by giving to the bobbins a disposition that will economise space, and save time in filling the creel and in keeping them at a suitable height adapted to the workers who attend this feature of the machine.

Figs. 12 and 13 illustrate a variety of methods of forming the creel, and plan views are also shown. In Fig. 12 the usual Bolton system is given; single rows of rails are

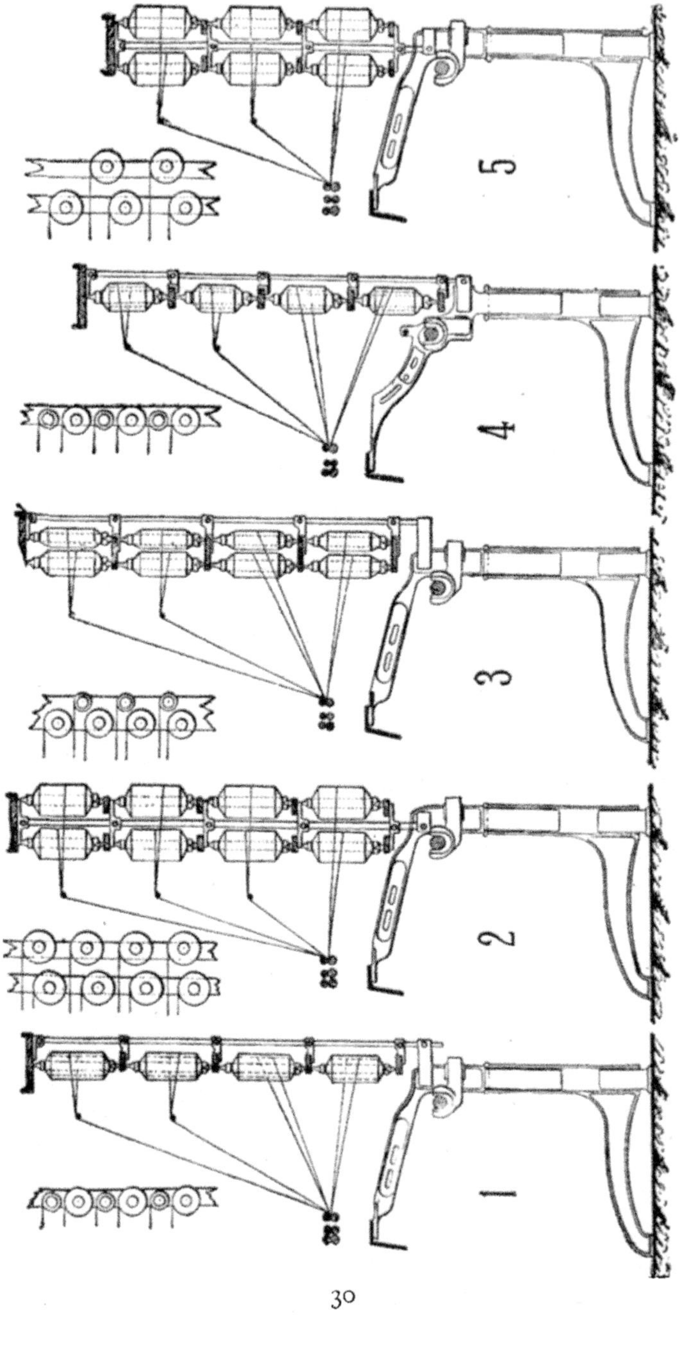

Fig. 12.

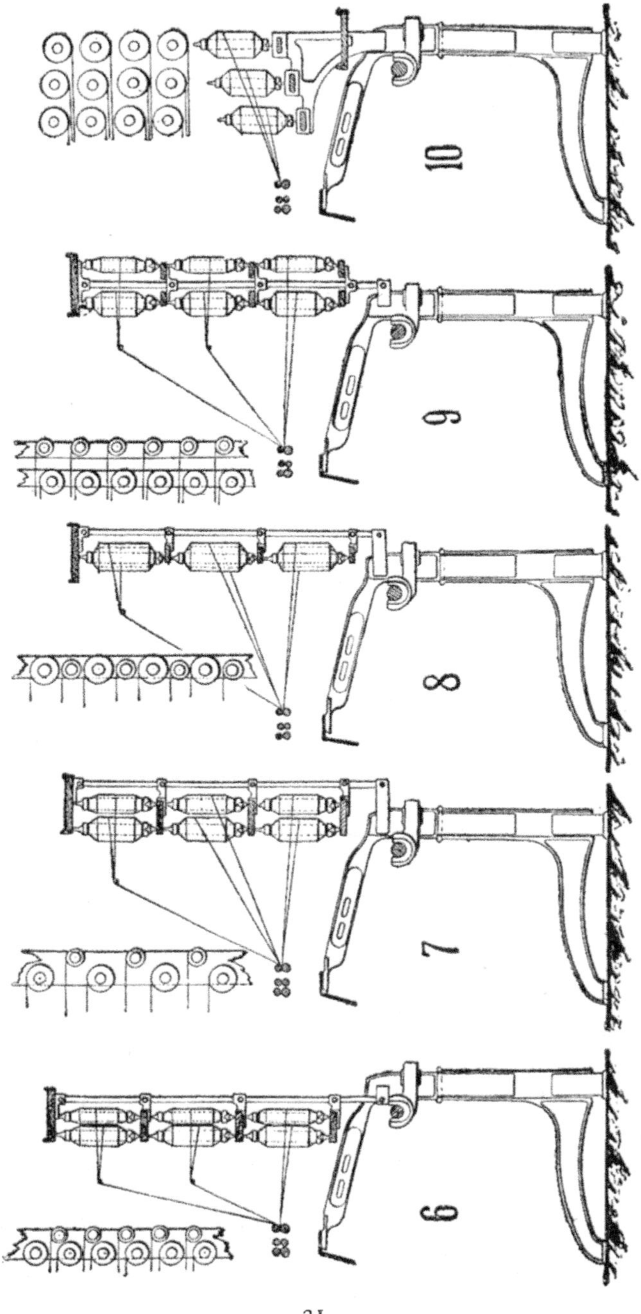

Fig. 13.

employed, which saves space, but it necessitates half and full bobbins with double rovings. Alternate arrangements for all full bobbins, with four heights, are shown, in which two rails are used, and also an arrangement with a broad single rail, the bobbins being arranged in zig-zag order. Three heights of bobbins for single rovings are illustrated.

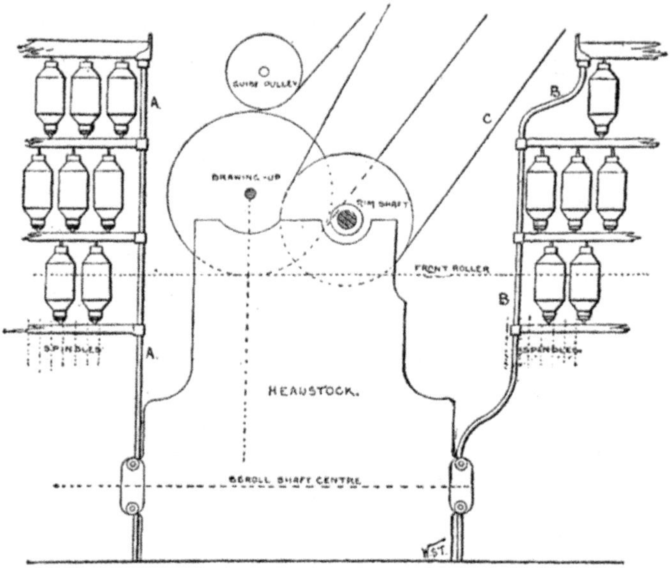

FIG. 14.

An exceptional method is illustrated which is only adopted when circumstances prevent the application of the other systems.

The creel itself is built up on a series of upright rods, firmly fastened to the spring pieces which carry the roller beam.

Driving the Mule.—Owing to the fact that the two headstocks of a pair of mules are always placed out of the centre of their respective lengths (see Fig. 9), the driving belt is often so much inclined as to necessitate a slight alteration in the arrangement of the end of the creel at the headstock. Such an alteration is shown in Fig. 14. It would clearly be impossible to have B straight up, as at A, on account of the driving belt C; therefore a method similar to that illustrated is usually adopted.

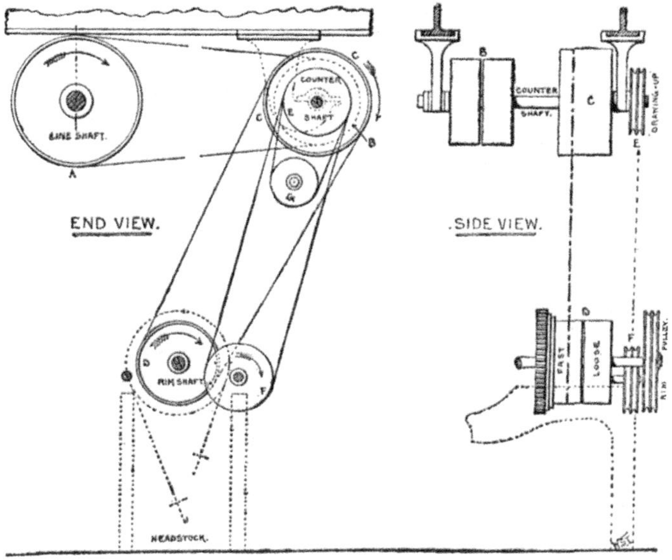

Fig. 15.

A general idea of the main driving of the mule can be obtained from Fig. 15. The end view is taken from the back of the machine, and shows all the shafts in section. It will be seen that the line shaft or main driving shaft is at right angles to the direction of the carriage length. In all new mills this shaft runs from end to end, and is driven direct from the engine; the various counter shafts for each machine are independent of each other, but all are driven from the line shaft in a manner similar to that shown in the drawing. The pulley A drives B on the counter shaft; a separate pulley C on the counter shaft drives D on the rim shaft. The driving pulley A is made double the width of the belt, so that to stop the mule all that is necessary is to move the strap on the loose pulley at B; this completely stops the whole machine. Owing to the alternate motions of the mule, it is necessary to continue the working of some parts whilst others are stopped; this is effected partly by means of a fast and loose pulley on the rim shaft, and also by the employment of clutch cones and wheels that are put into and out of gear at their correct times by other parts of the moving mechanism.

It has been remarked that the principal driving of the self-actor is performed through the driving belt from C to D. Formerly this belt supplied the entire machine with its motion, but within the last few years an important change has taken place by transferring a portion of its strain to a supplementary driving arrangement by means of a band. This is shown in the sketch, Fig. 15. E is a band pulley on the counter shaft, and drives the pulley F on the drawing-up shaft. Its principal function is to produce the inward run or drawing-up of the carriage; several other important actions are effected also by it, which ensure a more perfect working of the machine than in the old system, where the whole work was thrown on the driving belt. The liability of the drawing-up band to stretch is compensated for by means of a tightening pulley G, which ensures a regular tension. The above general description is given so that the more detailed descriptions of each action which follow will be better understood, and the illustrations also will be extremely useful for reference, as it is clearly impossible in illustrating such a complicated machine to show more than one or two motions in a single sketch.

Although Fig. 15 shows the line shaft at right angles to the carriage, and thus brings the rim pulley at the back of the headstock, it ought to be remarked that this is not invariably the practice. It sometimes happens that, owing to the formation of the mill or the necessity for having the shafting fixed in a certain position, the line shaft is placed parallel to the length of the carriage. When such is the case, the rim pulley is arranged at the side of the headstock, and by very little re-arrangement of gearing all the other motions work in the same way as when the rim is at the back.

Movement of the Carriage.—We will now consider the question of how the carriage is moved during its outward and inward run. The remarks previously made will have demonstrated that there are two distinct actions, namely, spinning and winding—spinning when going out and winding when coming in—and for each of these the motion of the carriage undergoes a change of speed. It is perhaps necessary to explain the reasons for such a change of speed. The motion of the carriage during the operation of twisting is clearly dependent upon the number of twists required to be put in a given length of the yarn; the quicker the twists can be put in, consistent with the character of the cotton and the perfect working of the automatic actions associated with it, will provide a foundation in obtaining the speed of spindle; and this speed, when decided upon, regulates the speed of the carriage. From these considerations it is an easy matter to reason in a general way that the lower the counts spun the quicker the speed of the spindle; and, as lower counts have less twist than the higher counts, it follows that the speed of the carriage is quicker for low counts than for high counts. It is also not difficult to understand from what has been already said that the twisting operation is necessarily slow. When, however, the spinning is completed, and winding on begins, there is nothing to prevent as quick a return as possible to the roller beam. We therefore find a wide difference between the two motions of the carriage, and moreover they are performed by two distinct actions of the mechanism.

To convey an idea of the difference of the time, an example is given as follows:—Suppose a mule is found to complete its whole cycle of operations three times over in 54 seconds, this would give 18 seconds for each draw, *i.e.* an outward and inward run. Of this 18 seconds there would be about 4½ seconds in which the mule would back off and run in, thus leaving 13½ seconds for the outward run during which spinning is taking place. In this time the carriage has travelled 64 inches, and, in order to put the right number of twists in the yarn, the spindles must run at the rate of 9000 revolutions per minute without allowing for slippage of the bands. This gives us a good conception of the comparative speeds of the chief working parts, so we can now proceed to examine the methods adopted for obtaining them.

In order to fully appreciate the methods adopted in moving the carriage, it is as well to thoroughly understand the reasons for their adoption. In the first place the carriage is very long, and consequently heavy; if it contains 1000 spindles of 1⅜ inch gauge its length will probably be about 120 feet. To move this long heavy mass, which includes the faller rods and all their connections, the spindles, tin drums, square, the framework of the carriage and its bowls, etc. etc., is of itself a difficult matter; but when this heavy mass keeps stopping and starting, it is still more difficult to regulate its movements so that it may commence smoothly, and also finish without any abruptness. The problem is solved, however, by the introduction of what are technically called "scrolls." These are a kind of drum in the form of a spiral, and of sufficient length to wind on the requisite amount of band for the stretch. The small diameter with which they commence enables a very slow motion to be given to the carriage on the commencement and finish of its stretch, whilst the intermediate portions of its movement are much quicker; abruptness of actions and its consequent strains are by this means avoided. The above remarks are general to the two movements of the carriage, but are specially applicable to the inward run. During the outward run the carriage moves very slowly, but the inward run being much quicker, both the commencement and finish are made as slow as possible.

The outward run is obtained direct from the front roller through a train of wheels to the back shaft. Fig. 16 illustrates this connection; it is an enlarged view of a portion of the general gearing plan given in Fig. 11. The motion in the first place is received from the rim shaft through the wheel J; from here it passes through the compound carrier K L, and to the back change wheel or speed wheel C. A bevel wheel R conveys the motion to the front roller bevel S. Connected to S by means of a clutch-box is a wheel T, and from this wheel through the wheels O, E, P, and Q, the back shaft is driven. The speed of the carriage is of course directly related to that of the front roller; any required change between the two speeds is readily obtained by changing the pinion P, and a further change, in which both front roller and carriage will be altered in speed, can be made through the wheel C, and sometimes by changing L and K. These speeds and the calculations connected with them will be dealt with under the head of "Calculations" when we reach that part of the subject.

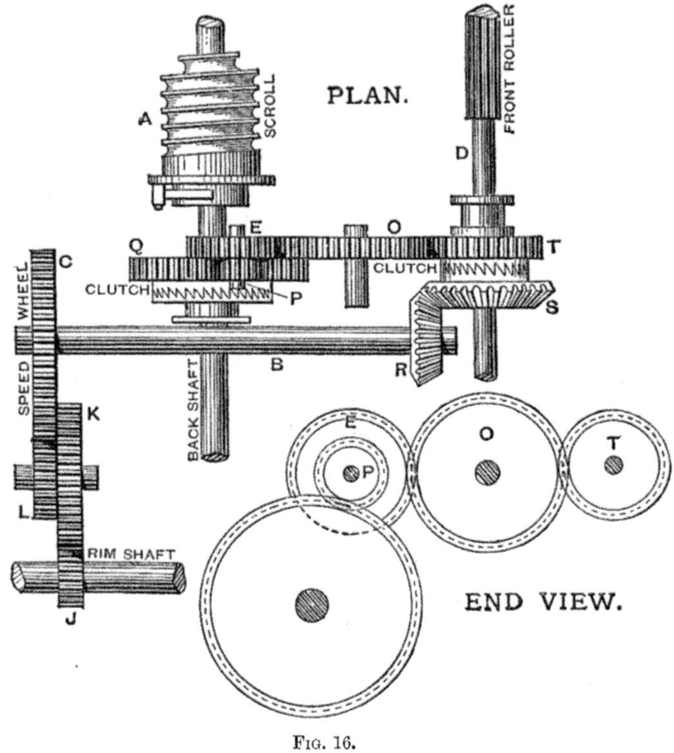

FIG. 16.

The arrangement for taking the carriage out by means of the back shaft is shown in the three illustrations: Figs. 17, 18, and 19. Bands passing over and around the scrolls are fastened to the carriage either at the back, front, or ends; the revolution of the shaft acting through the bands draws the carriage either outwards or inwards, as the case may be, this of course depending on the direction of the rotation of the shaft.

As a rule there are five scrolls in the back shaft; one, A, is connected by band to a large scroll 2 (see Fig. 17) on the scroll shaft. B and B are placed each about half-way between the headstock and the ends of the machine. One is also placed at each end, as at F and F. The method of connecting the bands to the carriage is shown in Fig. 17; but to make it more clear, drawings are given in Figs. 18 and 19, which show the attachment very

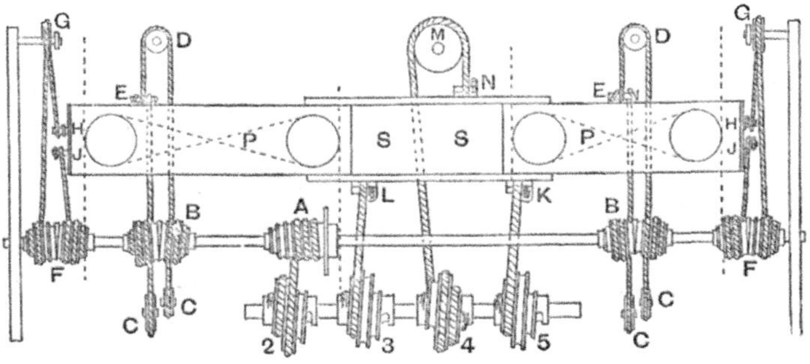

FIG. 17

distinctly. In Fig. 18 the scroll B is represented as drawing the carriage out; this it does by means of the band F, which passes under the carriage, over a guide pulley D at the front of the mule, and from here is fastened to the carriage at E; its motion in the direction of the arrow draws the mule out. The same drawing also shows that if the direction of motion of the scroll B is changed, the carriage can be drawn in through the band G, which is also fastened to the carriage. It must clearly be understood, however, that the motion of the back shaft for performing the "outward run" is obtained directly from the front roller, and the movement it gives to the carriage is a very regular one, except at its commencement, when the band is working on the small diameter of the scroll part at H, Fig. 18. The mule at this point is close to the

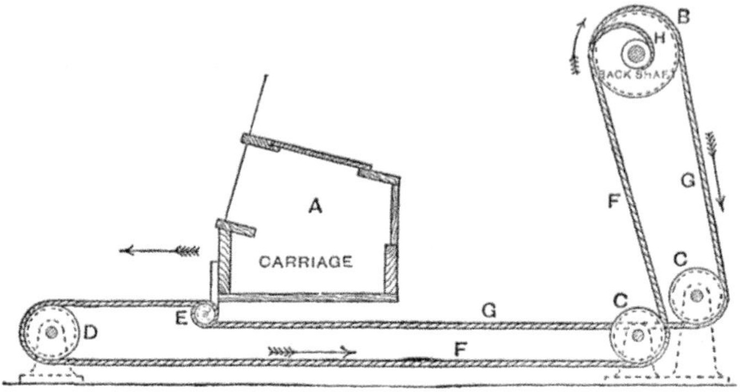

FIG 18.

roller beam, and stationary, and consequently the movement of the heavy mass must be brought about slowly. This is effected by making a short spiral at H for about half a revolution before attaining a maximum diameter at B.

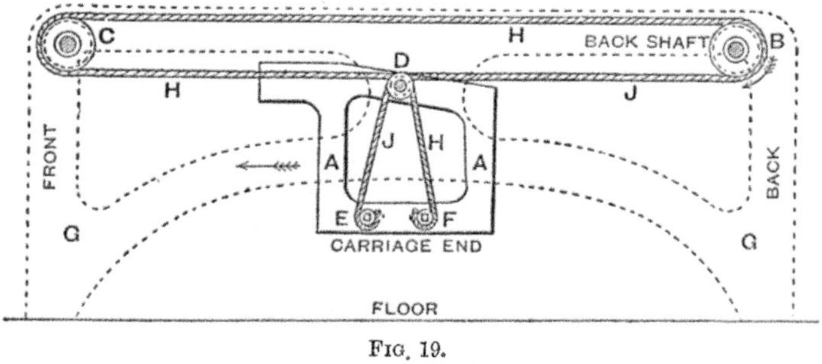

FIG. 19.

When this spiral portion of the drum is passed the remainder of the stretch is performed at a uniform speed by the straight portion of the drums.

The ends of the carriage are moved in the manner shown

in Fig. 19. B is the scroll, corresponding to F in Fig. 17. One of the bands H passes from B over a carrier pulley C, and a loose stud D, and is fastened to the carriage end at F; the other band, J, simply passes over D, and is then fastened at E. The revolution of B in either direction will produce a similar movement of the carriage. The direction shown by the arrows is the "outward run" during the spinning process.

When the carriage has reached the end of its outward run, an action called "backing-off" takes place, and immediately afterwards the inward run commences. As already described, this inward run is performed very quickly. The connection of the front roller with the back shaft is broken by disengaging the clutch, Fig. 16, which leaves the back shaft free to be driven from another source, namely, the scroll shaft.

The scroll shaft is driven through bevel wheels from the drawing-up shaft, see Fig. 11. On it are keyed four large scrolls, three of which are used in drawing the carriage in (Fig. 17), Nos. 3 and 5 are directly connected to the carriage to serve this purpose, while No. 2 is connected to the back shaft by a band on the scroll A. The whole back shaft is thus utilised for the inward run as well as for the outward run, its direction of revolution of course being reversed to enable the latter operation to be performed.

The fourth scroll, called the check scroll, is introduced in order, as its name implies, to check any irregularities of movement that may be caused through the varying and quick motion of the carriage during the inward run. Its effect will be better understood by comparing its position and action with the drawing-up scrolls 3 and 5. In Fig. 20 the scrolls Nos. 3 and 5 are shown attached to the carriage, being represented as drawing it in. When the band is on the smallest diameter the speed is slow, but on the large diameter it is quick, and attains its maximum speed on the largest diameter, and then begins to decrease. It is, however, quite possible that after the carriage has attained its quickest speed its momentum will compel it to continue at a slightly greater speed, for a moment or so, than that of the scroll. This is a contingency that must

FIG. 20.

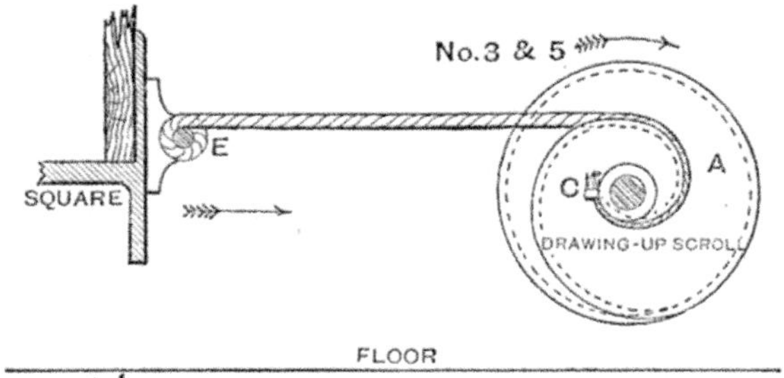

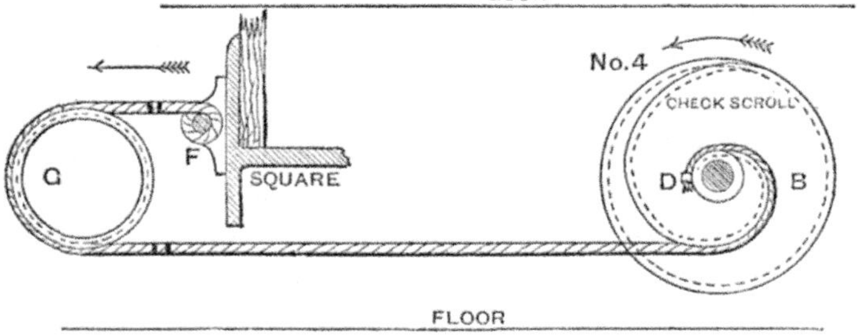

FIG. 21.

be avoided, as it might lead to disastrous results. The check scroll is therefore arranged to effect this, and Fig. 21 illustrates the arrangement. It will be noticed that its position on the shaft is opposite to that of the other scrolls, and that its band leads off from its lower side, and, passing underneath the carriage, is carried over a guide pulley G and connected to the front of the carriage. Now it is quite clear that any tendency of the carriage to overrun the scrolls 3 and 5 will be counteracted by scroll No. 4, because overrunning would result in tightening the band of the check scroll: in other words, this scroll serves the purpose of a drag on the carriage the moment it varies from the speed of the drawing-up scrolls.

There is a very important feature in connection with the various scroll bands that have been mentioned, which ought not to be overlooked. It will be observed that one very essential condition of the successful working of the mule is the necessity for maintaining the carriage perfectly parallel to the rollers. To maintain this requires in the first place a strong carriage to resist flexure, and the faller rods must be strong also as an aid to this condition; but the most important feature is the connection of the various bands to the carriage. Bands are at the best an uncertain element, so everything must be done in choosing only the very best bands and compensating in every way their tendency to stretch and to take up the extra length they acquire through the strain to which they are subjected.

The attachment of the bands to the carriage becomes therefore a very important factor in good work. Special ratchet arrangements are applied at the various points, so that an adjustment as fine as experienced judgment will allow can be attained. Frequent adjustment is necessary, for the bands are very irregular in their stretching qualities, and it is a serious matter if the carriage be allowed to vary from a straight line during its traverse. The yarn coming from the rollers will in such a case be irregularly stretched or drawn during the outward run, and on the inward run its winding on the spindle will consequently be unequal at various parts of the mule. When the carriage finishes its inward run, it ought to do this simultaneously throughout its whole length, coming against all the back-stops at the same moment with a smooth silent finish, and not abruptly, occasioning noise and shock, which would result in faulty yarn in the form of snarls or broken ends.

The extra long mules now made render close attention to the bands imperative. The carriage as now made is constructed on lines that reduce its flexure to a minimum; at the same time it is sometimes mounted on bowls that work on friction rollers, and the same feature is introduced in some cases for the faller rods and even for the back shaft. Everything, in fact, is done to prevent torsion and to preserve a perfectly straight line through the centre of the spindles, and also to maintain this line absolutely parallel with the front roller throughout the traverse of the carriage.

The adjustment of the bands just described is generally termed "squaring the mule," but "squaring band" is a name that is given to a special band which is used to obtain the movement of the end of the carriage equal to that of the middle part or square. It is illustrated in Fig. 17, but a detailed reference will be made to the diagram Fig. 22. Half the length of the carriage is shown, each half having its own bands, and the band is placed underneath it. Two bands are used, L and M. The band L is fixed at one end at a suitable spot E, and passes round the pulley C and D, the other end being fastened at F. A similar thing is done with the band M, but in the reverse order. Both bands are used for the same purpose, L for the outward run and M for the inward run; so we refer to L in the explanation. If the carriage be drawn outwards in the direction of the full arrow, a tension will exist in the band L as if it were being stretched in the direction shown. Now since the band passes from E to F, the same tension will exist in the band throughout its length, and by following it through we shall find that an effect is produced as if some force were pulling the band at F, in the direction shown. This has clearly the effect of pulling the end of the carriage out in the same direction as the middle, and with an equal force. The squaring band is therefore an important element in the "squaring" of the mule; but, like the other bands, it is necessary to keep a constant watch to see that it does not become defective for want of adjustment.

It will be interesting at this stage to devote a few words to a description of a "drawing-up scroll." The essential conditions to be fulfilled by such a scroll are—as slow a movement as possible at the commencement of the inward

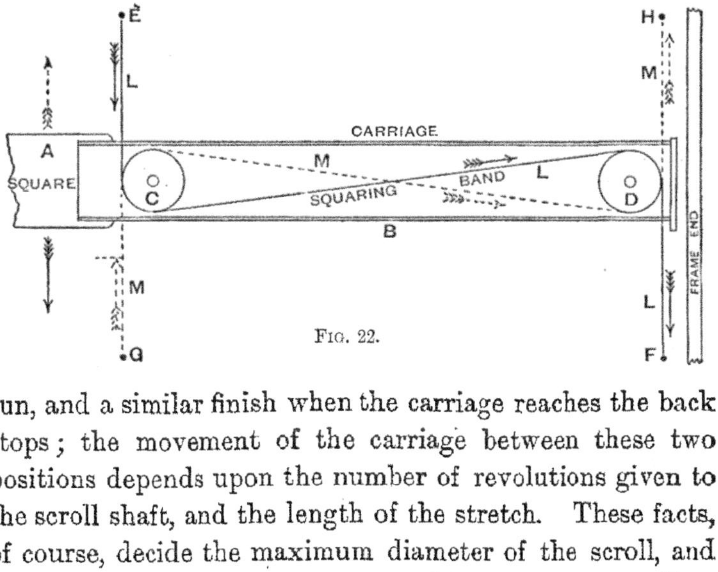

Fig. 22.

run, and a similar finish when the carriage reaches the back stops; the movement of the carriage between these two positions depends upon the number of revolutions given to the scroll shaft, and the length of the stretch. These facts, of course, decide the maximum diameter of the scroll, and the maximum diameter, in its turn, decides the intermediate speeds between the start and finish of the "run-in."

It is unnecessary to explain the method of obtaining the size of a scroll that will serve for any given stretch; but we will suppose the scroll has been designed, and that $2\frac{1}{2}$ revolutions of the scroll shaft are sufficient for the purpose. This means that during the inward run the scroll must make 2½ revolutions, and in doing so must wind on the band by which the carriage is drawn in. In order to obtain the commencing slow movement it is necessary to commence winding on a small diameter, as at B, Fig. 23; the diameter is then gradually increased by making the drum of a spiral form, until the largest diameter is obtained at C, where naturally the greatest speed is given to the carriage, which on examination of the diagram is found to be halfway in the stretch; from this point a reduction in speed takes place by a corresponding curve to the first half of the scroll, and it finishes on the same diameter as that on which it commenced.

To show the variation in the speed given to the carriage during its run, two portions of the scroll have been marked off. At B G a length is shown which represents the amount of band wound on during the first quarter of a second of the run in, while at the middle of the stretch the length wound on during the same time is shown at F E. The intermediate lengths could be easily shown in the same manner, but a better method is given in Fig. 24. The movement of the carriage for each quarter of a second is there shown; starting at A it would move to B in the first quarter of a second; each successive quarter would find the carriage at C, D, E, etc., until it had completed its journey at N. This diagram shows very distinctly the varying movement of the carriage; to those, however, who are interested in the matter, the diagrams in Figs. 25 and 26 will convey a much clearer idea of how the movement of the carriage is controlled. In Fig. 25 the straight lines D, C, B, show the development of the curve of the scroll, and the fact that straight lines represent such a development tells us that the carriage starting at B has a regularly increasing movement given to it until it reaches its greatest speed at C, from which point it at once begins to decrease to D. The point to observe in this diagram is that the change of speed, whether at the start, middle, or end, commences at once. Some authorities condemn this method and find much better results given by forming the scroll so as to give a movement as represented in Fig. 26. Here, instead of commencing to increase regularly, the initial slow movement of the carriage is continued a little

Fig. 23.

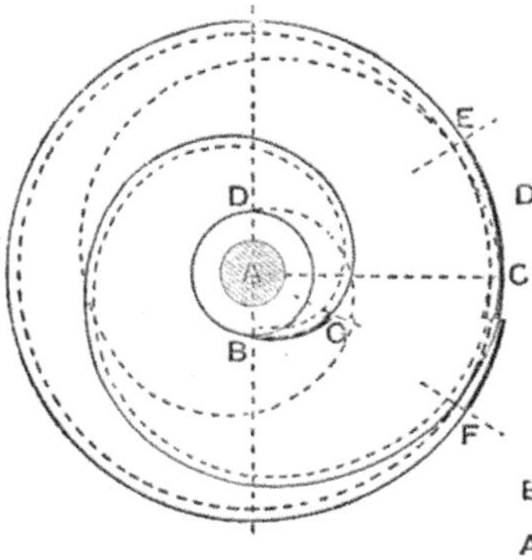

Fig. 25.

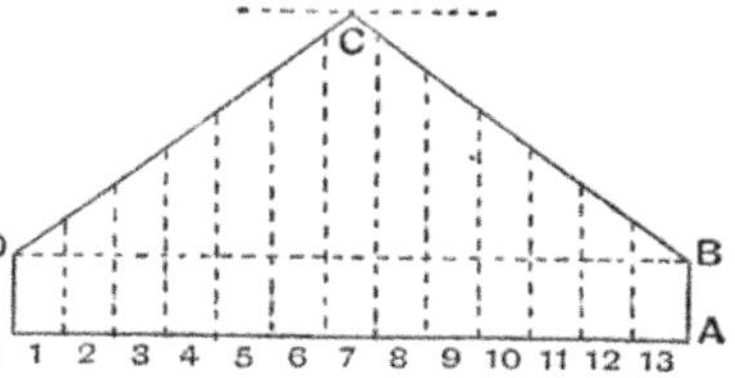

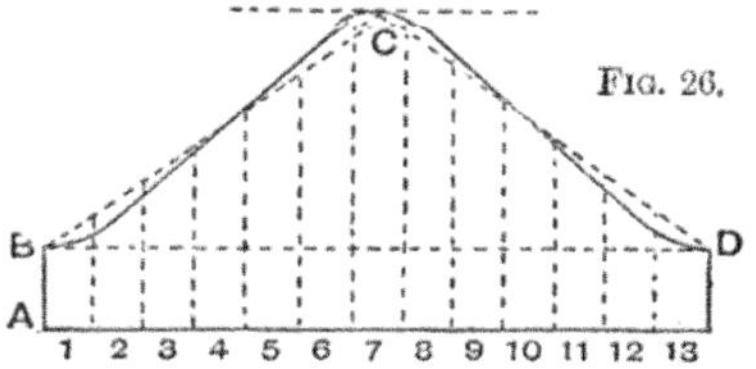

Fig. 26.

Fig. 24.

longer, and then gradually increased to a regular acceleration until near the maximum at C. Here the speed is maintained a moment or two longer, and then a more gradual reduction is made to the decreasing speed than in the case of Fig. 26 until D is reached.

In Fig. 17 was shown a system of arranging the scrolls which up to a few years ago was generally followed. At the present time, however, one or two important firms have arranged their systems on a slightly different plan. Instead of the scrolls Nos. 3 and 5 being placed so far apart and independent of each other, they are brought closer together, and one band only is used for the two. This band, instead of being fastened to the usual ratchet-tightening arrangement on the square, goes from one scroll and passes round a horizontal carrier pulley, or round fixing, on the square, and from there back to the other scroll. The object of this is to obtain exactly the same tension in the band of each scroll. We have seen how important a matter this uniformity of tension is, and it

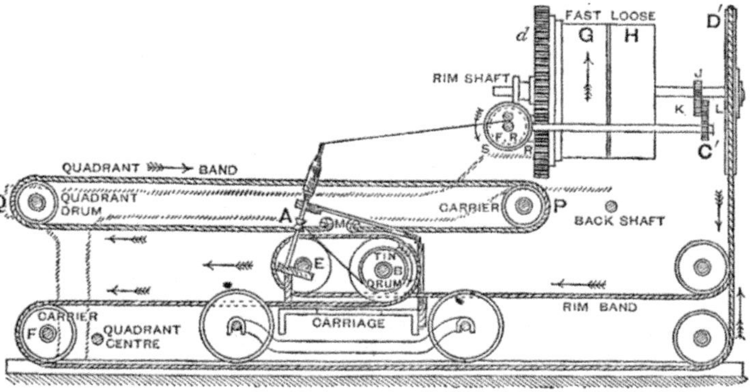

FIG. 27.

will be admitted that this method is an excellent one for attaining it. Of course it is necessary to keep the carrier pulley itself adjusted as the band becomes slack. Although having obvious advantages, it is open to question whether this method is superior to that illustrated. The pull, taking place at what is practically one point, is bound to be inferior in effect to that of a pull at two points as far apart as possible on the rigid part of the carriage called the square. Unequal wear that may take place in the band will lead to more waste and loss of time than in the old method, and the new one is under a distinct disadvantage when, as sometimes happens, the band breaks and the breakage is not immediately noticed; serious results in such case would certainly follow. Under the old system, when one band breaks the other band will prevent any mishap occurring until the minder discovers it and effects a remedy.

Driving the Spindles.—Fig. 27 illustrates the method

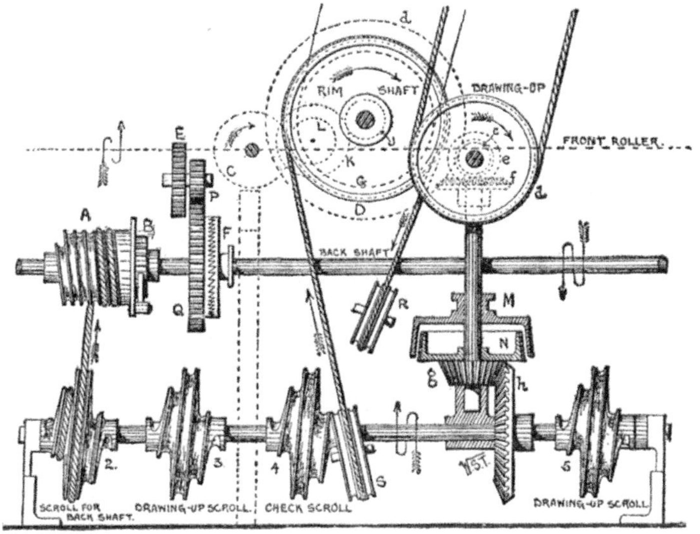

Fig. 28.

adopted in driving the spindles. They are driven from the rim shaft through a large band pulley D; the band passes down behind the headstock over a fixed back carrier pulley, and on to another carrier pulley E; from this it passes round a band pulley B on the tin drum shaft through which the spindles are driven. Continuing, it goes forward to the front of the headstock and over a carrier pulley F, by which it is guided on its return journey, and passing over another back carrier reaches the rim pulley. The back view of the mule is given in Fig. 28, and the positions of the back carrier pulleys R and S show the rim band guided in a direction at right angles to that in which it leaves the rim pulley.

The revolution of the rim shaft in the illustration is in the direction shown by the arrow, but it is not necessarily so in all makes of mules; some have the rim running the opposite way, and with an arrangement of the driving of the tin cylinder as represented in Fig. 29. There is practically no difference between the two methods, the wear, strain, and length being about the same in each case.

Although only a single grooved band pulley is shown in the sketch, this has merely been done to simplify the drawing. On the mule two or three grooved pulleys are used, and the band is consequently twice or three times the length represented in the sketch. A long length of rope, such as this, is subject to a considerable amount of stretching; and, especially when it is new, some attention must be given to it to keep it at a uniform tension. The carrier pulley F is fixed in a slide, which can be readily adjusted to compensate for any stretching that may take place. Unless the rim band be kept well to the grooves of the rim and the tin cylinder pulleys, considerable slippage is likely to occur; even under the best conditions some slippage is unavoidable, but neglect in keeping the band tight leads to very serious faults in the yarn. Every care should therefore be taken in attending to this feature of the mule. The tin drum or cylinder, extending the full length of the carriages, drives each spindle by means of a short length of banding, which passes round a small pulley on the spindle, called a wharve. The direction of rotation of the spindle can be varied by a change in the crossing of the band from the cylinder; a change in its speed is brought about by changing the rim pulley D and replacing it by a larger or smaller as the case requires, the end of the rim shaft being arranged so that this may be quickly effected.

An interesting point to observe in the two drawings, Figs. 27 and 29, is the effect of the movement of the carriage on the band. In both cases the band, when leading on and off, is running in the same direction as the carriage. To get an accurate idea of the revolution of the cylinder this must therefore be taken into account, for there is clearly a loss, which amounts to from $1\frac{1}{2}$ to 2 per cent in ordinary numbers. This loss of speed must not

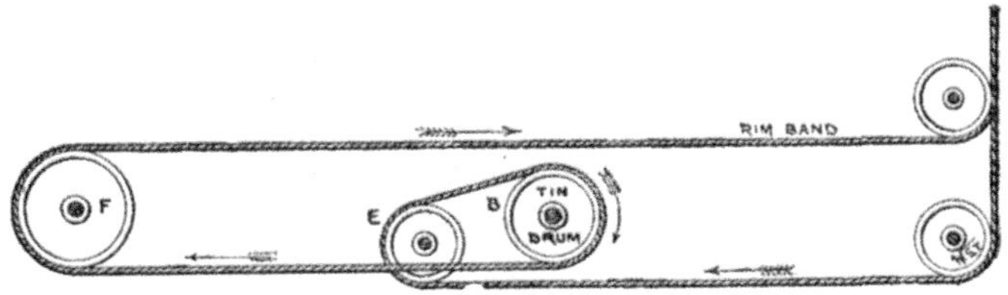

FIG. 29.

be confused with slippage, because it is due to an entirely different cause. As a rule, however, it is included in the term "slippage," such term including the difference between the calculated number of revolutions and the actual number. One chief reason for the employment of a three-grooved rim pulley is the desire to reduce slippage to a minimum, even when a slightly increased power is the result; and at the present time for good work the three-grooved pulley has become general.

Another interesting feature displayed by the spindles is the relative slowness by which they attain their speed on commencing the outward run. Theoretically the spindles are supposed to commence running immediately the carriage starts on its outward run, and at the same instant the roller also commences to turn. Careful observations extending over a large number of mules show that, starting from the beam, the spindles do not attain their maximum speed until the carriage has moved 10 to 30 inches away from its starting point. This accounts for much of the irregularities of twist, counts, and other conditions of mule yarn which affect its quality, and it ought certainly to be taken into consideration more than appears to be done in estimating twist, etc.

The explanation in a general way is that it is due to the enormously high percentage of power required to start the mule carriage and the spindles on the outward run. In a 1000-spindle mule the power required during the first half-second rises as high as 25 h.p., and this is developed immediately the strap goes on the fast pulley. Such a high power is undoubtedly due to the resistance of the carriage and spindles, both being at rest at the time. They yield gradually, and in doing so a large percentage of slippage must take place, especially on the rim band, and the spindle bands, and also on the driving belt. The carriage itself loses nothing, because slippage is almost impossible in its case, but it adds to the general disarrangement of the relative movements of itself and the rollers and spindles for the first second or so of the run out.

A very ingenious method of trying to overcome the difficulty just mentioned, that of starting the spindles at their full speed, has been introduced by a well-known firm of machine makers. It is illustrated in the accompanying sketch, Fig. 30. The variation in the relative motions has been overcome by what is practically driving the front roller and carriage from the tin cylinder. The tin cylinder is driven in the usual way, but by a special arrangement the same band transmits its motion to the front roller, which is therefore not driven in the direct manner by gearing, as is usual in other mules.

On reference to the drawing it will be seen that the driving pulleys E are mounted on a hollow shaft B, to which the rim pulley also is fixed. Within the shaft B another shaft A is placed, carrying at one end a band pulley H, and at the other end a wheel I from which the

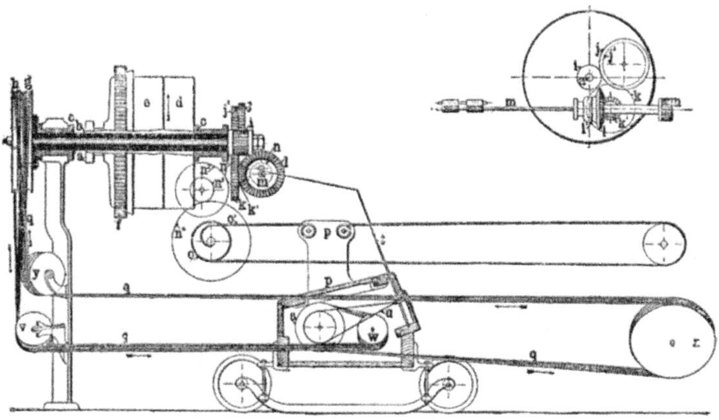

FIG. 30.

front roller is driven. Now as the outward run commences, the rim pulley G will be driven. Its band will drive the tin cylinder in the manner shown, and on returning to the back of the headstock is passed round the band pulley H on the inner shaft A A, which it therefore drives at a speed equal to E E, but minus any slippage that has occurred in the rim pulley G. The wheel I on the shaft A drives the front rollers, and from the front rollers the back shaft is driven in the usual manner by the train of wheels shown.

We are now in a position to see the peculiarity of this motion, and also its advantages. The usual method is to drive the front roller and back shaft direct from the rim shaft: consequently little or no slip occurs; but since the spindles are driven by band, a large percentage of slippage occurs, especially as the carriage starts out from the roller beam, and inequalities of twist of rather a serious character are therefore introduced. To neutralise these as much as possible the direct method of driving the carriage is dispensed with, and both spindles and carriage are driven by the rim band; any slippage that takes place in the band will now affect each motion, and if the spindles start slowly the carriage will also do the same, and in this way prevent any inequality of twist that would otherwise occur. It must be clearly understood that the "initial" slippage of the bands of the mule, which is very great, must not be confounded with what might be termed a "general" slippage, which must always exist throughout the travel of the carriage, and which is sometimes estimated to be as high as 5 per cent of the speed of the spindles. No band can be kept at such a tension, and in perfect contact with its pulleys, to an extent that would actually prevent slippage, so something must always be allowed for this when dealing with calculated speeds where bands are employed.

The Rim Shaft.—Before proceeding further in our description, it will be an advantage to illustrate and describe those parts of the mule from which the actions already mentioned receive their motion. The rim shaft is naturally the first point to which attention must be directed, and in order to show clearly the disposition of the driving pulleys, an illustration is given in Fig. 31, which represents in section this important feature. Reference may also be made to the sketch, which shows a section through the duplex system of driving.

The rim shaft is generally carried by two bearings, G G, which form part of the general framing of the machine. On the shaft between these bearings are placed the main driving pulleys. They consist of fast and loose pulleys B and C; the fast pulley B is keyed to the shaft, and through it the mule receives its chief movements. One

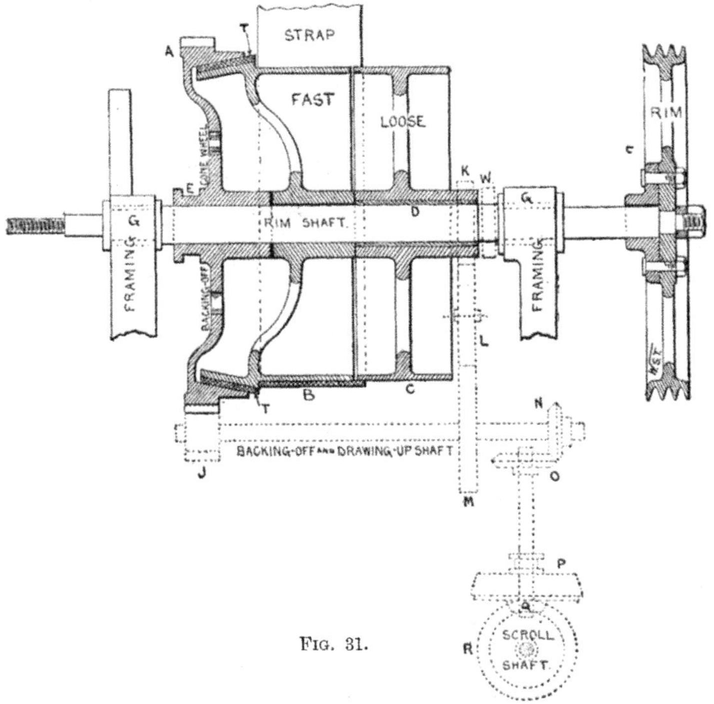

FIG. 31.

edge of this pulley is extended, and formed with a conical surface, upon which is riveted a layer of leather, T; a large wheel, A, called the backing-off cone wheel, also has its outer rim extended and its interior side recessed out in a conical form for the reception of the conical part of the fast pulley. The large wheel A is not keyed to the rim shaft, simply riding loose upon it; but by means of a fork, fitting in the grooved part of the boss at E, and levers, the backing-off cone wheel can be moved into or out of contact with the fast pulley. The loose pulley, to which the strap is moved when certain actions are at rest, rides loose upon a bush, as shown in the drawing. This bush may be either a separate piece or be formed as part of the brass bearing which fits in the framing and carries the rim shaft. The end of the rim shaft upon which the rim pulley is bolted is specially prepared to receive the rim and to effect a speedy change when a larger or smaller pulley is necessary; this detail is fully shown in the sketch. The place to which the pulley is bolted is, in some mules, forged on the shaft and case-hardened, by which means the possibility of breakage, owing to the sudden strains to which it is subjected, is reduced to a minimum; at the same time, the fact of its forming part of the shaft and being turned and finished therewith ensures more perfect running, and far smoother driving of the spindles. The three-grooved rim pulley is illustrated, as this form is now generally used, and is recognised as the best for driving purposes; through it the band, which is much longer in consequence, maintains a better grip in the grooves, and therefore reduces slippage.

Drawing-up and Backing-off, etc.—There are practically two systems in vogue on mules at the present time in regard to the "drawing-up" and "backing-off" arrangements. Owing to the great advantages that have been found to result from the "drawing-up" by means of a separate driving by band or strap, the older form is gradually becoming obsolete; but as a very large number of mules are working under the old conditions, a brief sketch of the arrangement will be given. As a preliminary, it must be clearly understood that the loose pulley on the mules is not used as a means to stop the mule: this is effected in the counter shaft; therefore the word "loose" is only used in a local sense. In the action about to be described, the loose pulley performs very important functions, to which reference will now be made. The drawing, Fig. 31, can be used to aid the description of the older form of "drawing-up" and backing-off," the special parts relating to it being shown in dotted lines.

When the strap is on the fast pulley B, the backing-off cone wheel A is out of contact with it, and therefore free on the shaft. Under the circumstances, all that is fixed on the rim shaft will revolve. Two important actions now commence, viz.—The turning of the spindles through the rim pulley F, which constitutes the spinning process; and the revolution of the rollers and outward movement of the carriage, which is effected through the wheel W fixed on the rim shaft. These actions continue as long as the strap remains on the fast pulley, but after the carriage has moved what may be considered the necessary distance, say 64 inches, its own movement, acting through levers, brings about what are technically called "changes": *i.e.* certain actions are made to cease and others come into operation. These "changes" will be described in detail; for the present purpose it is sufficient to mention that one of the changes causes the strap to be moved from the fast to the loose pulley C, which has the effect of stopping the rim shaft, and therefore the spindles, the rollers and the carriage.

During the time the strap is on the fast pulley, a small portion of its breadth is working on the loose pulley, and causing it to revolve. This movement is sufficient to make the wheel A revolve, because the loose pulley has on its boss a wheel K, through which the "backing-off" cone friction A can be driven. Gearing into A is a wheel on the cam shaft (not shown in the sketch); this latter shaft has a cone clutch driving arrangement, which is put into and out of gear by the carriage. At the termination of the outward run, one of the "changes" produced by the levers referred to above, puts the cone clutch on the cam shaft into gear, and enables the movement of the loose pulley to turn the cam shaft and by this means to put the roller and back shaft catch boxes out of gear, and thus stop the carriage, etc. Immediately the carriage and spindles have ceased working, the strap being on the loose pulley C, two other important actions are brought into play. One is called the "backing-off," its object being to cause the spindles to revolve a few turns in the opposite direction to that in which they revolved when spinning. This unwinds the yarn on the spindle, which is coiled between the cop and the spindle point. The action is brought about by certain levers forcing the backing-off wheel A into contact with the conical part of the fast pulley. As A is being driven at the time through the wheels K, L, M, and J, and in the contrary direction to the driving strap, it commences to turn the rim shaft in the opposite direction, and so gives the desired movement to the spindles. The other action is the "drawing-up" of the carriage during the inward run. This, as already stated, is the duty of the scroll shaft; it receives the motion enabling it to do this through the wheel K, on the loose pulley, acting through the wheels M, N, O, Q, and R.

It will be seen that all the movements referred to in this description are connected with one another almost directly; it is only by the careful adjustment in putting cone clutches in and out of gear that it is possible to bring about the several operations that have just been described, and in order to give a clearer idea of these complicated actions the following table may prove useful:—

When the strap is on the fast pulley, during the outward run:—

The spindles are revolving.
The rollers are delivering roving.
The carriage is making its outward run.
The "backing-off" cone friction is out of gear.
The "drawing-up" friction is out of gear.
The "backing-off" cone wheel A and the cone dish P on the upright scroll shaft are revolving, because a portion of the strap is on the loose pulley C which drives them through K.
The cone clutch on the cam shaft is out of gear.

When the carriage reaches the end of the stretch, changes take place which have the effect of:—

The strap is on the loose pulley.
- Putting the cone clutch on cam shaft in gear.
- Moving the strap on to the loose pulley C.
- Stopping the spindles.
- Putting the "backing-off" friction into gear with the fast pulley, and causing "backing-off."
- Stopping the carriage and back shaft.
- Stopping the rollers.

When "backing-off" has finished,

The cone clutch at P is put into gear, and the scroll shaft draws in the carriage.

The brief analysis just given does not by any means exhaust the actions of the mule during the period described; it merely presents in a concise form the chief points of the description already given; and much of it will of necessity be recapitulated as the mechanism is dealt with which is used to bring about the various "changes" referred to.

The modern form of the "drawing-up" and the "backing-off" can now be presented, and with this object Fig. 32 has been prepared, showing it fully in detail. It must be understood that other types of machine differ in the general disposition of the parts from that shown, but since the object is the same in each, one description will suffice. Advantage has also been taken in this sketch to illustrate what is now becoming a very usual practice in the driving of the mule, namely—two sets of fast and loose pulleys under the name of "duplex" driving. These pulleys are shown at H and G. Instead of a 5-inch strap being used, as seen in Fig. 32, working on a wide pulley, two narrow ones are now employed, generally each about $2\frac{1}{2}$ inches wide, working on a similarly reduced width of pulley; the direct object of the arrangement is to obtain a quicker change than is possible with a wide belt, and although special means are taken in most mules to assist the strap in moving from one pulley to the other, there must always be some little delay in doing it. The adoption of the "duplex" system results in a distinct saving of time, and although assistance in the form of a strap-relieving motion is not so necessary as before, it is still often employed, and usefully so, in helping to obtain the change in as short a time as possible. Slight objections are raised by some against the arrangement; such as the possibility of unequal tension in the two belts, which would throw most of the driving on to one strap and so cause breakages and also damages to the machine through entanglements, etc. These objections are of a practical character, which experience only can decide; but so far nothing has happened to prevent their very extensive adoption, and a large proportion of mules now made have the "duplex" arrangement applied to them. In order to obtain a clearer idea of the disposition of the driving in Fig. 32, reference ought to be made to a sketch already given in Fig. 11, where a full plan view is represented, the lettering in each, with few exceptions, being the same. A is the rim shaft containing the driving pulleys H and G; the fast pulley, as in the last example, Fig. 32, has a conical extension covered with leather for the purpose of forming a cone clutch with a corresponding recessed portion of the backing-off cone wheel D riding loose on the

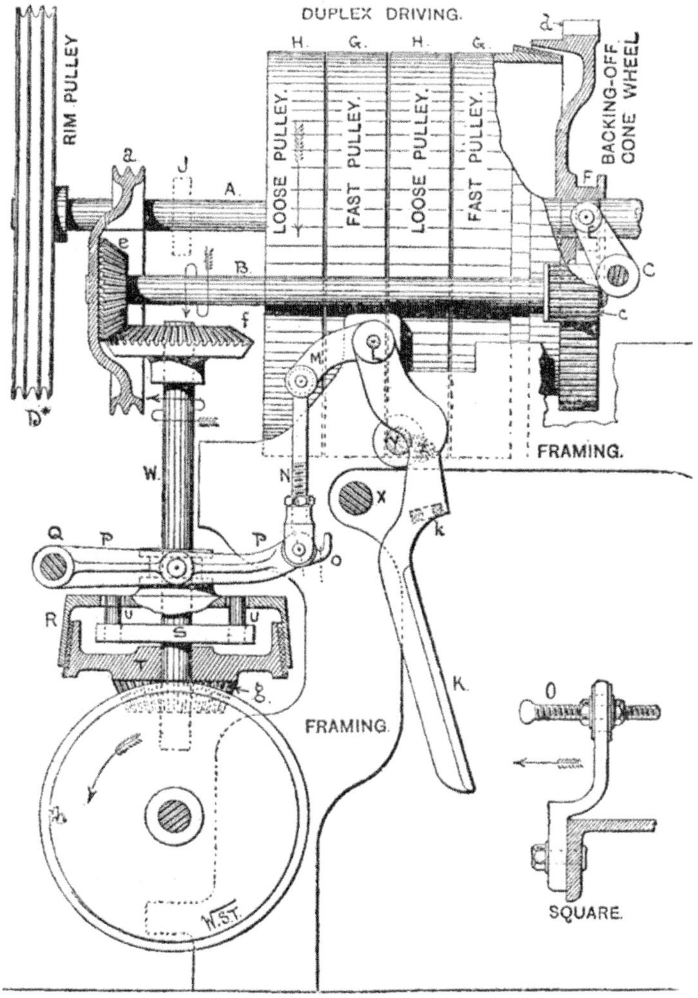

Fig. 32.

rim shaft. Situated on one side of the rim shaft is an extra shaft B, on which is keyed a band pulley "a," through which the "drawing-up" is effected; "a" is independently driven from the same counter shaft that drives the rim shaft, but its speed of course can be regulated to any extent required.. On the other end of the shaft B is keyed a small pinion "c," which gears into the backing-off wheel "d" and so drives it; so long as the driving pulley and backing-off cone are not in contact, the revolution of "d" serves no purpose, but directly the carriage ceases its outward run, "backing-off" must be performed by reversing the spindles; this is done, as in the previous case, by putting "d" and G into contact with each other, through the lever E centred at C, and so causing the fast pulley to be driven and consequently the rim shaft. It is only a momentary action, as will be seen when the subject is treated more in detail; it is mentioned here merely to show how it is effected by means of the separate driving through the pulley "a." On the side shaft B, close to the band pulley, is fixed a bevel "e," gearing into a large bevel "f" on an upright shaft W.

At the lower end of this shaft is the cone clutch and bevel necessary for driving the scroll shaft for the purpose of "drawing-up" the carriage during the inward run. The feature is given in detail so that it can easily be understood. Fixed on the shaft is a conical pulley T, on whose outer surface is firmly riveted a layer of leather; on its under side is fixed a bevel wheel "g," which gears into a large bevel "h" on the scroll shaft. Sliding on the shaft W and covering up the lower cone pulley T is a conical dish R, which rides loose upon the shaft; for the purpose of driving R the upright shaft is specially prepared by having forged on to it a plate S to which is fastened two pins U; these pins fit in holes in the cone dish R, and as the shaft revolves they carry the dish round with it.

During the outward run of the carriage the cone clutches R and T are out of gear, but immediately the run out is finished and the necessary change made, the cone clutch comes into gear; the scroll shaft is then directly driven from the band pulley "a" and the "drawing-up" commences and continues until the arrival of the carriage at the stops puts the cone clutch again out of gear. The drawing, Fig. 32, shows one method adopted for putting the cone clutch in and out of gear. The upper part of the cone dish R is prepared with a recessed boss for the reception of a forked lever P carrying studs fitting in the recess. The lever P is centred at Q and its other end is connected by means of an adjustable link N with one end M of the drawing-up lever K, centred on part of the headstock at L. The drawing-up lever hangs down and lies in the path of the carriage, so that towards the finish of the inward run, when the cone clutch is in gear, an adjusting screw O on the square moves the lever K forward and through its connections N and P lifts the cone dish out of contact with the cone pulley T and so stops the scroll shaft. As the cone clutch must be kept out of contact during the outward run, special arrangements are provided to prevent K from returning to its original position when the carriage and the adjusting screw O move away from it on their outward run; the details of the action will, however, be treated subsequently. On the completion of the outward run a "change" occurs which relieves the lever K, and a strong spring attached to the lever P at O forces the cone dish R into contact with T and so causes the scroll shaft to revolve and draw up the carriage.

Although the drawing-up cone friction is apparently a simple arrangement, and one capable of performing its work perhaps better than other methods yet tried, it has inherent faults which necessitate extreme care in using and setting it. Its whole action depends upon the friction between the external and internal conical surfaces; one surface is covered with leather, and this must be of the very best quality, firmly and evenly fastened on the lower cone and turned in the lathe before applying it to the machine. In the form of the cones several points must be taken into consideration; the chief are—Diameter, inclination and breadth of the surfaces in contact. In regard to the diameter it is clear that this depends upon the principle of leverages, and the economical use of power; small cones require much more power, and as a consequence the extra power and strain leads to a greater tendency to slippage of the surfaces and their speedy destruction. Large diameters are therefore a necessity, and within the limits set by the work they perform it may be said that the larger the cones are the better. All makers try to keep them as large as possible, and though local circumstances and individual opinions may cause one maker to have the diameter slightly larger than another, the difference is not now so great as to give more than a superficial advantage. Formerly much trouble was caused through diameters being too small and more especially when this was associated with a very narrow width.

The width of the surfaces brought into contact is highly important. Although friction is said to be independent of surface, it must be considered that in the case of a cone friction the wedge action is really the vital principle, and as such the ordinary idea of friction must be set aside, because surface under such conditions plays a very important part; the greater the surfaces bound together for the time being, the greater the force that can be transferred through them without yielding. In consequence of this, a large area of contact is obtained by large diameters and wide surfaces, and the difference in work is only too easily seen by a comparison of the work of a modern mule and the old narrow frictions.

The question of the inclination given to the conical surfaces is an extremely delicate matter. In the short space of probably three-eighths of an inch, the two cones must be brought together so firmly as to revolve as one and to convey in this condition force sufficient to bring the carriage in, and also when apart from each other to be perfectly free without the slightest tendency to touch. Instantaneous action is indispensable, and this must be effected with a minimum strain on the parts controlling it. If the angle is not sufficient, the smallest fraction of wear or permanent compression in the leather will prevent the grip of the surfaces, and even if the adjustment of the levers allow of a grip being obtained, the difficulty of separating the two cones, when once wedged together in consequence of a too slight taper, is so great that the strain is sure to result in frequent and considerable damage both to the machine and the yarn. On the other hand, if the angle is too great, the wedge action loses its power and the grip is not sufficient to draw up the carriage without an amount of slippage which practically destroys the value of the yarn that is being spun. It will be seen, therefore, that a strong element of success in the working of the mule depends upon perfect conditions in the formation of the friction cone.

Now, although a mule may be set to work with a friction cone practically perfect, its usefulness may be partially destroyed by carelessness in the setting of the parts that put it in and take it out of gear. Leather wears and is affected by the weather, so that constant attention must be given to it, to see that it is performing its work properly; and as all mules contain adjusting points this ought to be an easy matter, if care is taken to attend to it.

For fine spinning, say from 120's to 300's, some makers have found it an advantage to dispense with the drawing-up friction cone, and in the accompanying sketch is represented an arrangement of a very effective character adopted by one firm of machinists, who are noted for their attention to this class of work. Fig. 33 shows the chief points of the motion. The backing-off shaft G, instead of being driven by band, as in the last example, Fig. 32, has two pulleys on the end of the shaft. The drawing-up motion is effected when the strap is on the loose pulley B; it drives the scroll shaft through the usual bevel wheels C D and E F, the bevel C, of course, being fastened to the pulley B. The backing-off is driven from the fast pulley A through H and J, when the wheel J is put into gear at the proper moment with the fast pulley K on the rim shaft. The carriage, as in the previous case, moves the strap on to the fast backing-off pulley A. It does this at the termination of the inward run, by the adjusting stud M coming in contact with the drawing-up lever N, and this lever's connection with the strap-fork lever R produces the change. Means are taken to keep the strap on the fast pulley A during the outward run (see description and illustrations of "long-lever" mule), so that at the right moment for backing-off it is instantly performed by H driving J when J has been put into gear with the fast pulley K on the rim shaft. Means are also adopted to adjust the amount of strap on the drawing-up pulley B both by a stop-rod and screw.

In this way the speed of the drawing-up can be regulated to

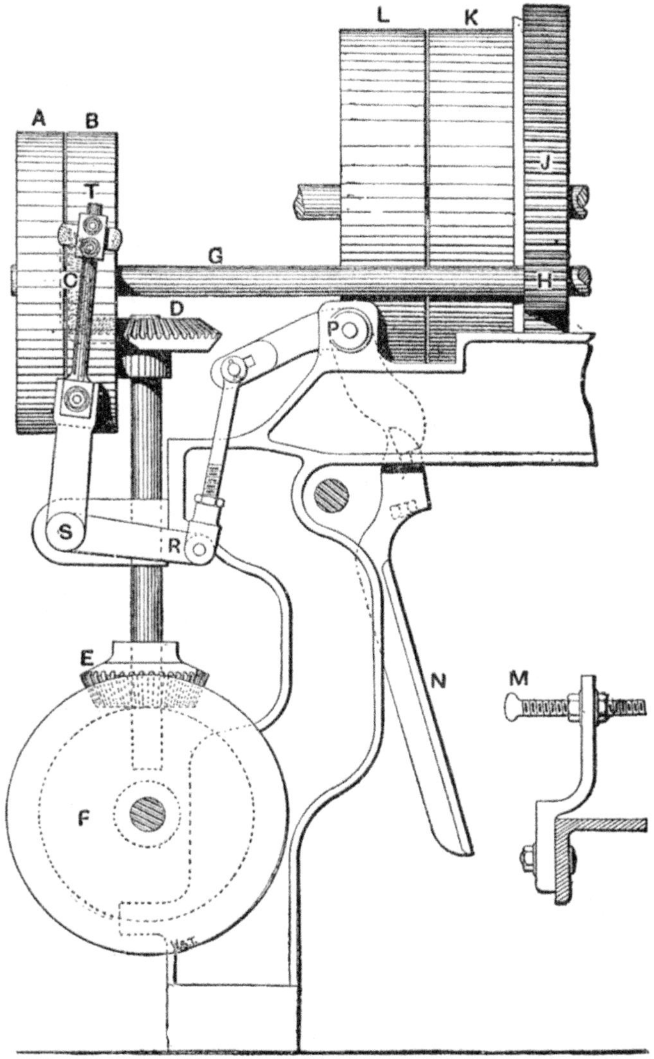

FIG. 33.

the amount considered necessary for the numbers being spun.

Changes on the "Cam Shaft" and the "Long-Lever" Mule.—It will be convenient at this point to describe how the various changes of action are produced, which give to the mule its characteristic motions. In doing this there will be an advantage in confining our attention to two well-known types of machines, known generally as the "cam-shaft mule" and the "long-lever mule." The first-named is so called because its actions depend chiefly upon certain important changes being brought about through the medium of cams; while the latter mule obtains similar effects almost directly through the regulated movements of a long lever. Both systems are good, and give excellent results for all classes of yarn, though there is a tendency in some quarters to consider the long-lever principle more applicable to fine spinning than to the production of coarse numbers. Such, however, is not the case; mules fitted up in either system give equally good results, whether for coarse or fine numbers. The application of the lever is becoming more general, on account of simplicity, easy adjustment, and certainty of action. The cam system of course also possesses these attributes, and it must be understood that it is only in a comparative sense they are indicated here, but the fact that the best fine-spinning mules are almost always built on the long-lever system shows that its advantages are fully recognised.

Cam-Shaft Mule.—As the cam-shaft mule is the one most generally known, this will be described first, and numerous illustrations will be used to illustrate the several features described. Fig. 34 presents a general view of the cam shaft as usually applied. In order to fully convey the idea of its working, a little recapitulation of what has already been said becomes necessary. The driving of the machine takes place through the pulleys on the rim shaft A.

The backing-off cone wheel C is driven continuously, either

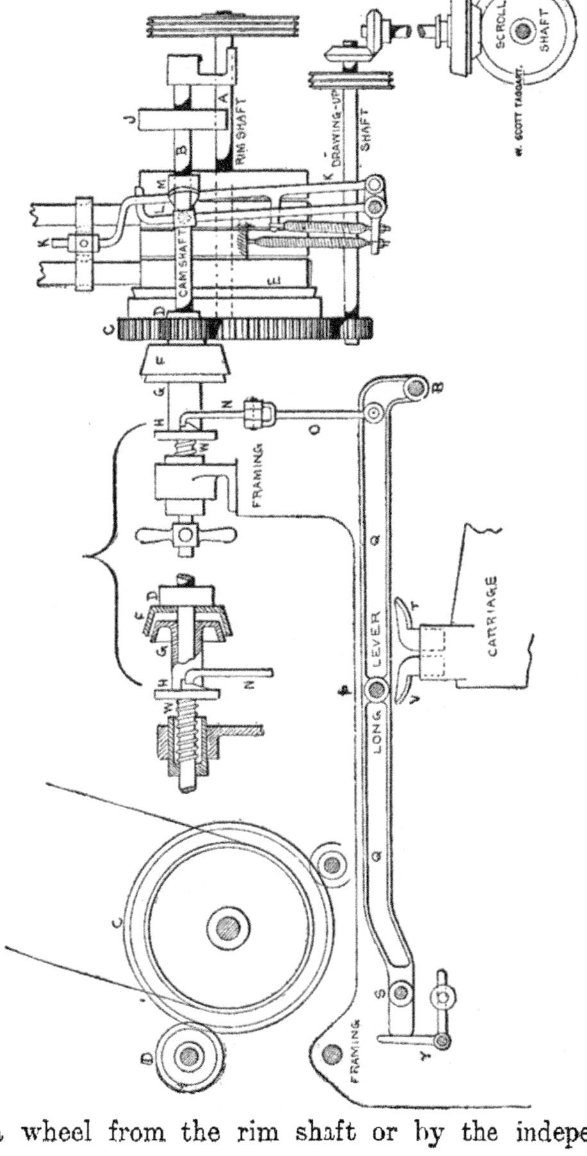

FIG. 34.

FIG. 35

by a wheel from the rim shaft or by the independent

system of driving by band or belt, as shown in the sketch. The front roller is driven from the rim shaft, and from the front roller the back shaft receives its motion.

When the carriage commences its outward run, the strap is on the fast pulley, driving the spindles, the rollers, and the carriage. On arriving at its outermost position, changes must be effected which will stop all these actions, and it is through the medium of the cam shaft that the necessary "changes" are produced. These changes may be summarised as follows:—The carriage must be brought to rest; the spindles must be stopped; backing-off must take place; the front rollers must cease to deliver the roving; and the back shaft must be disconnected from the front roller so as to permit the scroll shaft to bring the carriage in.

On reference to Fig. 34, the cam shaft B is shown alongside and parallel to the rim shaft. A wheel thereon, D, gears into the backing-off cone wheel C, and as C is always revolving, D, which rides loose on the cam shaft, will do the same. On one side of D is cast an internal cone dish F, into which can be made to fit a conical clutch G; G is made to slide on the cam shaft by means of a float key, and it is kept out of gear with F by a lever N pressing against it. So long as the cone clutch is not put into gear, the wheels D and F run loose on the shaft, but when, by the removal of the lever N, the spring at W forces G into contact with F, the cam shaft revolves and the desired changes can then take place. By the help of the drawing this action can be closely examined. A long lever on the inside of the headstock is centred at P; at each end are fitted pins R and S; on the carriage square an arrangement is made for carrying two inclines V, T, and by the motion of the carriage these inclines come respectively into contact with the pins S and R, and depress that end of the lever acted upon. The movement of the long lever raises or lowers the link O, which in its turn actuates the lever N, and in N we have the controlling movement, which puts in or takes out of gear the cone clutches G, F. In the drawing the cone clutches are out of gear, but directly they are brought into contact, the cam shaft revolves. On the cam shaft are placed several cams, which effect the necessary change; these are shown at M, J, and one on the back of the cone clutch at G.

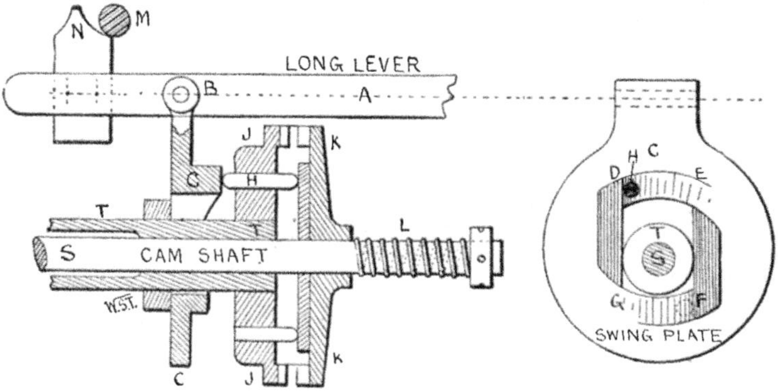

FIG. 36.

Their several actions will be considered in detail, as well as the special construction of the cam plate H, which controls the working of the cone clutch. A small end view in Fig. 35 of the two wheels C and D is given, which shows their relative positions to each other.

Although the cone-clutch arrangement on the cam shaft is the one generally adopted, there are other makes of mule in which clutch wheels are employed in preference to the frictional grip. One of the best known is illustrated in Fig. 36, where a partial end view is also shown. In this case the cam shaft is placed below the long lever; at each end of this long lever A is fastened an inclined bracket N. The carriage carries a bowl M, so disposed that it comes into contact with the inclined bracket and depresses that end of the lever. The movement of the lever A so produced lowers a specially constructed pendant plate C in such a way as to relieve the pressure of the spring at L, so that the two clutch wheels J, K are at once brought into contact. A view of the swing plate C is given in order to make it clear how this action is produced; but first it must be understood that the cam shaft S is continually revolving through a wheel thereon being in gear with the backing-off cone wheel. The revolution of the cam shaft S carries round the half clutch wheel K, which is connected to the shaft by a float key; the other half of the clutch wheel, J, is keyed to a loose shell T, which practically covers in a large part of the cam shaft. On the loose shell are fitted the various cams for producing the changes. These can only be driven when the two half clutches J and K are brought into gear. This is effected, as already stated, by the lowering of the swing plate C, in the following manner. The plate is arranged to fit loosely on the shaft, and also is made capable of rising and falling. On one of its faces, inclines are arranged directly opposite to each other, and these inclines, in a similar manner to those described in the previous example, serve the purpose of keeping K out of contact with J. A pin H, passing through the body of J, connects the clutch K and the plate C, and as long as the pin is on the highest point of the incline at D the two clutches remain out of gear. In the position shown in the sketch the long lever is on the point of being depressed; as it falls the plate C will be lowered, and will move out of the way of the pin H. Directly the pin is free from the incline the spring L, which is in compression, at once forces K into contact with J, and the shell T immediately commences to revolve. During this revolution the pin H is carried round by J, and is brought into the path of the incline G F, opposite to the one from which it has just been fixed; as it travels up the incline it forces K out of gear with J, and the cam shell instantly stops. Only half a revolution has thus been given to the cam shell, this being sufficient to produce the necessary changes. When the carriage finishes its run-in, the other end of the lever is depressed, with the effect that the swing plate C is lifted up, and the pin H is relieved from the high point of the cam at F, and therefore permits

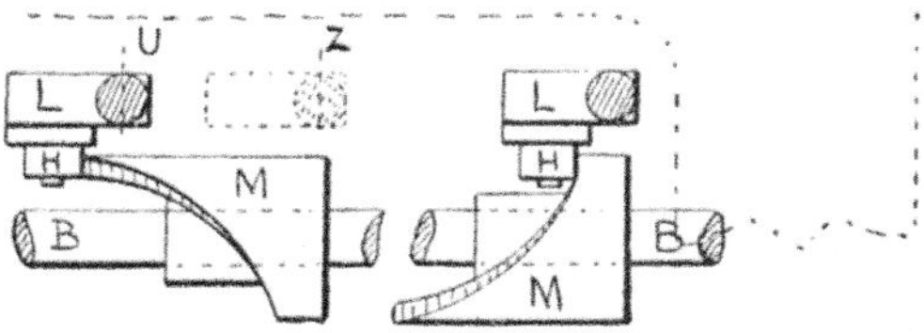

FIG. 37.

the clutch to gear again and perform another half revolution of the cam shell T.

Movement of the Strap-Fork.—The movement of the strap-fork from one pulley to another is effected through the cam M on the cam shaft. Its general position was shown in Fig. 34. Here a small plan view is given, in Fig. 37. Two positions are represented, showing the effect of the half revolution of the cam in moving the strap-fork from Z to U. Since the cam M is not double-acting, the strap-fork returns from U to Z by means of springs. By reference to Figs. 34 and 38 it will be seen that although L is referred to as the strap-fork, it is not so in reality, but is simply a kind of buffer or relieving rod between the cam

M and the strap-fork K itself. This feature is sufficiently important to warrant a more detailed description, which will now be given.

From the general plan of the cam-shaft arrangement we can proceed to consider it more in detail. The first feature to attract attention is the method of moving the strap from the fast to the loose pulley. Although the cam M is nominally spoken of as performing this function, it does not do so in reality; its chief duty is to move the strap from the loose to the fast pulley, and when the opposite effect is necessary the cam simply moves into a position which allows another action to bring about the change.

Fig. 38 shows sufficient of the parts to make the description clear. It will be seen that the cam M actuates the strap-fork lever K, not in a direct manner, but through another lever L, upon which the bowl H is fastened. This lever is centred on a short fixed shaft at Y, and its lower end is extended at Z in the manner shown, for the attachment of strong springs *a* and *b*. The strap-fork itself is centred at *k*, and simply works free in this position; it has a projection about the middle of its length, to which a spring *a* is connected, and by means of this spring the two levers L and K are kept together. For instance, if the cam M moves the lever L in the direction of the fast pulley, the lower end of it at Z will be depressed, and will exert a strong pull on the spring *a*, which will be sufficient to draw forward the strap-fork lever K. This depression of the lower part of the lever L will also bring into tension a spring *b*, which is attached to the framing of the machine, so that as long as the strap is on the fast pulley the spring *b* is in tension and always exerting its power to force the strap on to the loose pulley. It is prevented from doing so by reason of the lever L being locked in position by the twist lever Q, which is attached to it. This lever serves the important purpose of regulating the time at which the strap can be moved; the cam M may make its half revolu-

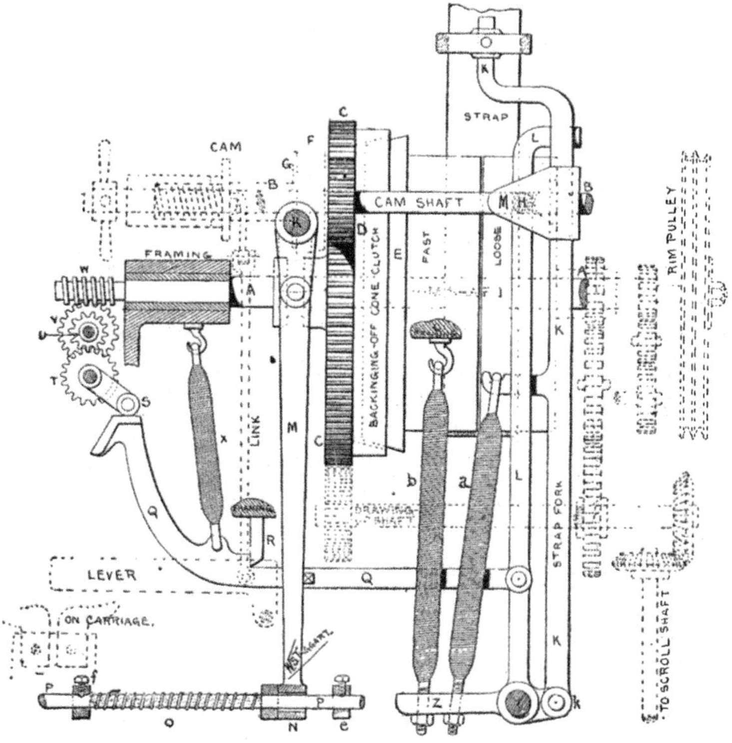

FIG. 38.

tion, but the strap-fork will not move on that account; the movement can only take place when the twist lever Q is relieved. The action of this lever is as follows:—One end of it is attached to L; at a certain part of its length a projection on it abuts against a stop R, which is fixed to the framing; the other end is brought into such a position that it can be acted upon by a lever S, which revolves by virtue of its connection to a wheel T. This wheel is driven from a worm W on the end of the rim shaft, through the worm wheel V and the pinion U. Now, since the relief of the twist lever is seen from this arrangement to be controlled from the rim shaft, it will readily be understood that the number of revolutions of the spindles is the dominant factor in deciding when the strap must be changed. It has already been shown that the movement of the carriage at the termination of its outward run brings about half a revolution of the cam M, and so leaves a clear space for the pin H to return and carry the strap on to the loose pulley; but unless the spindles have received a sufficient number of turns during the run-out, it is not necessary that the two actions should be simultaneous; it is frequently advantageous to continue to turn the spindles after the carriage has stopped—an action termed "**twisting at the head,**" and for this purpose the twist lever can be employed. This action is rendered very simple by the arrangement of the wheels; by changing the pinion U, the revolutions of T, and consequently the lever S, can be adjusted to cause the strap-fork to move on to the loose pulley simultaneously with the half revolution of the cam M, or to follow it at whatever interval may be considered necessary. During the revolution of S it comes into contact with the end of the lever Q and depresses it; this lowers it sufficiently to unlock it from the catch R, and when this happens the spring B pulls the strap-fork over on to the loose pulley. At the same time the spring X, attached to Q and also to the framing, is put into tension, so that when the cam M makes the next half revolution on the completion of the run-in of the carriage, the movement of the lever L pushes Q forward, and the spring draws it upwards and locks it again at R.

Of course it is often unnecessary to have "twisting at the head," and therefore an arrangement such as that described can be accurately adjusted to give that result, or it may be dispensed with altogether, and the twist be regulated from the gearing which drives the front roller. If the arrangement in Fig. 38 is absent, the only effect will be to cause the strap-fork to move on to the loose pulley at the same time as the cam M makes its half revolution; this is generally spoken of as "striking through"; but when a definite number of twists per inch are required, which it is not considered advisable or possible to put in while the carriage runs out, then this arrangement supplies the deficiency between the time the carriage stops and the backing-off takes place. It may be remarked that the change of the strap on to the loose pulley is not confined to the method shown, and frequently an independent system, called the "strap-relieving motion," is employed, which will be described presently.

Driving of the Cam Shaft.—The next point to be noticed on the cam shaft is the action of the long lever in bringing about the engagement and disengagement of the cone clutch on the cam shaft. One method has already been described in detail; the one now under notice refers to that illustrated in Fig. 34. Two other sketches are now given, Figs. 39 and 40, in explanation of the point. Therein B is the cam shaft in section, and H is the plate on whose surface are formed two inclines, E F and G J. The inclines are at different distances from the centre of the shaft, and each one at its highest point, E and G, falls abruptly, while the other ends, F and J, fall gradually to the level surface of H. The lever N, centred at A, is connected to the long lever by the link O; as the long lever is actuated by the inclines on the carriage, N will oscillate and be alternately brought into the paths of the inclines. Two positions of the cam plate are represented; the one in Fig. 39 shows the lever N on the highest point of the inner incline, in which position the cone clutch is disengaged and the spring W (Fig. 34) is in compression. When the end of the long lever opposite to O is depressed, O will be raised,

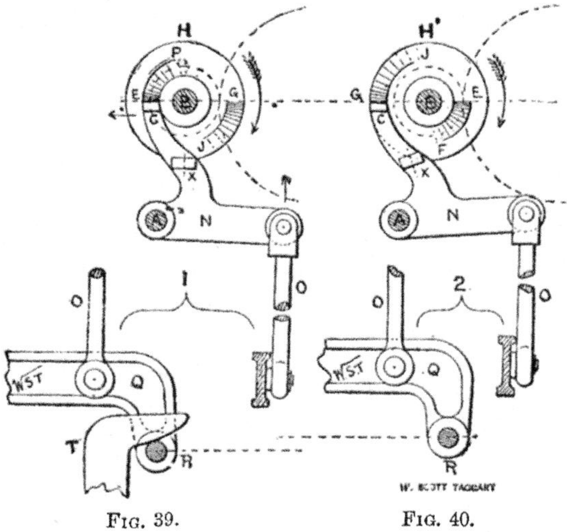

FIG. 39. FIG. 40.

and this movement takes the end C of the lever N away from the incline E F, so that the spring is free to force the cam plate forward and thus cause the two halves of the cone clutch to engage with each other and bring about the revolution of B, and consequently of the cam plate itself. Now it will be observed that, as H is pushed forward by the spring W, the surface of the cam plate is brought almost into contact with the end C of the lever, and therefore as the plate revolves it lies directly in the path of the opposite incline G J, which in passing tends to force the lever N on one side. In this effort, however, the lever remains rigid, the spring W at the back of H yields instead, and the cam plate is thus forced back and so disengages the cone clutch to which it is attached. This naturally stops the motion of the cam shaft after it has made half a revolution, and it remains stopped until the end of the long lever depresses O and moves the lever N from the highest point of the outer incline at G, and places it in a position nearer the centre to be acted upon by the inner incline during the next half revolution of the cam shaft.

It will be seen that the action just described is one of the utmost importance in the operations of the mule. Absolute accuracy must be sought for in the adjustment of the levers so as to obtain the greatest effect from the inclines. The precise moment for putting the clutch in and out of gear depends on a number of points that require careful attention, most of which are associated with the cam plate and levers. The long lever itself must be acted upon at the right moment, and of course adjusting brackets are always provided to effect this. The surfaces of the inclines are subject to wear, and although they are invariably case-hardened, still it is not unusual to find sufficient wear taking place to necessitate care in attending to them. They are likewise made separate so as to be easily replaced when required; and the same remarks can be applied to the end C of the lever N, upon which a great strain is thrown. The spring at the back of H must be strong enough to cause a thorough grip in the cone clutch, and attention must be paid to the two halves of the clutch to see that they are working correctly, and adjustment be made for any wear and tear that occurs.

Attention may be now paid to the other cams on the

shaft B. These are illustrated in Figs. 41 and 42, which represent the action of the cams in operating the front roller and the back shaft respectively. The front-roller cam G, Fig. 41, forms part of the back of the cone clutch on the cam shaft; it is double-grooved, so that the clutch wheel at F can be put into and out of gear without springs. A lever D, centred at A, works in the groove, and its other end fits in a ring groove on the half clutch E. In the position shown the carriage is performing its outward run and is spinning; therefore the clutch is in gear and the rollers are revolving. When the carriage stops, the cam shaft turns half a revolution, which brings the lever D from the highest point of the cam down to the lowest point directly opposite. This effects a separation of the clutch at F, and the rollers cease revolving, remaining stationary until the run-in is completed, during which winding is going on. (It may be remarked that this latter statement, though correct so far as the cam G and clutch F are concerned, has exceptions.) For certain classes of yarn of good quality and generally fine numbers, the rollers are subject to two other independent motions, namely, "winding delivery motion" and "jacking delivery motion"; these will be fully described subsequently.

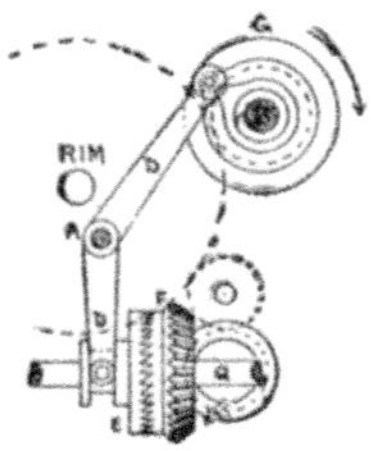

FIG. 41.

Outward Run of the Carriage.—The next cam to be noticed, J, serves the purpose of independently disconnecting the back shaft, and so stopping the carriage at the finish of the outward run.

As it performs other functions besides this one, the accompanying drawing, Fig. 42, taken from a recent machine, has been prepared, from which it will be seen that the putting of the back-shaft clutch box into and out of gear is not the only action it performs. To understand

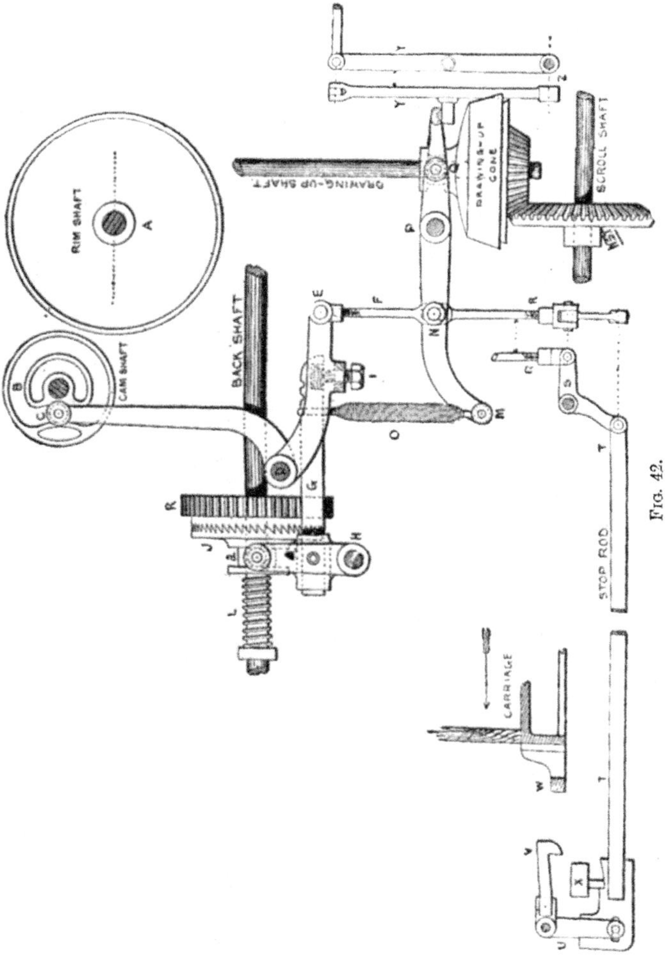

FIG. 42.

the various movements, great care must be taken to follow the description, and as it would be both difficult and inconvenient to show the different parts in their changing positions, the reader must picture to himself the clutches being put into and out of gear and the lifting and lowering of the levers.

In the drawing, Fig. 42, A is the rim shaft, from which the cam B is revolved half a revolution, just previous to or on the termination of the outward and inward runs of the carriage. Working in the groove of the cam is a bowl C, carried by one end of a bell-crank lever centred at D; the other end of this lever, at E, carries a link, whose lower end is slotted and slides on a pin fixed in the lever whose centre is at P. A projection on the bell-crank lever at V carries a set screw, capable of adjustment, which bears against the horizontal arm of another lever pivoted at H. The vertical arm of this lever carries a bowl A, which works in the groove on the back of the clutch wheel J; the drawing represents the clutch in gear, with the bowl C on the lower portion of the cam B. Confining ourselves for the moment to this action, it will readily be seen that the next half revolution of the cam will, acting through the stop-screw at I, lift the horizontal arm of the lever G and put the clutch J out of gear, thus stopping the back shaft and consequently the carriage. The spring at L serves the purpose, as already observed, of keeping the two halves of the clutch in gear during the outward run of the carriage.

Drawing-up.—We may now trace the action of B still further, for it performs the important functions of directly taking the drawing-up cone clutch out of gear and indirectly of putting it into gear; this it does in the following manner:—A lever centred at P has on one end Q a forked jaw, which fits in the ringed groove of the upper half of the drawing-up cone clutch; its other end M is connected by a strong spring O to the horizontal arm of the lever G. Also in connection with the lever P is the slotted link F, which is pendant from E; a direct connection is therefore obtained between the drawing-up cone and the cam B. As the cam B revolves, the end of the lever at E is raised; this action lifts the link F, but this has no effect on the lever at P, because of the slot at its lower end. At the same time the lever G is also raised, which puts a strong tension on the spring O, this naturally exerts a powerful tendency to pull the end of the lever at M in an upward direction; it cannot, however, effect this purpose at once, because the other end of the lever P is locked by the lever Y, and it is only at the moment of the finish of the run-out that the action of the carriage draws the lever Y on one side, and permits the tension in the spring O to lift M upwards and force the end Q downwards, thus putting the cone clutch into gear. Directly this happens the bevel on the scroll shaft is driven, and the carriage is drawn in. Now it will be noticed that the fact of lifting up the lever at M brings the pin at N to the top of the slot in the lower part of the link F; therefore, as the cam B makes half its revolution, when the carriage is on the point of finishing its inward run, the end of the lever at E is depressed, and the link F presses downwards on the pin at N and lifts up the other end of the lever at Q, thus taking the clutch out of gear and stopping the scroll shaft. At the same time the lowering of the end M puts tension on the spring O, which, together with the spring at L, forces the clutch-box at J into gear and enables the outward run to be made.

Locking Arrangements.—It will be readily understood that, although the actions just described are simple in character and are obtained by means free of complication, they are so highly important and depend on such a delicate adjustment, both in regard to the time of their action as well as the extent of their movement, that means must be taken to ensure accuracy and prevent any derangement happening through wear or accident, to the machine. Those practically acquainted with machinery will understand this fully, and in the mule the precaution of what is called "locking" each motion to guard against irregularities is an absolute necessity. We have already seen that the lever Y locks the lever P in position until the carriage itself moves it away and allows the cone clutch to drop into gear. (This feature will be more fully described presently.) There is, in addition, an arrangement at the front of the headstock by which an effective control of the machine may be obtained, which serves the purpose of locking the carriage at the termination of its outward run and during the period of backing-off. An enlarged drawing of it is given in Fig. 43. The rod T is connected by the bell-crank lever S and the link R to the lever P; any movement of P will therefore move the rod T. Now by locking T in any position it is possible to keep P from moving until the rod T is relieved. The locking is performed in the following manner:—A lever X, centred on a stud, is capable of being moved upward by the carriage as it is finishing its outward run. The rod T at this moment is out as far as it will move, and a projection on the lever X rests in the recessed part of the end piece F of the rod T, and keeps the rod from moving inward. It will be noticed on reference to Fig. 42 that until the rod T is relieved the cone clutch cannot fall into gear, so there have been two agencies at work during the outward run to prevent the scrolls being driven from the drawing-up cone clutch, namely: the lever Y and the stop-rod T. Directly, however, the carriage comes against the lever X, the projection at E is lifted up and the rod T is free to move when Y is released. Just previous to this, however, the carriage in moving outward has lifted up the catch V, and a projection W on the carriage passing under it allows V to fall and thus locks the carriage. For about the next two seconds the carriage is stationary, and backing-off takes place; when this action is completed the lever Y is moved aside and the cone clutch is freed from

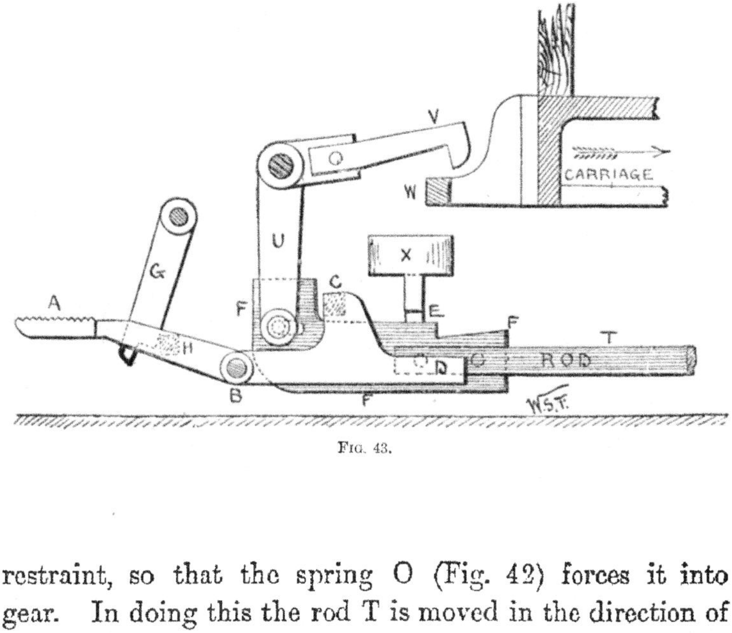

FIG. 43.

restraint, so that the spring O (Fig. 42) forces it into gear. In doing this the rod T is moved in the direction of the carriage, as seen in Fig. 43, and at the same time the lever V, through its connection with the rod, is lifted up and thus releases the carriage, which at once commences its inward run.

A lever A, centred at B, is devised to enable a stoppage of the carriage to be made in any position. The drawing shows the position of the parts as the carriage is on the point of running in; by pressing on A the projection at C will force the stop-rod T outwards, and lift the upper part of the cone clutch at Q out of gear, which at once stops the carriage. The lever A itself can be locked by the catch G when required. On the other hand, when the projection E is engaged in the recess at F, it can be raised out of contact by the projection D of the lever A, thereby relieving the rod any time during the outward run.

Backing-off.—"Backing-off" is the next feature calling for attention. At this point, however, only a description of the means adopted for obtaining it will be given. The reason and the effect will be dealt with fully when the spindle and cop are treated; the general idea of the action, already given, will meanwhile be sufficient to enable the reader to understand the following remarks:—

In Fig. 44 a full view of the arrangement is shown. The chief object in view, it will be remembered, is to put into and take out of gear the large backing-off cone wheel with the fast pulley, in order to reverse the direction of revolution of the rim shaft, and consequently of the spindle. The duration of the movement is scarcely more than a fraction of a second, but its importance necessitates extreme accuracy and promptness of action. The cone wheel contains a ring groove, in which works a forked lever, centred at I. This fork is connected to a lever H, whose lower end fits loosely on a rod that runs along the side of the headstock. This backing-off rod has one end E coupled to a bell-crank lever, pivoted at O, while the other end is joined to the upper part of the lever G, whose function it is to lock the drawing-up cone in position during the "backing-off." As the carriage moves out, the fast pulley and cone wheel are out of gear, as is also the drawing-up cone clutch. On the termination of the outward run

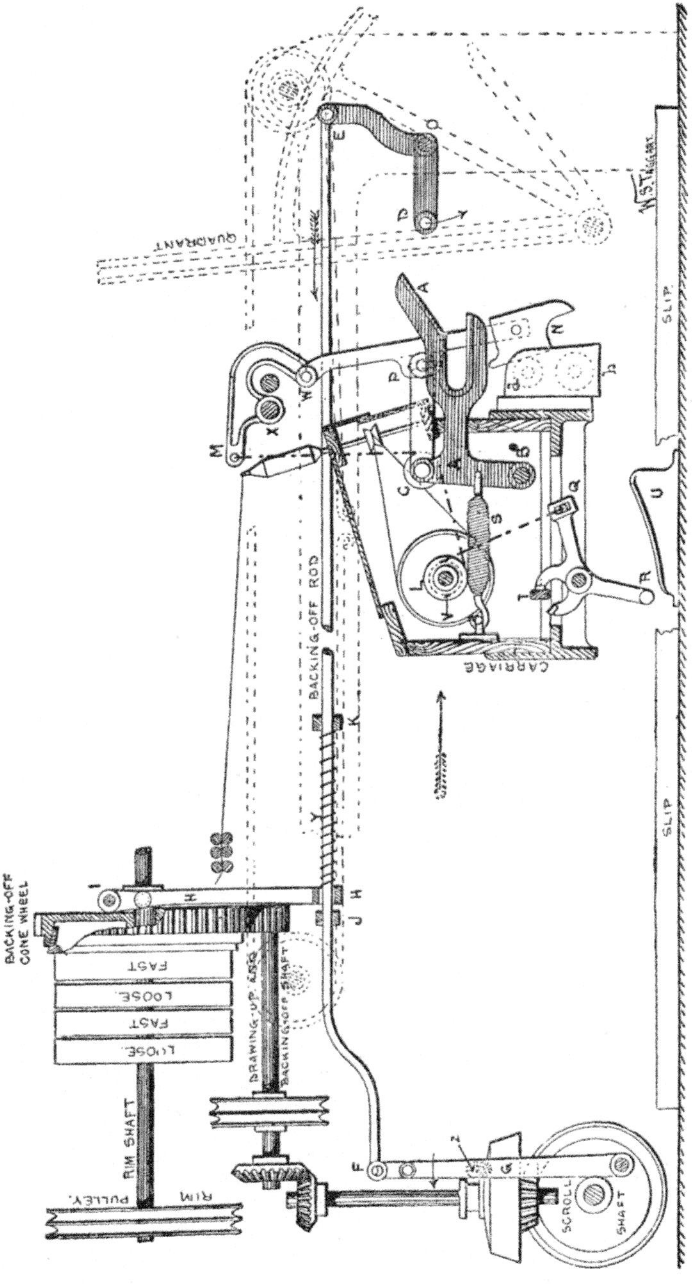
QUADRANT
BACKING-OFF ROD
CARRIAGE
SLIP
SLIP
BACKING-OFF CONE WHEEL
FAST
LOOSE
FAST
LOOSE
DRAWING-UP SHAFT
BACKING-OFF SHAFT
RIM SHAFT
RIM PULLEY.
SCROLL SHAFT

Fig. 44.

each of these must be put into gear—the first one the moment the carriage stops, and the second one on completion of backing-off. In the carriage square is pivoted at B a specially shaped jaw lever, the mouth of the jaw having an inclined projection. As the carriage approaches the finish of the stretch, the inclined portion comes into contact with a bowl D carried by the bell-crank lever, so that this end of the lever is depressed and the backing-off rod is moved in the direction shown by the arrow. The effect of this action is to move the fixed stop-washer K forward, and compress the spring Y, which bears against the lower end of the vertical lever at H; the compression of the spring exerts sufficient pressure to force the lever H forward and to bring about the necessary contact of the backing-off cone wheel and the fast pulley. At the same time the end of the backing-off rod moves the lever G outwards, and places the pin underneath the lever Z, thus preventing the upper cone clutch from falling into gear. Other actions have come into play as this occurs (as already described), by means of which the strap is moved on to the loose pulley, so that directly the cone wheel has turned through a portion of a revolution, it must be taken out of gear instantly, the moment before the scroll shaft commences to draw the carriage in. It is obvious that this necessary and rapid movement of release cannot be performed by allowing the carriage to move until the bell-crank lever is free; accordingly other actions are introduced to effect it. It will be sufficient at this point merely to indicate, rather than describe, the means adopted, as a fuller description will be given later.

On the faller X is fastened a lever, one end W of which is connected to a pendant arm, whose lower end N slides on a bowl *a*, carried by a slide, which is moved up and down by the shaper as the carriage moves in and out. As the carriage finishes its outward run, the position of the copping faller arm is approximately that shown in the sketch. Directly, however, the tin roller reverses its direction of revolution for backing-off, a small scroll (or "snail" as it is called) L winds on a chain, which passes under a bowl C on the lever A and on to the faller lever at M; this end of the lever will consequently be depressed, and so draw up its other end W, and along with it the locking faller arm N. It will lift this latter so high that the recess at N will come opposite the bowl *a*, and its natural tendency would be to fall forward and rest there. This is, indeed, what actually occurs, but to render this a definite action the lever A is connected to the locking arm by a link P, and, in addition, the lever A itself is connected by a strong spring S with the opposite side of the square. As long as the locking arm N simply rests against the bowl *a*, the lever A will remain fixed in spite of the strong pull of the spring S; but immediately the tin drum, through the snail L, draws M downwards and the locking arm N upwards, N becomes free from the bowl *a*, and the spring S draws forward with a quick action the lever A, together with the link P and the faller locking arm N. This movement of A is the one that releases the bowl D from the jaw; for as A is suddenly shot backwards, the jaw itself lifts up and carries D with it. The upward movement of D draws the backing-off rod forward, and in doing so the stop-washer J is brought against the lever H, and pulls the backing-off cone wheel out of gear with the fast pulley. At the same time the lever G is also moved forward, and its projecting pin is brought from under the end of the cone-clutch lever, and permits it to fall into gear, whereupon the carriage at once commences its inward run.

In Fig. 38 a drawing is given showing how the movement of the strap from the fast to the loose pulley is regulated from the revolution of the rim shaft, by means of the twist wheel. The strap can only be moved after the twist wheel has made a given number of revolutions, and by relieving the twist lever allowing the strap-fork to be pulled over by a powerful spring.

Strap-relieving Motion.—It is not always necessary to adopt this method of regulating the twist in the mule, and frequently, instead of employing it, a strap-relieving motion is used. An arrangement of this kind is shown in the drawing, Fig. 45. A few words as to the reason of its introduction are necessary before giving the description of its action. The application of a twist-wheel motion, as will be remembered, enables a very definite number of twists to be put into the yarn as the carriage runs out. If enough twists have not been put in by the time the carriage has finished the stretch, then, although the strap-fork cam has made its half revolution, the strap-fork cannot change until the twist wheel relieves it, and until this occurs the spindles will continue to be driven and so put extra twist into the yarn. Technically this is called "twisting at the head"; but it will be observed that when the twist is sufficient, their revolution must be instantly stopped. Such an action entails an enormous, though momentary, effort of those parts of the machine which perform it. There is an excessive amount of friction set up in the cone clutch, and difficulty is experienced in moving the strap quickly on to the loose pulley. It is, therefore, found that for some classes of yarn, lower numbers especially, and in many cases according to the opinion or experience of the spinner, the advantages of the twist lever are not sufficient to outweigh the advantages of a strap-relieving motion. In the first place, therefore, this motion displaces the twist lever. In its stead we have the arrangement shown in Fig. 45. The carriage is moving outwards, and, when within a few inches of its finish, comes into contact, through an adjustable stop A, with an inclined lever B, pivoted at C. The lever B carries a stud D, which fits a recessed part of E, so that the depression of B in the direction of the arrow draws the strap-relieving rod forward. This rod is attached at F to a pendant lever J, centred at K. On the stud or short

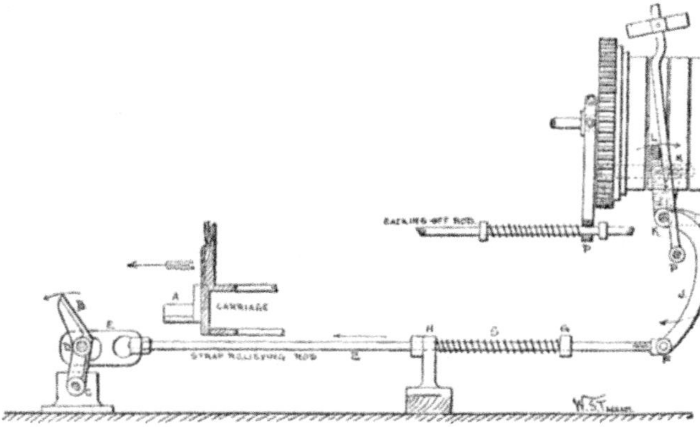

Fig. 45.

shaft K is fixed a lever L, whose upper end bears against the strap-fork rod. It will, therefore, be seen that the forward movement of the rod E will move the strap from the fast to the loose pulley, and it will do this gradually as the carriage is finishing the last four to eight inches of its outward run. By the time the carriage is at rest, and backing-off commences, the spindles have therefore lost a good proportion of their speed, and a great saving of power is effected, with its consequent reduction in strain, etc., in bringing them to rest and reversing them for backing-off. It must not be overlooked, however, that there may be a "slight" loss of twist through slowing the spindles; for although the rollers are affected in their speed in the same degree, the two are not driven in the same way, and it is possible for the proportion between them to be slightly disturbed. This cannot be regarded, however, as a disadvantage, because it has no practical value, especially in regard to the class of yarn it is used for.

As the rod E is moved forward, a spring S, threaded upon it, is compressed by the stop-washer G pressing it against the fixed bracket H. In addition, a spring O is put into tension at the same time, so that when the carriage commences its inward run the rod gradually returns to its original position, and leaves the strap-fork on the loose pulley until the cam changes it at the completion of the inward run.

A further point to observe is, that the backing-off lever P is locked by this motion, in a similar way to that adopted in the twist lever. In order to show this clearly, two detached and enlarged views are given in Figs. 46 and 47.

In Fig. 46 the arrangement is shown in position for the strap on the fast pulley and the backing-off out of gear. The strap-relieving motion keeps the backing-off lever locked by connecting a short lever M to the shaft K, and in turn connecting M to a link R; a projection T on R bears against the backing-off lever P, and until this projection is removed backing-off cannot take place. As the carriage acts upon the strap-relieving motion, the link R is drawn on one side (as shown in Fig. 47) and P is left free to put the backing-off cone clutch into gear with the fast pulley.

It will be seen from the drawing that adjustments can be made in several positions, and these are necessary. Sometimes it is only desired to begin moving the strap 4 inches before finishing the stretch, while in some cases a gradual movement through as much as 10 inches can be obtained. Means for obtaining this range of action are therefore provided; but when adjustments are made, an important point to be careful about is to see that the strap is clear of the fast pulley just as the carriage finishes its run-out.

Object of Backing-off.—Before proceeding further it will be advisable to give a general idea of what is meant by "backing-off," in order to explain certain irregularities which this action causes, and the mechanical methods adopted to compensate for them.

The spinning operation during the run-out of the carriage

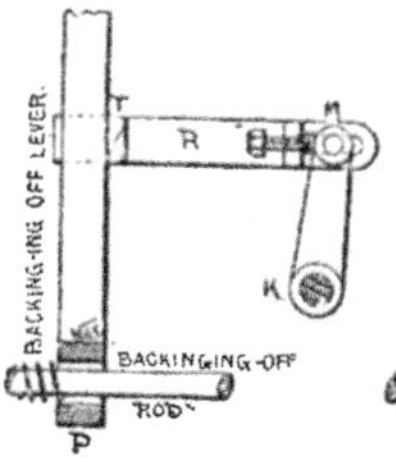

FIG. 46.

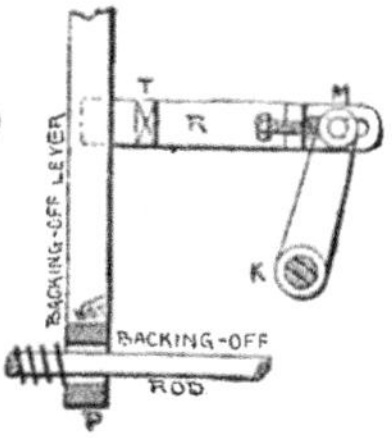

FIG. 47.

has already been fully explained and illustrated. Now one direct effect of this method of obtaining twist is that the yarn must be taken from that portion of the spindle on which the cop is being formed, and raised to the point: and, *vice versâ*, when spinning is completed, the yarn must be taken from the point of the spindle and guided on the cop in whatever part of the blade it happens to be. To make this quite clear, two diagrams are given, Figs. 48 and 49. Spinning is supposed to be taking place, as shown by the full lines. When winding takes place, the yarn must be taken from M and wound on the cop below N, and when that operation is completed it must be returned to the point of the spindle. Examination of the diagrams, Figs. 50 and

51, will very clearly show what the effects of these two operations are. In the first place, a wire C^1 running the full length of the mule is provided, and over this the yarn is guided on to the spindle during the winding process. It is carried by an arm centred on a shaft A, called the "copping" faller; this faller rod is actuated by levers from the shaping or copping mechanism, and by this means the wire C^1 guides the yarn on to the upper portion of the shaded part of the cop, and in doing so moves through the space between C and C^1. When the carriage arrives "in" against the rollers, the yarn must be transferred from the point C^1 to the point of the spindle, as in Fig. 50, and in effecting this we are brought into contact with one of the most interesting and characteristic features of the mule.

If the yarn were led on to the cop direct from the rollers, it is clear that the act of lifting it from C^1 to the top of the spindle would cause the whole of the ends to break, because of the longer length of yarn required in this latter position. And again, we saw when treating of the spinning process that the peculiar action of twisting in the mule necessitates a certain number of windings of the yarn round the spindle up to the point before spinning can commence; and this condition could not be fulfilled if the thread were guided direct on to the cop. To obtain each of these necessary elements, another wire at D is provided, carried by an arm working from a shaft B called the "counter" faller. Over the wire D the yarn passes on to the wire C; D is kept in such a position that the length of yarn from the cop to the rollers as it passes over the two wires is much more than the straight line between them, and consequently as the carriage gets in and before the spindles cease turning, the wire C rises up, and in doing so the spindle winds on the extra length in a series of turns, as seen in

Fig. 51. At the same time the wire D is lowered out of contact with the yarn, and the thread is free to be twisted as the carriage goes out.

When the outward run is complete, these extra turns on the spindle must be unwound before the winding can take place, and as they have been wound on in the same direction as the twist, it is evident that the spindles must be reversed

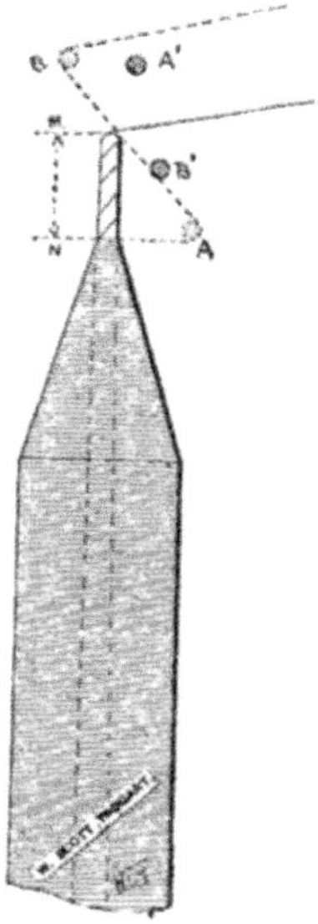

FIG. 48.

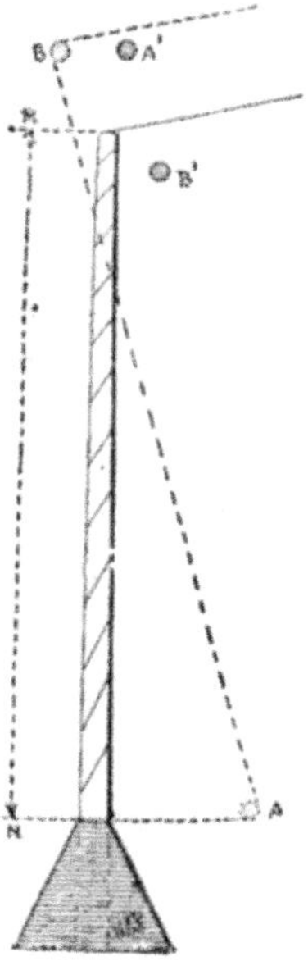

FIG. 49.

to unwind them; and also, since the unwinding means an additional length of yarn, something must be done to take up the extra length taken off the spindle.

The reversing of the spindles, in order to unwind the yarn from the bare portion of the blade between the cop and the point, is the special function of the "backing-off" process, already described. The action of the wires in compensating for the extra length unwound will be described subsequently, the object at present being merely to point out the necessity of backing-off and how it is effected.

Tightening the Backing-off Chain.—When the cop is in the early stages of its formation, the length of the bare spindle is considerable, and a good length of yarn requires to be unwound, as will be seen on reference to Fig. 49. As the cop gets larger, and gradually fills the spindle, the amount to be unwound comes less, until, at the finish, it is quite a small amount, Fig. 48. As we shall see presently,

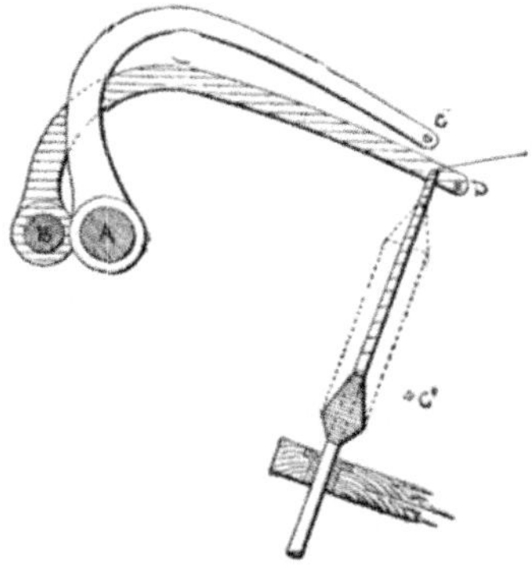

FIG. 50.

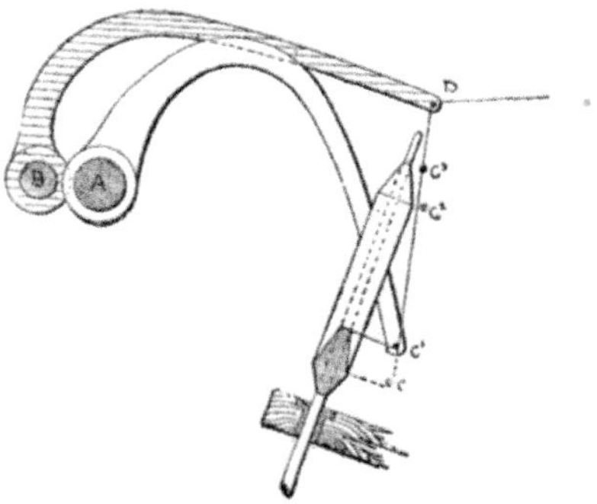

FIG. 51

the diminished revolutions of the spindles on reversal, which this necessitates, is very easily effected; but another point arises which requires a very careful consideration. The mechanism which causes the "copping"-faller wire to move from A^1 to A and the "counter"-faller wire to move from B^1 to B acts quickly, and therefore there is a danger that the downward motion of A will be much quicker than the rate at which the spindles unwind the yarn from the point to the cop. For this reason the movement of the copping-faller wire is, as it were, delayed, until the spindles commence to reverse, and by this means the likelihood of breakage is avoided; and if any slight slackness in the yarn results, the counter-faller wire has time to compensate for it. As the cop enlarges, however, the delay in the movement of the copping wire, as the spindles reverse, becomes a disadvantage; for there is less chance, owing to the shorter length of yarn to be unwound, of the wire overtaking the yarn, and therefore there is less necessity for the slight slackness in the yarn caused through the spindles reversing before the wire begins to move. On the contrary, this slackness of the yarn, in consequence of the lateness of the action of the wire, results in the making of very bad cops and snarly yarn. Several ingenious methods have been adopted to overcome this difficulty. Their object is that, while permitting the faller wire to be behindhand in its movement when the cop is beginning to be formed—because there is a distinct advantage in being so—it shall be so controlled that, at each layer added to the cop, its moment of action begins to approach that of the reversal of the spindle, until when the cop is finished the wire is brought to touch the yarn at the exact moment the spindles reverse. From this point, down to the cop, is so short a distance that there is no danger of the wire overtaking the yarn, and at the same time it maintains the thread at a tension that enables a perfectly solid cop to be formed.

Having explained the necessity for adopting some means of tightening the "backing-off" chain as the cop gradually enlarges, it remains to give an example of one method of doing it. For this purpose the drawing, Fig. 52, has been prepared. It is practically an enlarged view of a portion of Fig. 44, and, although showing a few variations in the arrangement and details, it can be used for reference in reading the remarks made when describing that drawing.

As the carriage comes out, the various parts are in the

positions shown in the drawing. The open jaw of the lever K depresses the bell-crank lever X, and so puts the backing-off cone clutch into gear with the fast pulley. In consequence of this, the tin cylinder Z reverses, and in addition to reversing the spindles in order to unwind the yarn from the bare part of the blade above the cop, it also winds on a portion of the chain L, and in doing so pulls down the faller arm C which is fastened to the copping faller A. The wire *f* is brought down by this action, and follows the yarn down the spindle as it is unwound; the rate at which it does this is regulated by the scroll surface on which the chain L is wound. A slight slackness of the chain L during the earlier part of the cop is not of much consequence, as already explained, and therefore the wire *f* need not touch the yarn the exact moment it begins to unwind from the spindle. As, however, the cop enlarges, the action of *f* must be brought earlier into operation, and this necessitates the use of some arrangement similar to that shown in the drawing.

Attached to a kind of boss of the scroll M is a chain L, whose other end is connected to a lever centred at N. As the carriage moves outward, one end R of this lever is so arranged that during the time the cop is having its first layers formed, it just comes into contact with an inclined plate S. This plate is connected to the shaper-plates by the rail T, and as these shaper-plates move during the building of the cop, the incline S is also moved, so that, instead of the end of lever at R just coming into contact with it the moment the carriage stops, the advance of the incline causes R to come into contact a little earlier after each layer is added. The effect of this is to cause the lever to yield and pull down the chain M, which in its turn moves the scroll on which the chain L is wound, and this action draws L tighter and gradually takes out the slackness, so that directly backing-off commences, the chain responds a little earlier after each draw, to the backward turning of the tin cylinder. In order to present these features of the self-actor as fully as possible, it will be necessary to give other examples of most of the arrange-

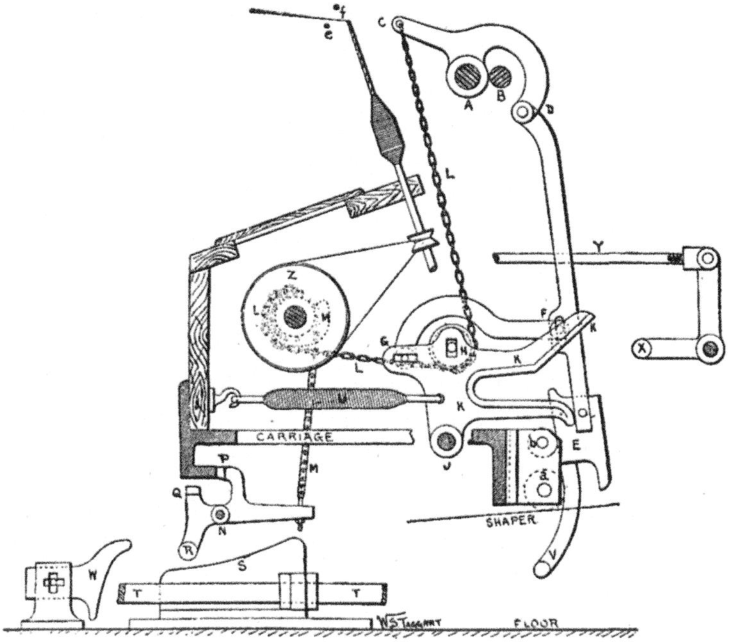

Fig. 52.

ments previously illustrated; but it is also necessary in order to prevent complication to present the subject in a consecutive form as far as possible, and with this object in view it is advisable to proceed with an explanation of the building and winding mechanism, after which reference to and further description will be given of other methods of performing the actions already so far described.

The Mule Cop.[1]—The only way to thoroughly under-

[1] See the author's book, *Quadrant and Shaper*, for a more detailed description of the Mule Cop.

stand the operation of building the cop and winding the yarn upon it is to make a complete examination of the cop itself, and from it to deduce the reasons for employing the special mechanism by which these results are obtained. In this way much of the description that follows will be less difficult to understand, and a better understanding of the problems will follow from the careful reasoning which it will involve.

It has already been pointed out that the spindles are carried by a long wooden structure, called the carriage. The portion of the carriage which does this is shown in Fig. 53. The spindle is supported at two points B and C, and the wharve is placed between them, its position being nearer the upper or bolster-bearing C than the footstep-bearing B. Above the bolster-bearing there projects the part of the spindle upon which the cop is built, and it is to this feature that our chief attention will be given. An enlarged drawing of the cop F is shown in Fig. 54. Its general shape is that of a cylinder, with conical ends, one end having usually a longer taper than the opposite end. The reason for adopting this shape in making a cop is not far to seek, and may be summed up in the words, solidity, and facility in being unwound again.

Let us now see how this peculiar shape is obtained, and ask ourselves various questions as to what is necessary in fulfilling the conditions of its structure. To begin then, the yarn must be first wound on the surface of a steel spindle, say ¼ inch in diameter. Frequently this surface is slightly enlarged by using tubes as a foundation; but for the present purpose it will be preferable to confine ourselves to the most usual course of winding the first layers on the bare spindle. That part of the blade on which the yarn is first wound is practically parallel, and we might almost say that the whole of the cop bottom is wound on to what might be termed a perfect cylinder. Above the cop bottom, however, the blade gradually tapers to the point T, where its diameter is made as small as possible consistent with

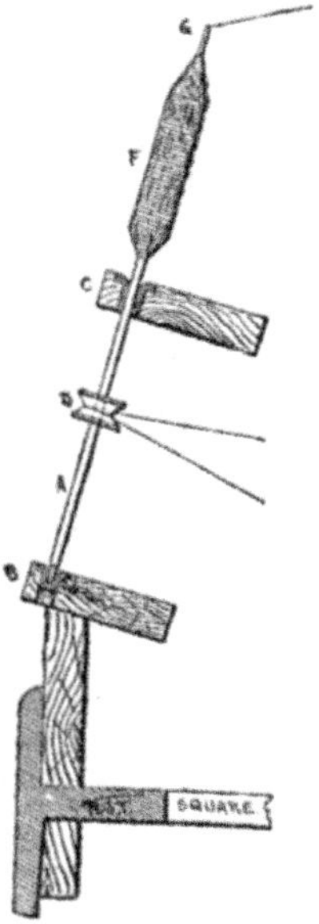

FIG. 53.

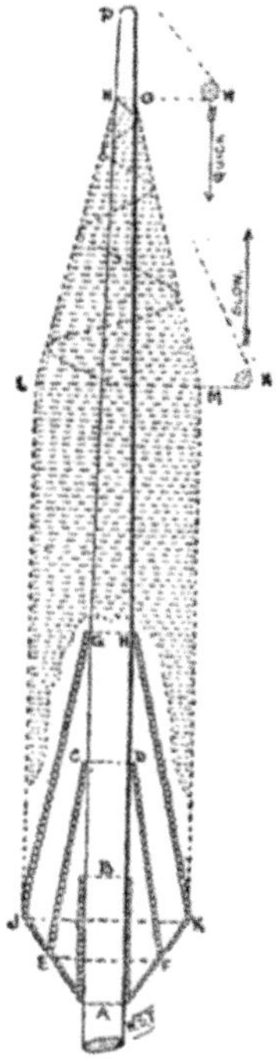

FIG. 54.

strength and with the yarn it is spinning. The reason for this has already been given in an earlier part of the book.

As winding takes place, during the return of the carriage to the roller beam, it will be necessary to revolve the spindles constantly at such a speed that at each inward run they will wind on the 64 inches of yarn that has been delivered by the rollers and twisted during the outward run of the carriage. Readers will understand that it is not desirable to complicate matters by mentioning the gain of carriage, etc.; therefore, for our present purpose, the delivery of the rollers, of 64 inches, must be accepted as tentative, being only for the simplicity of illustration. The first layer of yarn will therefore consist of 64 inches, and it will be wound upon a solid cylindrical surface. The length of that portion of the spindle upon which it is wound is, of course, arbitrary, but, as will be shown presently, it is made as short as possible, so that the layers are compact and close together; $\frac{7}{8}$th of an inch, or 1 inch, is the usual length. Subsequent layers are added, and mechanism is employed which gradually causes the cop to assume the shape shown in the lower part of the diagram, Fig. 54. The first layer is represented at A B. From A each additional layer has its commencing point raised in such a manner that an inclined surface is produced along the line A E J. At the same time, the surface or "chase" upon which the yarn is laid is also lengthened; this lengthening of the traverse or "chase" is shown by the lines E C and also J G. When a diameter has been obtained, as at J K, which is considered large enough, a cessation of some portion of the mechanism, and a slight modification of other portions, cause the commencing point of each layer to be raised, but this is done in such a manner that instead of giving a conical form, as it did from A to J, it begins to rise vertically, and in this way it continues to L, so that a cylindrical shape is given to the body of the cop.

It is clear that, no matter what diameter may be decided upon as large enough, the yarn must always finish winding on the spindle, so that the conical form is continued throughout the cop in the same condition practically as it had when the foundation A J G H K, or "cop bottom," as it is termed, was finished.

It has already been remarked that the first layer on the spindle from A to B is wound on the bare spindle, and is practically a parallel layer. To do so it will be necessary to revolve the spindle a certain number of times—a number readily calculated. For instance, a $\frac{1}{4}$ inch spindle must turn $\frac{64 \times 4 \times 7}{22} = 81\frac{5}{11}$ times to wind on 64 inches. Now when the next layer is added, it will begin on a larger diameter, represented by the extra layer of yarn; but it will finish on the same diameter as the first layer did. It will readily be seen that the speed of the spindle, for the second layer, will require to be altered; but this alteration must only take place at the commencement, for since the end diameter remains the same, so also must the speed. Succeeding layers increase the diameter of the cop at the bottom, but finish at the top with the same diameter, until we get to the full diameter, as at J K, and a long conical surface, as at J G, where the alteration in speed, in order to wind yarn on this surface, must undergo a considerable variation from that necessary at the commencement.

While the speed of spindle during the winding of the first layer was uniform, because of the cylindrical surface on which it was wound, the speed during the winding of the last layer, J G, must be ever varying, simply because the yarn is wound on varying diameters. The same length is wound on and in the same time as the first layer A B. To do this and at the same time maintain an equal tension on the yarn, it is clear that the speed of the spindle, when the yarn is passing on at J K, must be slow; and, correspondingly, when it travels up the cone the diameter becomes less, and the speed increases until it reaches the smallest diameter at G H, and here we must have the quickest speed. One revolution of $1\frac{1}{4}$ inch diameter at

J K will wind on $\frac{5 \times 22}{4 \times 7} = 3 \cdot 92$ inches, while one revolution of the small diameter, $\frac{1}{4}$ inch, at G H, will only wind on $\frac{22}{4 \times 7} = \cdot 7854$ inches; that is, the small end must revolve five times quicker than the large end. This increase of speed must therefore be gradual, and of such a nature that it corresponds as nearly as possible to the gradual decrease of diameter. From this reasoning in regard to the last layer of the cop bottom, we can see that a variation of speed must exist in each layer after the first one, and the only difference is that the variation between the first speed and the last one is not so great, this, of course, depending on the relative sizes of the cop at its various points. For instance, when the cop is 1 inch diameter, the variation in speed between the bottom and the top is as 4 to 1, and so on for the different diameters. The gradual variation in speed during the winding of any single layer, as well as the variation of speed between the different layers, can easily be shown by means of a diagram, and this we shall proceed to show.

On the assumption that the bare spindle is $\frac{1}{4}$ inch diameter, it has already been shown that a little over 80 revolutions will be required to wind on the 64 inches of stretch; and, moreover, since the first layer is wound on a parallel surface, the 80 revolutions must be made without variation in speed during the winding.

After the first layer, a new set of conditions arises, and each successive layer afterwards necessitates a change in position from the previous one, and also a complete change of the variation of speed which was required for the last layer put on.

When dealing with the building of the bobbin on the

fly-frames, we saw that each layer required a different speed as the diameter increased; the same necessity also arises in the case of the cop, for as the diameter enlarges from A to J K, Fig. 54, the speed of spindle must be altered, in order to wind on the yarn at this point in the same time as when wound on the bare spindle. But here the similarity ceases; in the fly-frame bobbin a parallel form is built throughout, while in the cop a conical form is

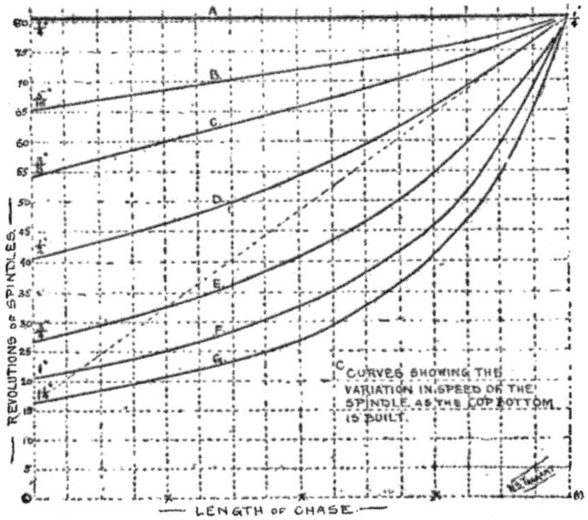

FIG. 55.

made, which tapers from a larger diameter to a smaller one; and, in addition, the proportion between the two diameters varies with each layer. This continued change of shape renders necessary a change of speed to suit each new set of conditions.

In building the conical form of cop it will readily be seen that the speed of the spindle must vary, from being slow at the large diameter to quick at the small diameter, and that this condition must hold good from the first to the last layer. It must not, however, be assumed that, because the cop has a "straight" taper, the variation in speed is a uniformly increasing one; this will clearly be seen as each layer is carefully examined and its speed found. As an aid to making this examination, the diagram in Fig. 55 has been prepared to show in a graphic form the variations of speed for different parts of the cop. (For the sake of simplicity the length of the "chase" is assumed to remain the same throughout the cop bottom; this assumption makes no difference to the "character" of the curves, but to those who desire it, it is an easy matter to realise that the length of chase for A is 1 inch, and for G 2 inches, all the others, of course, lying between these extremes.) The horizontal lines of the diagram represent the speed of the spindle; on the first line we can, therefore, mark off the number of revolutions that any given diameter will require in order to wind on. In this way we find that $\frac{1}{4}$ inch diameter requires 81·5 revolutions; $\frac{5}{16}$ inch diameter commences to revolve at the rate of 65·2 revolutions; and so on for the other diameters as shown in the table:—

$\frac{1}{4}$ in.	diameter	commences	at the rate of	81·5	revs.
$\frac{5}{16}$ in.	,,	,,	,,	65·2	,,
$\frac{3}{8}$ in.	,,	,,	,,	54·3	,,
$\frac{1}{2}$ in.	,,	,,	,,	40·75	,,
$\frac{3}{4}$ in.	,,	,,	,,	27·16	,,
1 in.	,,	,,	,,	20·4	,,
$1\frac{1}{4}$ in.	,,	,,	,,	16·3	,,

These initial rates of speed give us the starting-points of the curves. The other points are not difficult to obtain; but first let us notice what character the curves must have, before drawing them. The line representing the speeds for the first layer will naturally be straight, as representing a uniform rate the full length of the chase; this is drawn at A and shows the same speed throughout. Layers are added until the diameter becomes $\frac{5}{16}$ inch. Starting at

65·2 revolutions, it finishes at the same rate of speed as the first layer, namely 81·5. It is readily seen that the slight difference in the end diameters necessitates a variation in speed, but not sufficient to show clearly the character of the variation, so the line B is almost straight, though it will be observed that the end of it takes an upward curve at a little quicker rate than at its commencement.

To emphasise the characteristics, the larger diameter of $1\frac{1}{4}$ inch will be taken as an example. Here we begin with a rate of 16·3 revolutions, and finish at 81·5 revolutions. As the yarn travels upwards along the line G 7, Fig. 56, it will reach a point that is 1 inch diameter, and, continuing, will pass the $\frac{3}{4}$ and $\frac{1}{2}$ inch diameters. The question is now—At what rate must the spindle run in order to wind on the yarn evenly, so as to maintain the same tension in it at these various diameters? This can readily be answered; for we simply have to remember what was clearly explained in reference to the flyer bobbin (see Vol. II.), that the rate of speed must vary inversely as the diameter of the bobbin. For instance, if the spindle revolve at 16·3 revolutions for $1\frac{1}{4}$ inch diameter, then at 1 inch diameter it will run at $\frac{5}{4}$ of 16·3 = 20·4 revolutions, and so on for the other speeds. A table will show this better:—

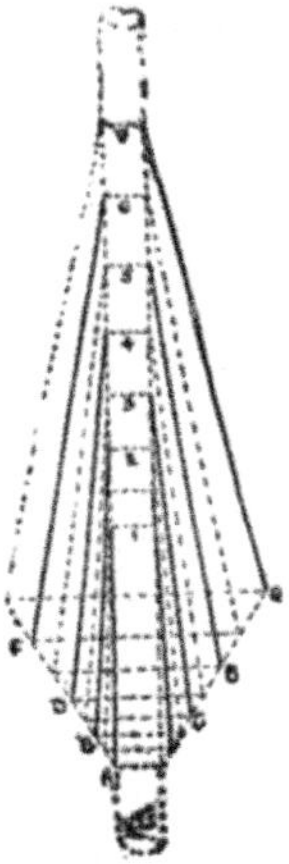

FIG. 56.

$1\frac{1}{4}$ in. requires a rate of revolution of				16·3
1 in.	,,	,,	,,	$\frac{5}{4}$ of 16·3 = 20·4
$\frac{3}{4}$ in.	,,	,,	,,	$\frac{10}{3}$ of 16·3 = 27·16
$\frac{1}{2}$ in.	,,	,,	,,	$\frac{10}{4}$ of 16·3 = 40·75
$\frac{1}{4}$ in.	,,	,,	,,	$\frac{5}{1}$ of 16·3 = 81·5

It is to be observed that the speeds in the above table vary inversely as the diameter, for, on comparing the speed at $1\frac{1}{4}$ inch and $\frac{1}{4}$ inch, we find that while the $\frac{1}{4}$ inch is one-fifth the diameter of $1\frac{1}{4}$ inch, the speed is five times quicker; and the other speeds follow the same proportion. It only remains to add that from this consideration we recognise at once that the characteristic curve of the hyperbole will represent the true variation in the revolution of the spindle while winding on a conical surface. Any diameter similarly treated will give the same characteristic features, so we are now in a position to represent graphically the information obtained from the table.

By marking off on the line O M points representing the 1, $\frac{3}{4}$, and $\frac{1}{2}$ inch diameters, and on the vertical lines measuring the number of revolutions corresponding to those diameters, we obtain points through which a curve may be drawn. This is shown at G, and from it we see at a glance the full character of the variation. Starting at 16·3 revolutions a gradual increase takes place; instead of being uniform, however, the increase occurs at an irregular rate, and as it approaches the smaller diameter it rises very rapidly, until it finishes on the bare spindle five times quicker than at its commencement. This irregular increase of speed must be thoroughly understood, for the principle of the "quadrant" entirely depends upon it; and it must not be confounded with a uniform increase in speed, which would be represented by the dotted line joining the two ends of the curve G. Such a variation differs greatly from what should be the real variation, as shown at G. If this increased but irregular acceleration of the speed of the spindle, as the yarn is wound from the base to the apex of the cone, be completely realised and comprehended, the understanding of the quadrant will be a comparatively easy matter.

Thus far we have assumed the diameter of the spindle

to be $\frac{1}{4}$ inch, but this refers only to the part on which the cop bottom is built. From this point to the end, it is tapered, and therefore each additional layer finishes on a smaller diameter, and consequently at a quicker speed. This problem will be dealt with at a later stage, as will also the question of guiding the yarn on the cop as winding takes place.

Another feature to be noticed in regard to the cop is the method of obtaining as solid and compact a form as possible. We have spoken of laying the yarn on the conical surface, from J to G, Fig. 54, but before it can be brought from above G to J it must pass over the conical surface. This is taken advantage of by causing the faller wire W to fall very quickly as the carriage commences its inward run, which has the effect of winding the yarn on the cop in several spiral turns, which binds together the layer below. On reaching M, the wire X commences its upward movement. It is this special movement that we have been considering, and it is this which is generally understood when "winding" is mentioned.

The Mule Quadrant and its Action.[1]—Having given an explanation of what is required in regard to driving the spindles at a correct speed while building the cop, we proceed to examine and explain the means adopted to obtain it. A rough outline only of the mechanism will be given at this point; fuller details will follow as we proceed with the examination of its action.

As already described, the spinning or twisting process takes places as the carriage moves out and the spindles are driven from the rim shaft. During the drawing-up, the tin cylinder is disconnected from this source, and receives its motion for winding purposes from an adjacent drum to which it is geared. This will be observed on reference to

[1] See the author's book, *Quadrant and Shaper*, for a more detailed description.

Fig. 57; the tin cylinder *u* is seen to be geared, by the wheel *x* and *z*, to a drum, round which a chain is wound.

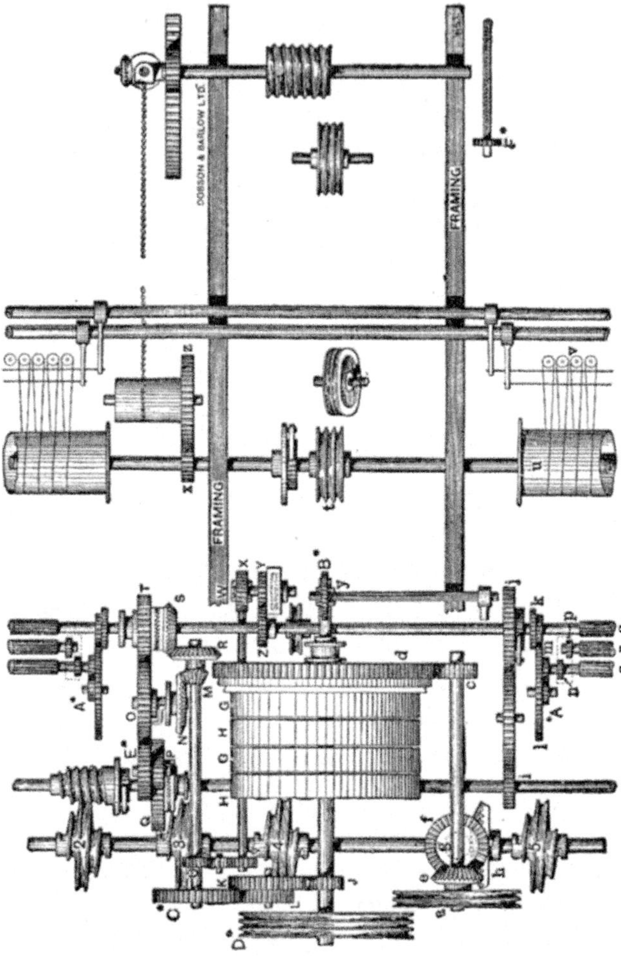

GEARING OF S.A. MULE

FIG. 57.

This chain is firmly fastened to the drum, and after passing round it several times it is connected to an oscillating arm called the "Quadrant." It is from this quadrant that the spindles receive their speed, or rather they are controlled and regulated by it as the cop passes through its various stages.

A good idea of its position and proportions can be obtained from the drawing, Fig. 58.

An enlarged view of the winding chain and drum is given in Figs. 59 and 60. One is a plan view, and shows the chain A passing round the drum B and connected to a hook D. If the hook is fixed, and a horizontal movement be given to the drum in the direction of the arrow E,

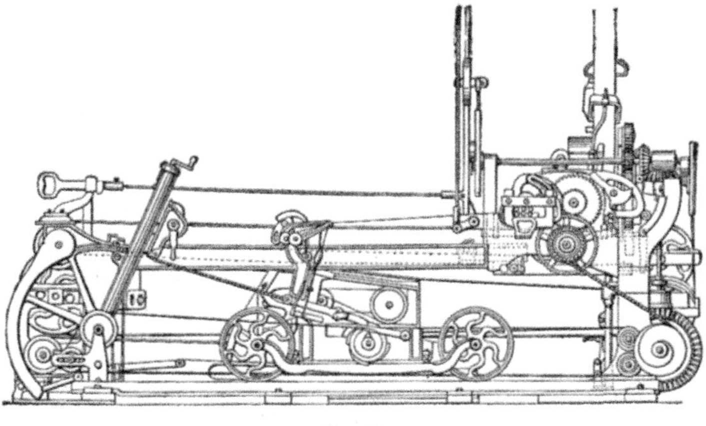

FIG. 58.

the drum will be compelled to yield by turning on its centres; this it will do by revolving in the direction of arrow F, and so unwinding some of the chain as the distance from its first position increases. In this apparently simple method of producing rotation there lies the germ of the mule quadrant, and we shall try by reasoning, to follow out the course which led Roberts to devise a mechanical arrangement that takes rank as one of the most remarkable and ingenious inventions of the last century. In passing, it may be as well to point out that readers are occasionally met with who look on the mule quadrant as the "differential motion" of the self-actor. It is scarcely possible for a reader, who has followed what has already been said, to labour under this impression, for it was emphatically shown when dealing with the fly-frame (see Vol. II.) that a differential motion is simply a convenient method of combining two distinct motions, through the medium of which a variation in one or the other can be effected. It possesses no variable element in itself, nor has it any part in either building or winding in the fly-frame. The variable motion of the bobbin in this later machine is entirely brought about by the cone drums, and the differential motion has

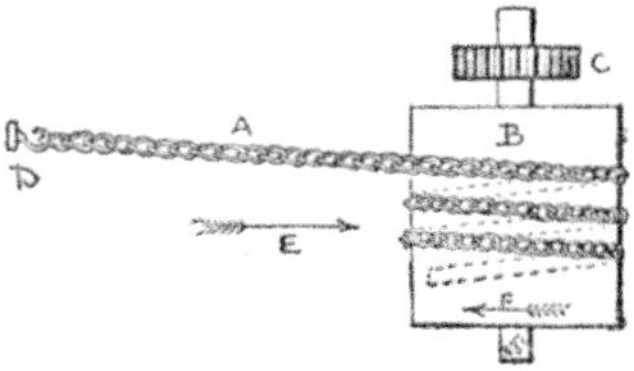

FIG. 59.

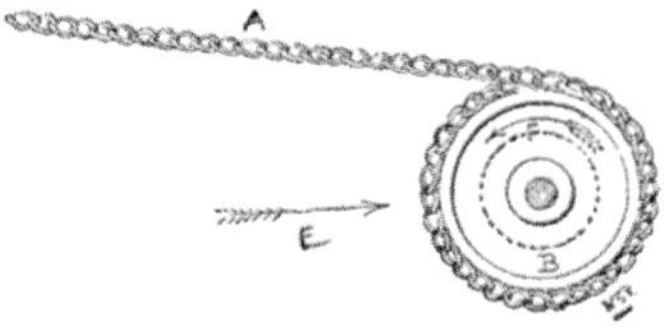

FIG. 60.

nothing whatever to do with it, except as an arrangement of wheels which assists in transferring a variable motion already given.

If the quadrant can be compared to anything, it is to the cone drums that it bears a resemblance—but only to the extent that they are both the direct means of giving a variable speed to whatever they drive. They do this, however, by such entirely different methods and principles, that a similarity exists only in the "name" of their purpose, namely—winding. Readers are therefore warned against falling into the error pointed out above, for it denotes failure in the attempt to understand the principle and purpose of either arrangement.

In the following explanations of the various phases of the action of the quadrant, the illustrations are mainly of a diagrammatic character, the chain, quadrant, and cylinder being represented as simple lines, free from details.

Although the explanation will be made as thorough and comprehensive as possible, and from it an almost complete understanding of the subject can be obtained, it must not on any account be considered a "theory" of the quadrant. It is rather a practical demonstration of the action of the quadrant drawn out to scale and shown in diagrams, a mere fringe of the theory being introduced in order to explain some of the results brought to light by these drawings. This is stated in order to prevent readers from falling into the error of ascribing to a brief explanation the term "theory." A theoretical consideration of the problem would be entirely out of place in these pages, chiefly because the subject requires a degree of knowledge for its comprehension which is totally beyond the average reader. The practical view here given, is designed to give the required information in the simplest manner, and also to dislodge some of the peculiar ideas which many hold on the subject.

Our first attempt will be confined to noticing the effect of the chain on the winding drum, when the point of its attachment is fixed, during the whole of the period of the run-in of the carriage. The accompanying series of diagrams will illustrate the remarks. In Fig. 61 the chain is fastened at H, and the other end is wound round the drum, which in its outermost position is shown at A. As the run-in takes place the drum will travel from A to G, and by dividing the stretch into equal parts, say six, we get seven different positions as occupied by the carriage whilst winding, these being shown at A, B, C, D, E, F, and G. Now, since the end of the chain is fixed at H, the motion from A to B will cause a certain length of the chain to be unwound from the drum, and, as before explained, this will cause the drum to revolve, the amount of the revolution of course depending upon the length of chain unwound. On account of the position of H in relation to the drum (which, it will be observed, is in the same horizontal line with the movement of the upper diameter of the drum), the chain unwound equals the distance moved by the carriage, and as each distance moved is exactly equal to the last, we get, for each of the divisions shown in the diagram, equal lengths of chain unwound. The chain unwound from A to B is equal to I J, and from F to G it is equal to N P, and so on for the other lengths, all of which are equal to each other. The movement of the carriage under these conditions clearly produces an equal rate of revolution in the winding drum in each division, and therefore a "uniform" rate of speed is obtained throughout the stretch. This equal horizontal movement of the drum, producing a uniform revolution, must be specially observed to depend on the position occupied by the fixed end of the chain at H. If this position is changed, another set of conditions arise which totally destroy all ideas of uniformity; and to emphasise this important point an illustration will be given. Let it be supposed, as shown in Fig. 62, that the point of attachment is raised vertically over the position H^1 to H; the chain would then pass from H to the drum A, and its point of contact there, would be at I (the unused part of the chain is shown in dotted lines throughout). The drum moves equal horizontal distances, as in the upper figure, so we may readily compare the effects of the two sets of conditions. In Fig. 61 it was found that the length of

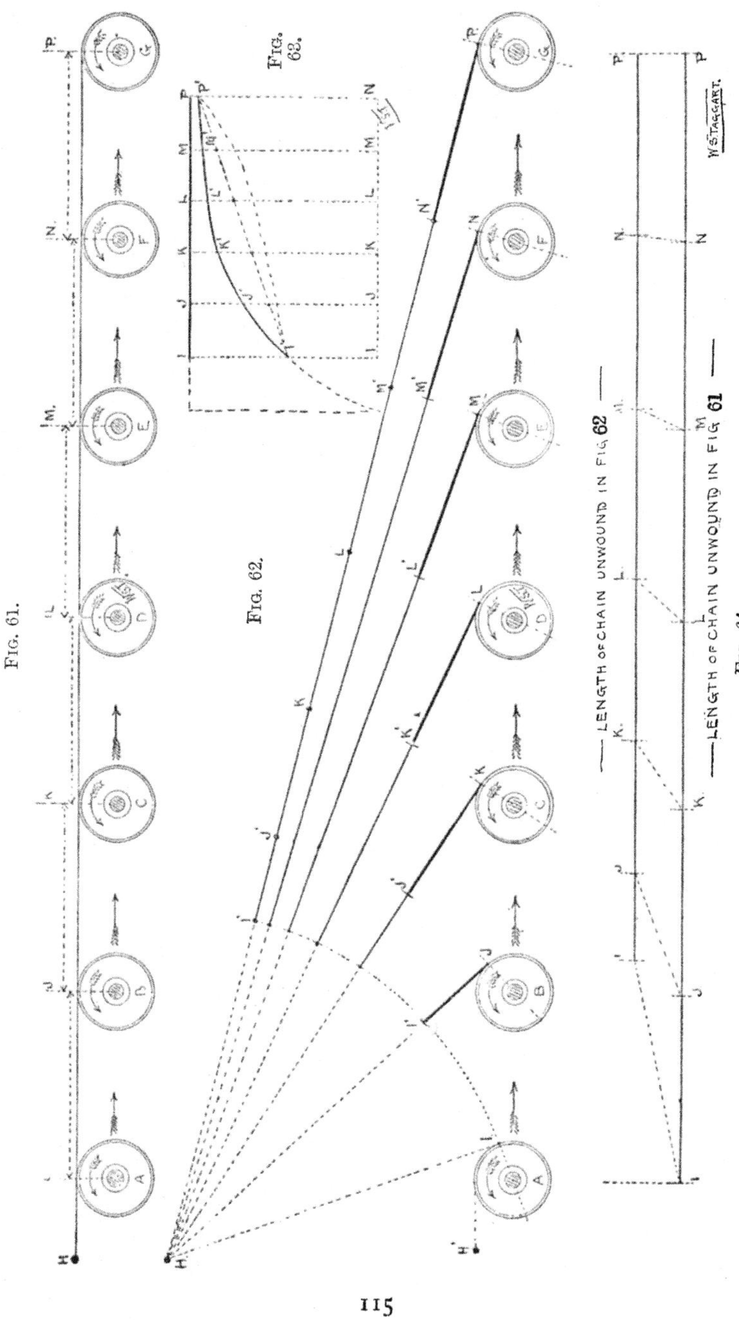

FIG. 61.

FIG. 62.

FIG. 63.

FIG. 64.

chain unwound was exactly equal to the horizontal distance moved by the drum, but in Fig. 62 the chain unwound is very far short of the distance from A to B. This length is shown by a thick line at I^1 to J, and a glance will show the great difference between the two. A further movement from B to C will cause another length of chain to be unwound, which is shown in thick lines from J^1 to K; this is a greater length than was unwound during the first movement from A to B. We shall also find on following out the other movements of the drum that each successive length unwound is longer than the previous one, and when we come to the last one, from F to G, the length N^1 P unwound is over twice that unwound during the first movement, from A to B. We thus find that by altering the position of attachment and making it fixed we destroy the uniform motion of Fig. 61, and obtain a gradually increasing and varying one in its place.

At the first glance it might appear that these results would produce the variation in the speed of the spindle required in making the cop bottom; and, as a matter of fact, in a limited sense, a conical cop could be built by this arrangement. The first layer would be wound on a parallel spindle, when the chain was at H in Fig. 61, and by moving the point of support vertically the various layers of the cone would be added until the point H was reached in Fig. 62.

This may be made much clearer in a diagram showing by means of a curve the relative variation of speed for each position. Fig. 63 has been prepared with this object. The upper line I to P represents uniformity of the motion in Fig. 61, and corresponds to a similar line in Fig. 55. The curved full line I^1 to P^1 represents the variation as produced in Fig. 62, and we can readily see that it has all the characteristics for giving the variable motion necessary for a conical form of cop. By comparing this curve, however, with the corresponding one in Fig. 55 it will be immediately observed that, although the two are allied in character, they are the reverse of each other. In Fig. 55 the curves increase slowly at first, and finish rapidly. In Fig. 63 the opposite is the case; we get a rapid increase at the beginning and a slow finish. In other words, Fig. 55 is the curve for building a conical form with the larger diameter at the bottom, while Fig. 63 is a curve of speeds for a cop "upside down," with the smallest diameter at the bottom. The dotted curve represents the variation required for the actual conditions of a mule cop, and we can clearly see a reverse order of their characteristics.

Two lessons can be learnt from this illustration. The first is that a statement which makes out that with a fixed point of attachment for the chain, and equal horizontal movements of the drum, a uniform motion is produced, is entirely wrong in principle; the second, that statements in connection with the quadrant, which point out that certain variable results in motion are produced, is not sufficient to explain, even from an elementary point of view, the principle underlying such an important piece of mechanism. A comparison is made in Fig. 64 between the total length of chain unwound from Fig. 61 and Fig. 62, and corresponding points in each length are connected by dotted lines to emphasise the difference between them. We see that in addition to the variable motion of Fig. 62 a shorter length of chain is used, and consequently the total revolution of the winding drum, and therefore the spindles, is less than in the case of Fig. 61, which winds the first layer on the spindle.

We shall now consider the question as it actually presents itself in the mule. The point of attachment for the chain, instead of being fixed, is carried by an arm, which is made to oscillate round a fixed centre. The point of attachment at the commencement of the cop is as near this fixed centre of the lever as possible; and as the cop enlarges, the nut to which the chain is hooked is raised up by a screw working within the arm of the lever. The new positions of the point of attachment, in conjunction with the movement of the arm itself, brings about the required degree of variation in the unwinding of the chain, and therefore in the speed of the spindles.

Examining the action of the quadrant in bringing about this result, let us first take the case when the point of attachment is near the fulcrum of the quadrant arm, Fig. 65. The arm, centred at H, is caused to move in unison with the carriage, through a quarter of a circle. It is arranged to commence from a line which is a little back from a vertical through the centre H, probably about 15°, as at H J; from here it moves through 90° to H Q. As the carriage moves from A to G, the quadrant moves through this quarter of a circle. An important feature must be noticed in this connection: during the earlier portion of the inward run of the carriage, the copping faller wire is depressed quickly, and lays some yarn on the cop in a few coarse-pitched spirals—an operation called "crossing"; the carriage has moved a little distance, 10 or 12 inches, before this operation is finished, and during this time the quadrant arm has also moved forward. When "crossing" is complete, the essential part of the winding commences, and it is for this feature that the quadrant serves its real purpose, and to which we are now drawing attention. Generally speaking, the quadrant arm is vertical when "crossing" is finished, and, relatively, the winding drum is in the position at B. The movement of the quadrant from J to K, and of the carriage from A to B, has nothing to do with the problem of winding, except that "crossing" takes place during this period. From B onwards, however, the spindles must be revolved to wind the yarn from a large diameter, which gradually tapers, until the bare spindle is reached. The movement of the carriage during winding is divided into five equal divisions, giving six positions of the drum; by dividing the path of the quadrant nut into the same number of equal divisions we get the position the nut occupies for each position of the drum, and we can then, by measurement or otherwise, find the lengths of chain unwound as the carriage moves in. These respective lengths are shown in thick lines from 3 to 4, from 5 to 6, from 7 to 8, from 9 to 10, and from 11 to 12. The difference between them is very slight indeed, and while theoretically they correspond to a conical surface, it is so little as to be almost imperceptible. During this movement of the carriage from B to G, the point of attachment of the chain has moved forward in the small arc of a circle from K to Q, and by doing this has prevented the unwinding of a little of the chain which would have been unwound if K had remained fixed. We get the first layer wound on the bare spindle during this period. As the layers are added, the nut is caused to travel up the screw of the quadrant until the cop bottom is complete, and its position at this point is shown at K in Fig. 66. The quadrant arm never varies in the angle it describes; so with the nut at K it still traverses the same angle, but as the circle is much larger, the length of the arc K M Q, which the nut travels along, is much greater than K Q in Fig. 66; consequently the amount of chain unwound is considerably less, because the nut moves in the same direction as the carriage to a greater extent than when the smaller arc of a circle in Fig. 66 is being traversed.

When the quadrant arm is vertical the nut is at K, Fig. 66, and the chain passes from this point to the winding drum B, which it touches at 2. The length of chain between K and 2 is unused chain. As the carriage moves inwards to C, the quadrant travels from K to L; and as this movement is almost a horizontal one, the difference between the lengths B C and K L represents nearly the amount of chain unwound from the drum. The amount unwound is shown by the thick line 3, 4; it is relatively a short portion of chain, and from it we see that the spindles are revolving slowly, because at this time the yarn is being wound on the thick part of the cop bottom. By measuring off or calculating the length of chain unwound as the carriage traverses each of the divisions C to D, D to E, E to F, and F to G, we get for each of these movements respectively a length equal to each of the dark lines at 5 to 6, 7 to 8, 9 to 10, and 11 to 12. These lines represent the amount of chain unwound, and it is clearly to be seen that the drum is revolved very slowly at first, and much quicker at the termination of the run-in. They represent very graphically the varying speed given to the spindles during the winding of the last layer on the cop bottom.

In order to present the results in the same way as those given for the speed of spindle in Fig. 55, a small diagram is given in Fig. 67, for the purpose of comparison, so that an idea may be formed as to whether the quadrant turns the spindles at a correct speed for winding. It is generally assumed that the quadrant does wind correctly, and therefore we find writers dismissing the subject by pointing out a variation in certain lines, and

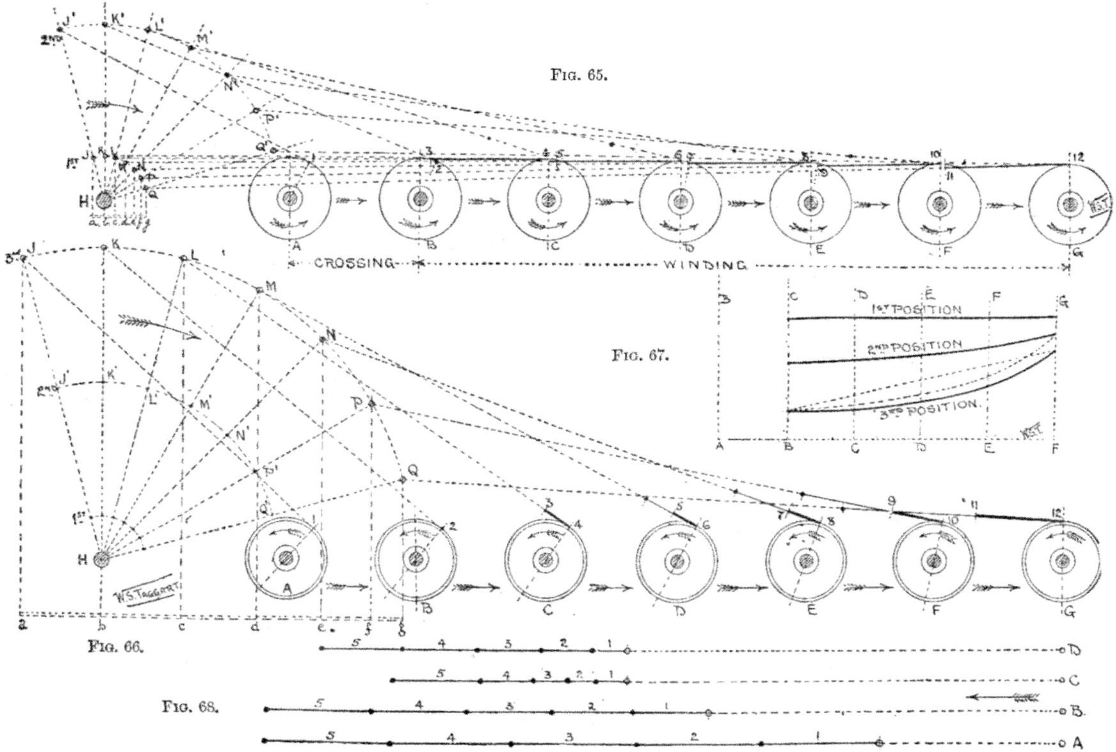

Fig. 65.

Fig. 66.

Fig. 67.

Fig. 68.

saying this variation explains the action of the quadrant. We have warned readers against this kind of explanation, and it would scarcely be consistent for the writer in this case simply to point to the thick lines in Fig. 66, and say these represent the necessary variation in the speed of the spindle for building a conical cop. Fig. 67 is therefore prepared to show why an oscillating arm, as we have it in the quadrant, gives results in winding of an opposite character to those produced by a fixed arm, and which approach most nearly to the actual conditions of speed required.

In the diagram, Fig. 67, the upper dark line represents the variation in speed produced when the quadrant nut is in its lowest position, as in Fig. 65. It is practically straight, and from this fact we see that an almost uniform motion is given to the spindles during the winding of the first layer on the bare spindle. The lowest curved line shows the variation in the speed of the spindle as produced when the nut occupies the highest position on the quadrant arm, during which time the full conical form of the cop bottom is completed. The dark lines transferred from Fig. 66 to Fig. 67 give the curve for the third position; its character corresponds closely to the similar curve in Fig. 55. It will be noticed that it rises very slowly at first, and afterwards the acceleration is greatly increased. This is what we know ought to be the case for a conical form of cop, but a very important point must not be overlooked: it ought to be asked whether this curve is actually similar to the one required for the speed and spindle. If any variation exists, then the quadrant is not performing its work perfectly. It would be impracticable to enter into the question fully, so it must suffice to point out that the two curves do not correspond. The dotted curve shown in Fig. 67 represents approximately the variation of speed the spindle *ought* to have, while the thicker curve underneath shows us the speed it actually has given to it by the quadrant. There is a very perceptible difference between the two curves, and it represents a considerable percentage of variation, which extends throughout the "stretch." The quadrant is therefore by no means "perfect" in giving the correct speed for winding; the difference just pointed out must be compensated for in some way, in order that proper winding can take place. Fortunately this can be effected very simply in the mechanism employed to put the yarn on the spindle, so that by means of the "shaper" the errors of winding, produced by the quadrant, are practically eliminated. In Fig. 65 a middle position of the nut has also been taken, and from it the second position curve in Fig. 67 has been drawn.

Another method of showing the length of chain unwound during each horizontal movement of the carriage is given in Fig. 68, A, B, and C representing the 1st, 2nd, and 3rd positions respectively; we see, in the full parts of each line, the varying portions of the chain unwound. The total length of chain used for turning the drum gets shorter as the cop builds, and from this we gather that the total number of revolutions made by the spindle becomes less and less as the cop bottom nears completion.

It is the practice, sometimes, in explaining the action of the quadrant, to draw a diagram somewhat similar to Fig. 66, and to drop vertical lines from the points J, K, L, M, N, P, Q. The horizontal and varying distances between these lines, as at *a b*, *b c*, *c d*, *d e*, *e f*, and *f g*, are then considered to represent the variation in speed produced by the quadrant, because it is said the quadrant delivers chain, as it were, in these proportions to the drum as the carriage moves in. It need scarcely be pointed out that such an explanation is entirely wrong, and the use of a pair of compasses in measuring the diagram will at once prove how totally at variance it is with the actual conditions. Another point in the explanation is the statement that the amount of chain delivered forward as the carriage runs in is equal to the horizontal distance *a* to *g*. This can also be so easily tested and found to be wrong that it is strange the above explanation, with all its errors and the wrong conception of the principle of the quadrant, should be so persistently repeated. A point also to be carefully guarded against is that on no account must the movement of the quadrant from J to K be allowed to enter into the question of the building of the conical part of the cop.

Having shown how the quadrant produces, approximately, the necessary variation to the speed of the spindle, during winding on a conical surface, there remains another feature to be pointed out and explained. The description so far has been confined to demonstrating how the above variation from a large diameter to a smaller one is brought about. We shall now describe how the initial speed for each new layer is produced. Every fresh layer makes a new conical surface, and while the smallest diameter of the cone practically remains the same throughout the cop bottom, the base of the cone is continually enlarging; and this necessitates a different initial or starting speed for each additional layer. For instance, the bare-spindle diameter will wind on 64 inches by revolving a little over 80 revolutions during the run-in. (NOTE.—It has not been considered necessary in this remark or in the previous ones to subtract the amount of yarn used during crossing from that actually wound on after crossing, as it makes no difference at all to the reasoning employed or the character of the curves deduced from it.) When the base is enlarged to $\frac{1}{2}$ inch diameter the initial speed must be at the rate of a little over 40 revolutions, and for $\frac{3}{4}$ inch diameter a corresponding reduction in the initial speed is produced equal to about 27 revolutions. For 1 inch diameter the starting speed becomes a fraction over 20 revolutions, and on being enlarged to $1\frac{1}{4}$ inch diameter a slight reduction on this (to about 16 revolutions) is necessary. By incorporating these results in a diagram, Fig. 69, a curve can be drawn which represents very distinctly how the starting speed for each new layer varies from a quick speed on the bare spindle to a slow speed on the $1\frac{1}{4}$ inch diameter. This variation in the initial speed, although not previously mentioned, can be clearly noticed in the diagram, Fig. 55, which shows the full variation for several parts of the cop. The curves in that diagram, if transferred to Fig. 69, would start from the points A, B, C, D, and E, and would follow the directions shown by the lines F, G, H, J, and K. From Figs. 67 and 68 the same information can also be deduced.

It was made clear in describing Figs. 65 and 66 that the movement of the point of attachment for the chain, up the quadrant arm and away from its fulcrum, enabled us to obtain the desired condition of winding. The question arises—What position must the nut to which the chain is connected occupy, for the various layers as they are added, in order to wind correctly? Only a relative answer can be given here to this question; to deal with it fully would require a number of very carefully-drawn diagrams, or a complicated system of calculation, which would scarcely be of use, at present at any rate; so we will simply give a practical example.

It was seen in Fig. 69 that a very great reduction takes place in the initial speed of spindle during the time the first ¼ inch increase of diameter is added; in fact, it falls to one-half. It was understood from Fig. 66 that the initial speed becomes slower as the nut travels up the quadrant. From these deductions, therefore, we can conclude that the first ¼ inch increase of diameter necessi-

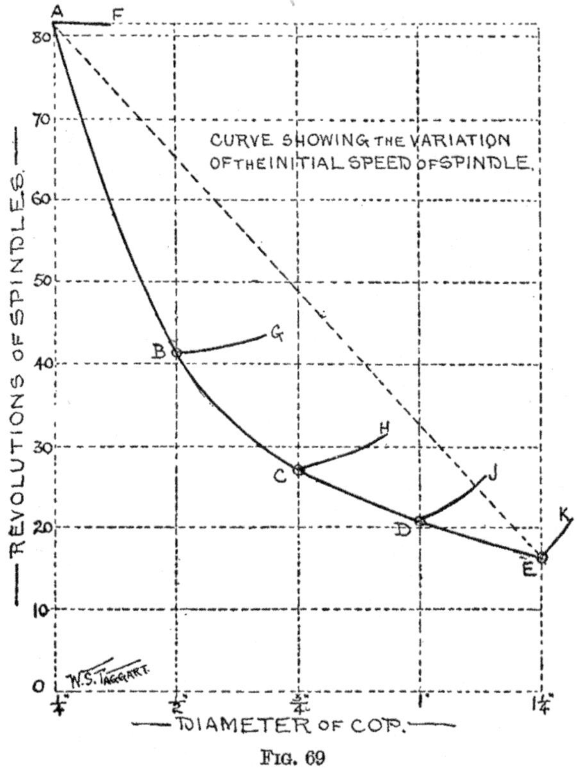

FIG. 69

tates a considerable movement of the nut up the screw to correspond to the great reduction in speed of the spindle. Now let us notice the reduction of speed when the last ¼ inch is added, as from D to E, Fig. 69. It is comparatively little, and therefore, as before, we conclude that only a slight movement of the nut up the quadrant screw will produce the necessary change. Between the two extremes the movement of the nut gradually lessens, and at first sight it might be said that the curve in Fig. 69 if reversed would represent the rate of movement. This conclusion, however, would be wrong; the curve gives us a "clue" to the rate of travel of the nut, but it by no means represents the actual rate.

In order to present to the reader an actual practical

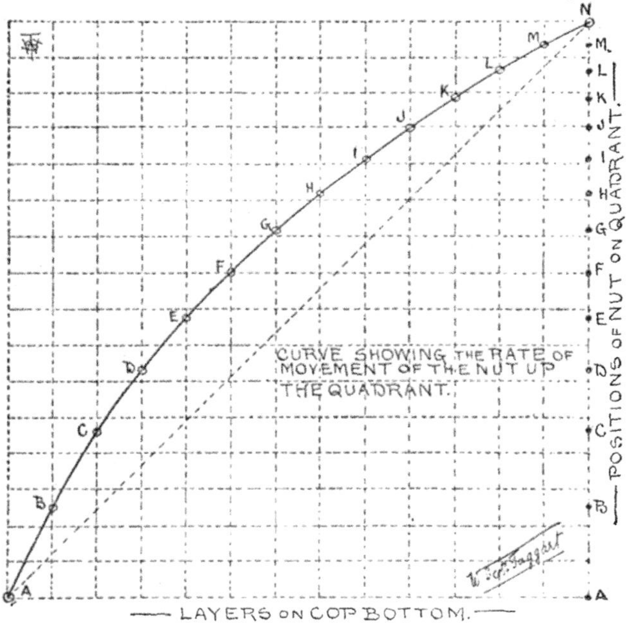

illustration of the movement of the nut up the quadrant, the diagram, Fig. 70, has been prepared. It was taken under ordinary working conditions. A good minder was chosen, and was permitted to "govern" the quadrant just when he thought proper; notice was taken of each movement of the nut and its amount, as well as the number of draws in the cop bottom, and the intervals between each movement.

Number of draws showing at which number the quadrant was actuated.

1	**44**	87	130	173	216	259
2	45	88	131	174	217	260
3	46	89	132	175	218	261
4	**47**	**90**	**133**	176	219	**262**
5	48	91	134	177	220	263
6	**49**	92	135	178	221	264
7	50	**93**	136	179	222	265
8	51	94	137	180	223	266
9	52	95	138	**181**	224	267
10	53	96	139	182	225	268
11	54	97	140	183	**226**	269
12	55	98	**141**	184	227	270
13	56	99	142	**185**	228	271
14	57	**100**	143	186	229	**272**
15	**58**	101	144	187	230	273
16	59	102	145	188	231	274
17	60	103	146	189	232	275
18	61	**104**	147	190	233	276
19	**62**	105	148	191	**234**	277
20	63	**106**	**149**	192	235	278
21	64	107	150	193	236	279
22	**65**	108	151	194	237	280
23	66	109	152	**195**	238	**281**
24	**67**	110	153	196	239	282
25	68	111	154	197	240	283
26	69	112	**155**	198	241	284
27	**70**	**113**	156	199	**242**	285
28	71	114	157	200	243	286
29	72	115	158	201	244	287
30	73	116	159	202	245	288
31	74	117	160	203	246	289
32	75	**118**	161	**204**	247	290
33	76	119	162	205	248	**291**
34	77	120	**163**	206	249	292
35	78	**121**	164	207	250	293
36	79	122	165	208	251	294
37	80	123	166	209	252	295
38	**81**	124	167	210	253	296
39	82	**125**	**168**	211	**254**	297
40	**83**	126	169	**212**	255	298
41	84	127	170	213	256	299
42	85	128	171	214	257	300
43	86	129	**172**	215	258	

The results when drawn out in diagram form yielded the curve shown in Fig. 70; and it is striking as showing most clearly what is readily deduced from the foregoing descriptions. The vertical lines, Fig. 70, represent equal intervals of layers on the cop bottom, and the horizontal lines represent inches on the quadrant arm. The first few layers required a movement of the nut from A to B, about 2½ inches; the next few layers necessitated its moving from B to C, 2 inches only; and the last lot of layers (equal to the first lot) required only a movement of a little over half-an-inch. The intermediate positions of the nut are shown on the last vertical line, and to those unable to understand the curve, this line will show how the nut moves less and less as the cop bottom increases in size.

The straight dotted line joining A and N represents the uniform movement of the nut up the screw, and it is easy to compare the two lines and from them understand how the movement is quick at first and slow at the end. It is almost needless to add that a curve similar to Fig. 70 would be given if the results were based on an investigation of the quadrant itself. When the mule is not fitted with some automatic arrangement for actuating the quadrant screw, the "minder" attends to it himself, and it requires a considerable amount of skill and attention to so move the nut as to give good results. When dealing with the subject of automatic "governors" (or, as they are sometimes called, "strapping" motions) further reference will be made to this subject.

To avoid complications, no attention has been paid to the effect which the tapered spindle has upon the question of winding. The diameter of the spindle where the cop bottom finishes is larger than the part where the full cop is complete. To compensate for this taper, some additional arrangement is necessary to help the quadrant; the mechanism employed is usually termed a "nosing" motion, but, as it is generally actuated from the "shaper" or fallers, its consideration will be deferred until a full examination of both these features has been made.

In the accompanying sketch, Fig. 71, a view is given of the quadrant and its connections. Only the essential features are shown, the chief ones being drawn in full lines. The driving of the quadrant is obtained in an indirect manner, and is an example of a rectilinear motion producing a circular one. Previous pages have described how the carriage receives its inward motion through a large scroll on the back scroll shaft, and it will be remembered how the carriage by this means had an irregular movement given to it. Now it is quite clear, from the foregoing explanation of the quadrant's action, that the forward motion of the quadrant must correspond to the motion of the carriage; therefore the irregularity of the one must be reproduced in the other. The best way to obtain this result is to drive the quadrant from the carriage itself, either directly, as shown in the drawing, or indirectly through the back shaft.

On reference to Fig. 74 it will be noticed that a band is fastened on the carriage square at J, whence it passes towards the back of the headstock and over a loose pulley H, or in some cases over a pulley on the back shaft. From this point it returns to the front of the headstock, and after passing round the quadrant drum G several times, it is attached to the carriage square at K. In whichever direction the carriage moves, it will, by means of the band, drive the drum G. On one end of the shaft that carries the drum is keyed a small pinion F, which gears into the toothed portion of the quadrant E; the drum and wheel F are so proportioned that one complete draw causes the

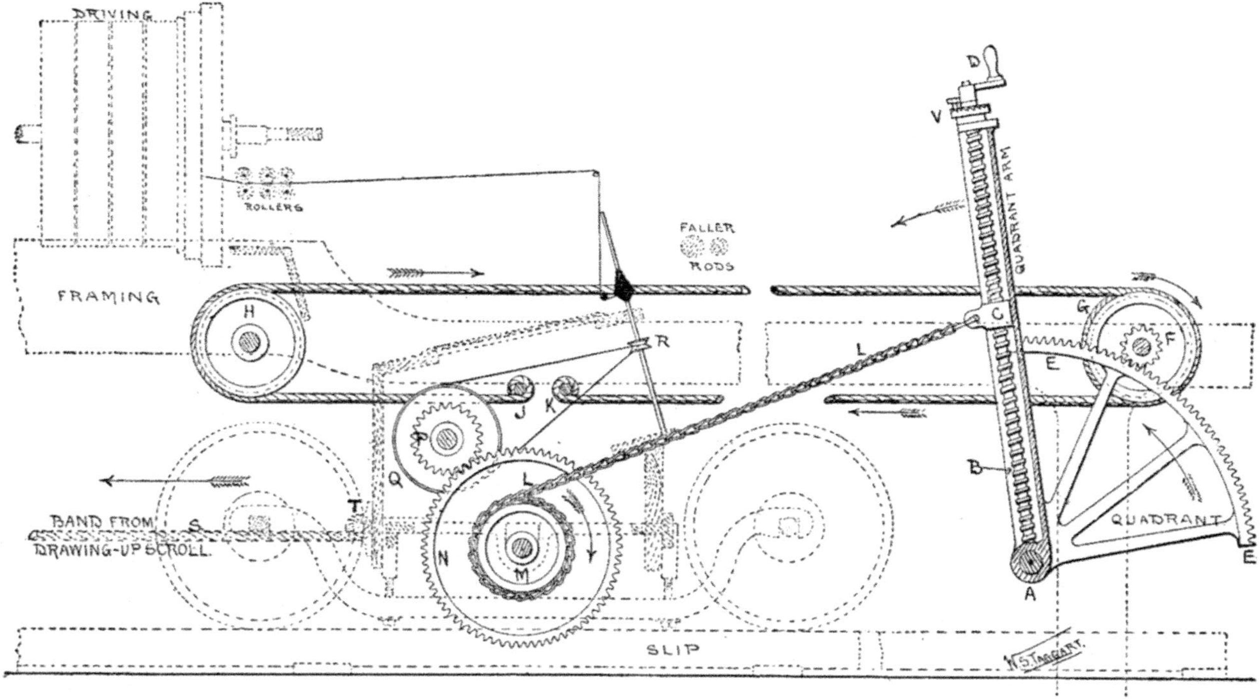

DRIVING
ROLLERS
FALLER
RODS
FRAMING
QUADRANT ARM
QUADRANT
SLIP
BAND FROM
DRAWING-UP SCROLL
W. S. TAGGART.

FIG. 71.

quadrant to move backwards and forwards through a right angle about the centre A.

The screw B is carried in the hollow box part of the arm by bearings at each end; its upper end at V has fitted to it a ratchet wheel, into the teeth of which a pawl engages. A handle D enables the screw to be turned in either direction, so that the nut C can be raised during the building of the cop bottom, or lowered to its starting point for the commencement of a new set of cops. The chain L passes from the quadrant nut on to the winding drum M; the end of this drum carries a large wheel N, which gears into a smaller wheel P on the tin roller. In this way the spindles receive their motion as the chain is unwound from the winding drum. The precise action of this connection between the winding drum and the tin cylinder can now be explained; a large view of the arrangement is shown in Fig. 72.

Winding Drum and Tin Roller.—We have already described how the spindles receive a quick speed from the rim shaft during the spinning process, and a drawing was given in Fig. 27 illustrating the driving arrangement. When winding takes place, with its comparative slow speed, some method must be adopted to disconnect the tin cylinder from the rim shaft driving, so that the spindles can be driven independently from the winding drum.

The sketch, Fig. 72, fully explains the means adopted. The chain L passes round the winding drum M; the end wheel N from this receives its motion, which it transfers to the wheel P, which runs loose on the tin roller. This tin roller wheel P is formed with a disc Q fixed on its boss, or cast in one piece with it; on the disc is fastened a stud, which carries a catch or click C; this catch can, when occasion requires it, be put into gear with a ratchet wheel A, which is keyed on the tin roller shaft. This occurs when winding takes place, so that the revolution of N, brought about by the unwinding of the chain from the drum, causes the tin roller to be driven through the catch C and ratchet wheel A.

Details are given in the sketch, which enable the action to be easily understood. When the winding drum receives motion, immediately the carriage commences its inward

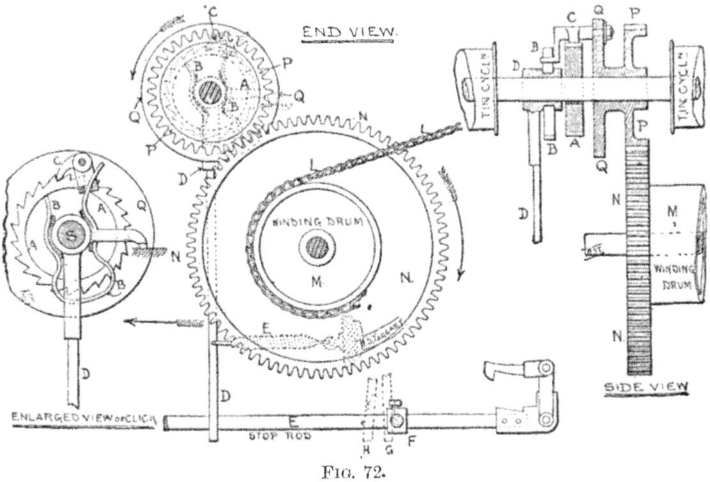

Fig. 72.

run, the disc will begin to revolve, and in doing so will cause the click C to engage with the teeth of A, and so rotate the tin roller. This is the chief element of the arrangement, but there are two considerations to be taken into account: in the first place, the click C must be kept out of contact with the teeth of the click wheel A during spinning; both noise and damage would result if this were not done; secondly, we must recognise how important it is that winding begins immediately the carriage starts in. In order that this shall happen, the click C must be in a position to engage instantly with the teeth of A; otherwise a slight interval will elapse; for instance, when the click is just on the point of one tooth, it must pass over the pitch before engaging with the next. In this time, although it is slight, the carriage will have moved, and as no winding has taken place the yarn becomes slack and snarls are formed. To obtain both of the above necessary conditions a pendant lever D hangs loosely from the tin roller shaft; and on a groove in its boss there is placed a spring B, so shaped as to grip the boss firmly; a leather lining on each side of the spring gives the necessary resistance in the form of friction to the free movement of the spring. One end of the spring is extended, and fits in a slot made in a projection of the click C. Now let it be supposed that the carriage commences its inward run; the rod D hangs vertical, and its lower end is in contact with a stop fixed on the rod E (the purposes of this rod have already been fully explained in connection with Figs. 42 and 43). When backing-off is finished the rod E shoots back, and the stop F moves the rod D on one side, as shown, for example, by dotted lines, from D to H. This oscillation of D carries with it the spring B, which clips its boss; the spring in its turn acts on the click C and forces its other end into engagement with the teeth of the ratchet wheel A; winding can therefore take place instantly, because the click being already in contact with a tooth can at once commence to turn the tin roller shaft.

When the inward run is completed, the strain on the click is taken off, the disc begins to revolve in the opposite direction, because the winding drum must wind on the chain during the outward run; and the ratchet wheel A continues, after a moment's stoppage, to revolve at a high rate, inasmuch as twisting is now in progress and the outward run has commenced. The click is therefore inoperative during the run-out and backing-off, and only comes into action when the run-in commences. A spring E ensures the clip being kept out of contact with the teeth of the ratchet wheel, and brings the rod D back into position after it has been acted upon by the stop on the rod E.

Shaper or Building Motion.—At this stage we can enter upon the discussion and description of the building of the cop, a subject so closely allied to the "winding" process that neither can be perfectly understood without reference to the other. In the analysis of the quadrant it was shown how the spindles were made to revolve whilst winding on the yarn, so that in spite of a continual variation of conditions the same length was practically wound on during each inward run.

In treating of the building operation it will be shown how the yarn is guided on to the spindle in a manner suitable to the revolutions produced by the quadrant. In considering this question the important point must not be overlooked that the quadrant, in its fundamental action and principle, is, within narrow limits, practically an unadjustable piece of mechanism. The building of the cop therefore must be performed in entire accordance with the conditions set up by the action of the quadrant, as far as the conical form is concerned.

On referring back to Fig. 54 it was pointed out that the cop underwent several changes in form from the first layer put on the bare spindle to the full size of the cop bottom. It is the duty of the building motion to guide the yarn on the spindle in such a manner as to produce as perfect a shape as possible during these various changes of form. The subject is not by any means a simple one, and to those having charge of the mule, it proves to be one of the most delicate and intricate features of the machine, requiring, when necessity arises, a considerable degree of skill and attention in its management and manipulation.

A general description of the mechanism will first be given, in which its essential features will be pointed out and the principal effects outlined. A detailed examination will then be made, which will probably be of great practical use to those most interested in the subject.

Fig. 73 presents all that is necessary at this stage to enable the action to be understood. The yarn coming from the front rollers is guided on to the spindle by passing over a wire V. This wire is carried by a series of levers or sickles fixed to the copping faller rod W. To give motion to the copping wire, this faller rod must be actuated from the building motion, and it will partake of a species of oscillating action, producing an upward and downward movement of the wire. A special lever O, usually called a "sector," is fixed to the copping faller W; one end carries the wire V, while the other is hinged at U to a pendant arm T, called the "faller leg." The lower end of this faller leg rests upon a stud R, which carries a bowl P. The bowl, during the travel of the carriage, runs upon the upper edge of a specially formed rail, called the "copping" or "shaper" rail. It will readily be seen that if this rail is inclined in any way, the bowl will rise and fall as the carriage moves in or out. During the outward run, the faller leg is quite free from the slide that carries the bowl; but, as already explained, the backing-off brings about a change, which causes the leg to be raised, until a recess on it passes over the upper end of the slide at Q, and in this position it rests upon the slide Q, and is said to be locked. When the inward run commences, the bowl P occupies the first position, as shown on the sketch, and as it passes over

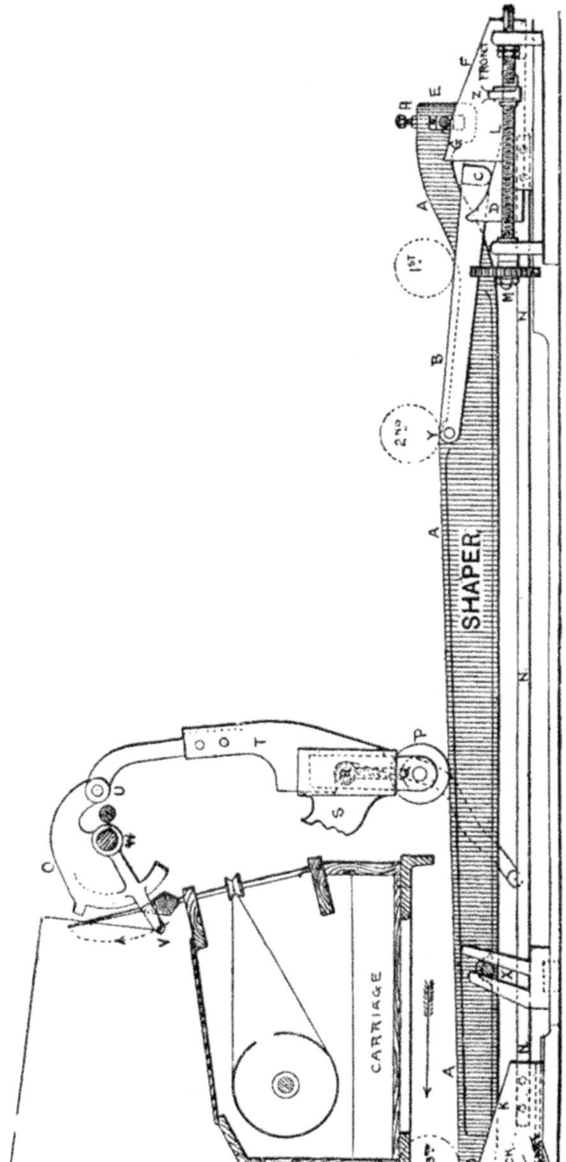

FIG. 73.

the surface of the rail at B it rises up until it reaches the second position at Y. This of course lifts up the whole of the faller leg T, and causes the wire at V to be correspondingly depressed; "crossing" takes place during this period, and a few coarse spirals of yarn are wound on the spindle. As the bowl passes from 2 to 3, it descends the inclined rail A, and the faller leg falls. This causes the wire V to be raised, and while this takes place the yarn is being guided in close spirals upon the cop. At the end of the inward run the faller leg is freed from its connection with the slide Q R, so that the wire is raised out of contact with the yarn, and the outward journey of the carriage is made without the shaper in any way affecting the copping faller.

Crossing.—The preceding remarks will have conveyed the essential idea of the builder motion. We can now go a step further, and point out the difference between the two inclines on the main copping rail. The earlier portion on which the bowl travels as the carriage goes in, is short compared with the later portion, although the vertical height through which the faller wire passes is practically the same; this means, of course, that the wire falls, and therefore puts the yarn on the cop much more quickly during the downward movement than during the upward movement. In doing so, the result is that the yarn is laid in a coarse series of spirals one way, and in a closely arranged series the other way, thus producing a foundation that is strong and not easily unravelled through carelessness. The proportion of length that the shorter incline bears to the longer one does not, however, give any idea of the respective rates of motion of the downward or upward movement of the wire; this can only be obtained by a careful consideration of the speed of the carriage during the traverse over the two parts of the shaper rail. By such observation it will be found that the downward motion is performed considerably quicker, compared with the upward movement, than a comparison of the two lengths would lead one to expect. This point is mentioned because there is a tendency to use simply the length of the inclines for a comparison of the two movements, thus conveying an altogether erroneous impression.

The word "crossing" is generally used to describe the quick downward movement of the wire; and when it is completed the bowl Y is on the point where the two inclines meet. When in this position it is the usual practice to have the quadrant screw vertical; but of course other considerations of a practical character may lead to a variation from this practice, and it may remain a factor to be regulated according to the requirements of the machine; therefore no hard-and-fast rule can be laid down in regard thereto.

We will now refer again to Fig. 54. The drawing shows what other conditions must be fulfilled by the shaper mechanism in order to build a cop. In the first place it will be noticed that the cop commences with a short layer at A B, and each layer afterwards is made longer, as at J, G. The shaper must therefore be adapted to produce this result; in other words, the inclines of the rail must be altered in such a way, that for each inward run, the vertical height between the highest point of the rail and the lowest must be made to increase.

Again, it will be noticed that in addition to the increased length of the traverse of the faller wire ("chase" it is generally called), the finishing point of the downward movement, and consequently the starting point of the upward movement, is not quite so low after each layer. This is quite a small difference when each layer is considered by itself; but taking the cop as a whole it results in a conical end being formed on the lower part of the cop as at A, E, J. The shaper must be arranged to produce this result, for the strength of the cop depends equally upon this lower end being well formed as upon the upper conical portion. On reference to Fig. 73, the complete arrangement is shown by which the above conditions can be fulfilled. The shaper rail is represented as resting, by means of pins or small bowls at E and K, upon short inclined surfaces; these are termed front and back inclines, and are connected by a rod N, so that any movement is produced equally in each. It will readily be understood that if the front and back inclines are moved, the ends of the shaper rail will be raised or lowered as the case may be, and will thus alter the position of the path along which the bowl travels. This will cause the yarn to be put on the spindle in a new position. To have this position correct, it is necessary to move the inclines in a special way, and also to have the inclines so formed as to give the required shape. The movement of the inclines is brought about by means of a screw L working in a nut carried by the front plate F. Supports from the floor fixing prevent the screw having any horizontal movement, so that any motion given to it produces a movement in the front plate, and this is transferred to the back plate through the connecting rod N. The screw is actuated from the carriage during each draw, through a ratchet wheel M fixed on one end of it. This wheel plays an important part in the working of the shaper, and attention will more fully be drawn to it at a later stage.

It may then be briefly stated that the inclines have their surfaces F and K, as a rule, unequally inclined to one another to produce variations in the chase; the earlier, or higher, portions differ in form from each other, in order to produce the bottom conical surface called coning; the inclines are moved, in order to lower bodily the whole of the rail and thus make the cop longer; the shorter incline B on the copping rail is made loose, so that it can be guided on the short incline D in such a manner that the faller leg can be locked always about the same distance from the "nose" of the cop, no matter whether the cop is short or long.

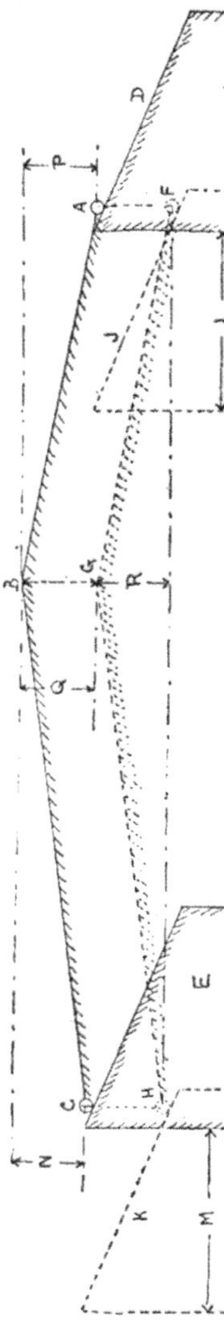

Fig. 74.

The above brief statements will now be enlarged upon, and by a series of diagrams it is hoped to make the matter perfectly clear to the reader. In the first place, while dealing with the principles, it will not be necessary to make the diagrams proportionate to actual condition; and in the next place, it will be assumed for a short time that the two inclined surfaces of the copping rail are in one piece.

Although all the requirements for building the cop are carried out in an apparently very simple manner, yet the difficulty experienced by most people in thoroughly understanding the mechanism, proves the necessity of a little more than the usual description being given. The motion will, therefore, be analysed as completely as possible, consistently with the object of this book.

In the first place, let it be noted how the bodily movements of the back and front inclines alter the length of the cop. This is illustrated in the diagram, Fig. 74. For the present we will assume the shaper rail A B C to be in one piece, and the front and back inclines to be equal to each other and quite straight. According to the diagram, the length of the layer of yarn put on the cop would be proportionate to the vertical distance P between A and B, or to N between C and B. If the inclines D and E are now moved forward to J and K, a distance equal to M and L, the shaper rail A B C will fall bodily to the position F G H; and since the front and back inclines are equal, every point of the shaper rail will fall an equal amount—which in the drawing is shown at Q. The effect of this "lowering" of the rail is to "raise" the faller wire so that the yarn is wound on to a higher part of the cop and thus lengthens it. From this diagram it is an easy matter to understand the general principle underlying the method of lengthening a cop; and as the cop is built up by additions to its length, the principle remains the foundation of the mechanism; but owing to the necessity of increasing the length in a special manner, variations must be made in the arrangement in order to fulfil the required conditions.

A drawing is given of a portion of a cop in Fig. 75. From it we shall quickly see what the building motion must do in placing the four layers of yarn shown in the diagram. The first layer F G is wound on the bare spindle, and is a comparatively short one; other layers are added until the layer K H is reached; and here we notice that it is longer than the first layer. It is, however, from the last layer C J of the cop bottom that we shall be able to observe the changes that have been effected in the form of the cop and the layers. In the first place, C J is considerably longer than F G, though it is well to bear in mind that in spite of this there is practically the same length of yarn

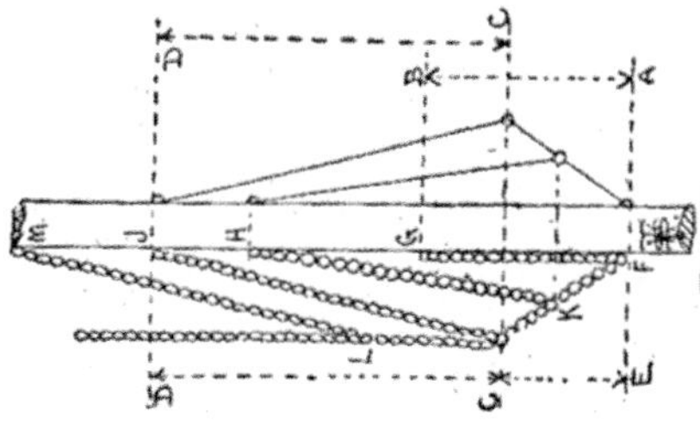

FIG. 75.

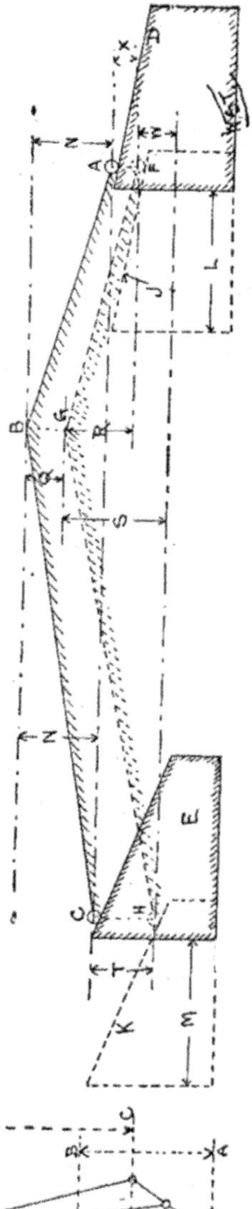

FIG. 76.

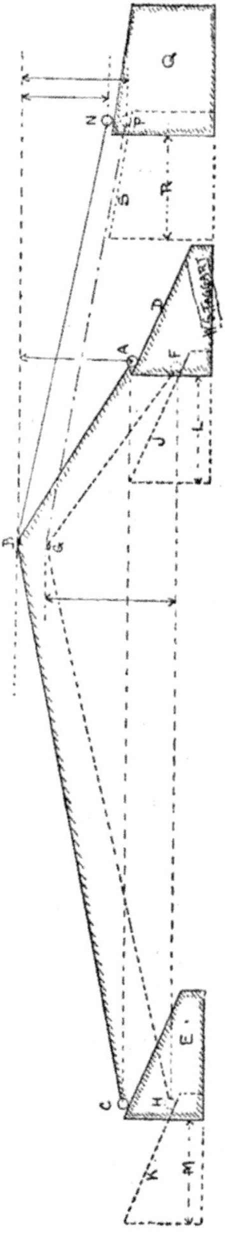

FIG. 77.

wound on in both cases. In the next place, we notice that the point on the spindle where the downward movement of the faller wire, or "crossing," commences, has been raised from G to J; and at the same time we observe that the point for commencing the upward motion of the faller wire has been raised from F to C. A comparison of these two lengths will show that the finishing point of the "building" layer has risen at a quicker rate than the commencing point, and consequently the length E C is much less than G J. It is this fact that causes the layer to be lengthened; therefore in arranging the shaper mechanism, one of the chief considerations is to so adapt the shaper rail that the point on which the faller leg bowl rests, when the "building" layer commences, shall not be displaced to the same extent as the point which represents the finish of the layer. This opens up a very interesting question, and it will be profitable to fully discuss it. Fig. 76 represents in a diagrammatic form the simplest arrangement; A B C is the rail all in one piece; both ends A and C rest, as in the previous case, upon inclines, but these inclines, instead of being equal to one another, are made with different angles: for instance, the front incline D is more horizontal than the back incline E; when, therefore, the inclines are moved forward to J and K, the rail A B C will be lowered; but owing to the difference in the inclines, the ends A and C will not fall to the same extent, and the angles of the two portions of the shaper rail will consequently become changed, and a variation will be introduced in the length of the layer of yarn put on the spindle. The diagram, Fig. 76, shows the extent of the alteration occasioned by making the front and back inclines of different inclinations. The vertical distance between A B or C and B is equal to N; this, we will presume, gives the first layer, as at F G, Fig. 75. If the inclines are now moved a distance equal to L and M, the end A will fall to F, B will fall to G, and the end C will fall to H. A comparison of these distances as marked at X, Q, and T respectively, shows that the end C has been lowered much more than B or A. Now, since the point B represents the beginning of the upward or "building" layer (for instance, from F to G, K to H, or C to J), the lowering of B to G will be shown on the cop by the change of the commencing point from F to C; and also, since C represents the finish of the same layer, we find the termination of the layer is much higher up the spindle at J than when the shaper rail occupied its first position, which gave the terminating point at G. The distance G to J produced by the lowering of the rail from C to H is much greater than the distance E to C, which is brought about by the lowering of the point B on the rail to G. It will be observed that only sufficient of the front and back inclines D and E have been used to make the cop bottom; and the variation in their inclination has had the effect of raising the point F to C (Fig. 75) and so forming a conical end on the bottom of the cop; it has also had the effect of lengthening the layer as the cop bottom enlarged. The operation just described is generally termed "coning"; and it must be carefully noted and understood that the chief essential in producing it is in the difference of the inclinations of those portions of the front and back inclines on which the shaper rail rests while the cop bottom is being built.

In connection with the diagram, Fig. 76, it will be noticed that A and C are in the same horizontal line. This means that the yarn in Fig. 75 commences at G, and comes down to F, and back again, exactly the same distance, to G. It is very desirable that the same thing should occur in the last layer also; but according to the diagram this is impossible, for it will be observed that, in consequence of the end A not having fallen to the same extent as C, the beginning of the downward "crossing" movement does not correspond with the finish of the upward "building" movement. The result is that crossing commences below the actual nose of the cop, and since the position of the bowl and its carrier, which travels along the shaper rail, regulates the locking of the faller leg, we get the locking operation performed a little later than is desirable.

The method now adopted of overcoming this fault of the shaper rail being in one piece, is to make the front short incline loose, so that its inclination can be so regulated to give both the crossing and building layers the same exactness, in order that locking shall take place always at or near the nose of the cop.

Fig. 77 has been prepared to illustrate graphically the essential features of the shaper with loose front inclines. The shaper rail is A B C; the part A B, instead of being in one piece with B C, is loose, and hinged at B, so that it can alter its inclination irrespective of the inclination of B C; by this means it is possible to lower the point A to the same extent as the point C, and in this way the locking of the faller at the termination of the backing-off will always take place at the nose, or, in other words, at the highest point of each layer of yarn.

Let us now carefully examine the arrangement as shown in Fig. 77, and see how by its means we can build the cop bottom. To sum up the conditions: it is necessary that B C should be capable of altering its inclination; also it must be so arranged that the point B will fall to a less extent than the end C; and at the same time the portion of the rail at A B must be so arranged that the end A will be on about the same horizontal level as the end C.

In order to fulfil the first condition, the long portion of the shaper rail C B is lengthened out to N, and this end rests upon the front incline Q. This incline, so far as the cop bottom is concerned (and it is this part of the cop with which we are now dealing), has a different inclination from the back incline E, so that any movement of the two inclines, for instance to S and K, will lower the rail C B to G H. This differential lowering fulfils the conditions for lengthening the chase of the cop bottom, as was shown in connection with Fig. 76, and by its means the bottom conical end of the cop is obtained.

The second condition is fulfilled by the end A of the loose incline A B resting on a separate incline D, which moves forward to the same degree as the back and front inclines. The short loose incline is therefore lowered quite independently of the position of B, and we can, by making a suitable profile on D, lower the end A to any required extent necessary for locking the faller leg correctly.

From the sketch it will be an easy matter to make a comparison of the various distances moved; A and C are seen to be on the same horizontal distance, and in this position the first layer is put on the spindle. When the rail is lowered, F and H are still on the same level, and the last layer C J, Fig. 75, is put on in this position. The drawing will also show that the point B has been lowered much less than the ends of the shaper rail, and so we conclude that while the "chase" has been lengthened the ascent of the point F has not taken place at the same rate.

When the cop bottom is finished, all the inclines, Q, D, and E, partake almost of the same inclination, and continue to build the cop to whatever length is required. The next sketch will show the three inclines in such a manner that they can easily be compared, and a few remarks will be made on the character of their profile.

It must be understood clearly that in the preceding descriptions the principle only of the copping motion has been dealt with, and for that purpose only those portions of the front, middle, and back plates have been used to illustrate in diagram form the building of the cop bottom. The inclinations of the three plates were also denoted by straight lines. In practice, however, it is found beneficial (and indeed necessary) to avoid the sharp angle where the cop bottom finished; and moreover the bottom conical part is very seldom straight, but slightly convex in outline. These considerations necessitate the use of curved inclines specially shaped to produce the desired result. The inclinations of these curves, however, differ from one another, so that our explanation is not affected.

In order to show actual conditions, and so that a comparison can be made of the inclinations, the front, middle, and back plates, taken from a mule, are shown in the drawing, Fig. 78. Comparing the front and back plates, we find a great difference in their inclination at those parts used during the building of the cop bottom, the reason for which has been so fully explained, that there ought to be no difficulty in thoroughly understanding it. In regard to the middle plate, the curve up to F is almost similar to that of the back plate, the reason for which is almost obvious, since it has been shown that the end of the loose front incline of the rail must fall almost in the same degree as the back end of the shaper rail.

There remain yet the other and longer portions of the plates to be mentioned. These portions B C, D E, and F G are used for building the remainder of the cop after the bottom is completed. They are, as a rule, perfectly straight in profile, but circumstances may possibly arise necessitating a very slight variation from the straight line. The inclinations of the middle and back plates are seen to be practically alike, but a difference between the front and back plates requires some explanation. We have seen that a long chase is being made when the cop bottom is complete. This may be carried through to the finish of the cop, but

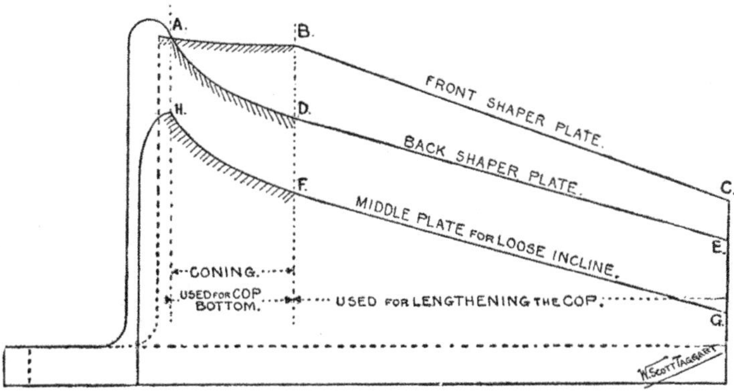

FIG. 78.

frequently it is caused to shorten, the object being to gain compactness and weight. The difference of inclination of the two plates brings about this result, as was pointed out in diagram, Fig. 75. In that illustration, however, the difference caused a lengthening of the chase; but in Fig. 78 the opposite effect is produced, because the back plate has less inclination than the front one.

Long Incline of Shaper Rail.—So far, it has been assumed that the long incline of the shaper rail is an inclined surface on which the shaper bowl travels to and fro; but a little consideration will show that this incline must be made a special shape in order to lower the bowl in such a manner that the faller wire guides the yarn on the cop in a series of regularly spaced spirals. This point will now be examined. It has already been shown that each revolution of the spindle, during winding, winds on unequal lengths of yarn. We also know that each length of yarn wound on represents the distance moved by the carriage. A further condition known is, that the faller wire must rise equal distances for each revolution of the spindle, this being necessary if the spirals of yarn are to be spaced equally.

Having these facts as our guide, it is an easy matter to find the outline or shape of the rail required. A cop has been carefully unwound and the length of each turn of yarn on the cop measured. These various lengths have been marked out on the line 1 to 22 in Fig. 79, so that in reading the drawing it must be understood that as the carriage moves from 1 to 2 the spindle has revolved once and wound on a length of yarn equal to the distance moved by the carriage, viz. from 1 to 2. The same thing occurs in the carriage moving from 2 to 3, and so on through all the positions marked up to 22, when the stretch is completed. It is to be noted that 21 revolutions of the spindle have been made and that each revolution has wound on a less length than the preceding one, so that a long length has been wound on in the first revolution and a short length in the last revolution. But it is readily seen that although all the lengths wound on are unequal, we still know that they differ from each other by a practically equal amount.

Now that these lengths have been measured off on Fig. 79, the long incline of the rail is drawn in, and the distance between the highest point or shoulder and

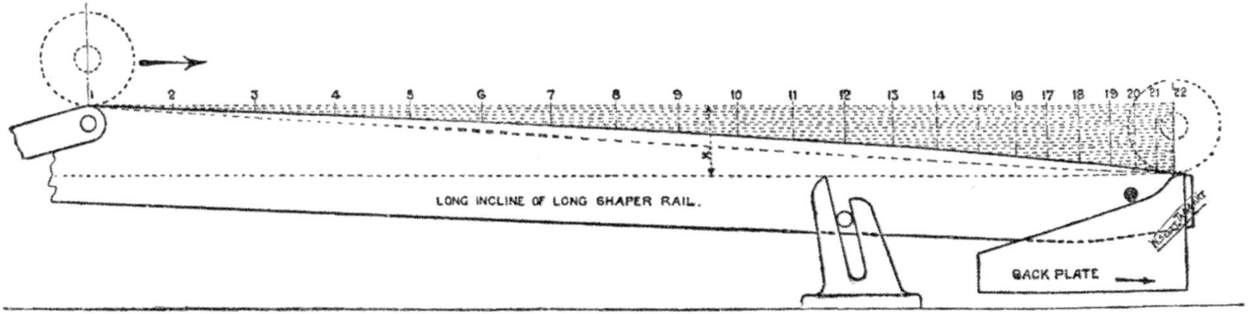
1
2
3
4
5
6
7
8
9
10
11
12
13
14
15
16
17
18
19
20
21
22
X
LONG INCLINE OF LONG SHAPER RAIL.
BACK PLATE

FIG. 79.

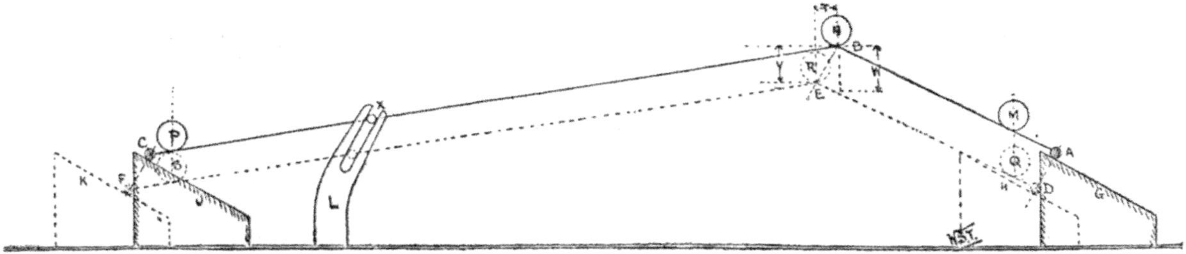
P
C
S
F
K
J
L
X
Y
R
E
B
W
M
Q
A
D
H
G

FIG. 80.

the lowest point is divided up into *equal* divisions corresponding to the number of revolutions of the spindle during winding. (This can be calculated, but it is much easier to count the turns by using a well-made cop.) This equal division is done because of the equal pitch of the spirals. By drawing vertical lines through the carriage positions and horizontal lines through the shaper bowl positions we obtain intersections through which a curve can be drawn, and this curve is the shape required on the long incline of the rail in order to move the faller wire in such a way as to guide the spirals of yarn on the conical surface at equal distances apart. The curve on the rail is parabolic in character, and it clearly cannot be correct except for a certain layer or a series of layers all of which are equal. The layers in the length of the cop after the cop bottom is finished are the most numerous, and therefore the rail is made to suit these layers. The cop-bottom layers are not so regular in their spacing, and, moreover, it so happens that the action of the quadrant not being absolutely perfect, the two qualities counteract each other somewhat and thus enable a fairly perfectly-shaped cop to be made.

It must also clearly be understood that the question of absolutely equal spacing of the coils on the cop is assumed to some extent. Careful measurements of cops show differences from different types and makes of mule, and when it is remembered that practically all shapers as at present used are the result of cutting, carving, and filing, such differences must be expected and allowed for; they are compromises in adjustment both for the quadrant and the curved edge of the long incline of the shaper plate.

Another feature of the shaper motion that calls for

some explanation is the use of the small inclined floor bracket X, as shown in Fig. 72. This bracket is generally known as the "steadying" bracket. A diagram has been prepared in Fig. 80 in order that its action may be explained, but in this connection it must be observed that the reasoning applies only to the old form of shaper. Loose incline shapers do not require an inclined steady bracket, and although they are generally used the inclines are formed to compensate for them. If a vertical slotted bracket was used it would simply mean altering the shaper ratchet wheel. The rail A B C rests on the two inclines G and J, and a pin X in the long rail fits in an inclined groove of the bracket L. As the inclines are moved to H and K the rail is lowered, but instead of falling vertically the groove of L causes it to fall in the direction of the incline and to take up the position shown by dotted lines at D, E, and F. The chief effect of this lowering of the rail is to give a horizontal movement to it, so that the relative amounts of the short and long inclines of the rail are altered, and the highest point of the rail is moved inwards to the extent shown at T. Several reasons are assigned for this action. The principal one may readily be understood when we point out that a longer time is taken up in crossing and a shorter time in winding; the yarn is consequently relieved of considerable strain as the cop lengthens, because since backing-off unwinds less and less off the bare spindle as the cop lengthens, it is advisable to do the crossing a little more gradually, and this can be done by taking more time to do it in. "Crossing" in any case induces an unusual strain in the yarn; but in the earlier stages of building there is so much yarn over and above the actual length of the stretch that the strain is not of any moment. As this surplus yarn becomes less, it is almost necessary to adopt a relieving action to prevent the rapid winding, which takes place during crossing, from breaking the yarn; and this is obtained by lengthening the time during which the crossing is performed. Of course a guide bracket of some kind is necessary to guide the rail in its descent, and for this purpose a straight bracket would be effective. The inclined bracket, however, serves another purpose in giving greater stability to the rail; for, since its inclination is opposite to the inclined plates on which the rails rest, it prevents vibration and keeps the shaper steady; hence its name.

Shortening the long part of the rail has the effect of winding a little more tightly, and thus helping in compensating for the diminishing diameter of the spindle. The positions of the faller-leg slide bowl are shown at M, N, and Q; and as the start and finish are always the same throughout the cop they begin and end at the same place, Q and S, at the finish; only the position at the highest point of the rail is altered from N to R; but as previously observed, this alters the relative lengths of the short and long portions of the rail.

Defective cops and their remedies.—To the practical reader the foregoing description of the shaper, and the explanation of its principle, will be of great assistance as an aid in solving many of the problems associated with the formation of the cop. It is in connection with the shaping mechanism that perhaps the greatest amount of intelligence and skill is required in managing the self-acting mule of to-day; and, as good and bad results of its working concentrate themselves to a large extent upon the shape of the cop, the subject of defects in its formation may be made the occasion for investigation. A very great number of imperfections arise in the shape of the cop either locally or generally. To enumerate them all would unduly extend the limits of this book, but the importance of the subject necessitates that some attention should be given to the most characteristic faults that arise in connection with the cop.

Badly-formed cops are not always the result of a faulty shaper, so that when a decision has to be given upon the cause of any particular irregularity in the cop's condition or shape, a very careful investigation into the matter is required before fixing on the exact point for correction. In the following notes, therefore, it must be understood that the remedies pointed out are only suggestions of what may be the cause. We shall restrict our attention first to defective cops resulting from other causes than those directly connected with the shaper rail or its inclines.

(1) Cops, instead of being perfectly parallel, may be very ridgy on their surface. Several causes are capable of producing this result. For instance, the bowl that runs along the shaper rail may be worn flat in one or more places; it may be loose on its stud; or it may be badly mounted and work on its edge instead of its full width. The bowl also on which the faller leg locks, if faulty in the same way as the rail bowl, will cause ridges. The fault can be corrected by returning the bowl and replacing the studs, or so mounting the bowl that it runs level on the rail. The ridges may be produced by the copping faller rod not working smoothly in its bearings through having play in the faller stands, especially in those near the head-stock. The shaper screw may not be perfectly true, and its irregular movement will give unequal advances to the inclines; a good screw will remedy this fault. If the collar which fits the screw binds, it may also cause a ridgy appearance. One frequent cause of ridginess is due to

the tumbler not taking the teeth of the ratchet wheel regularly; this gives an irregular movement to the inclines, and the irregularity is reproduced on the cop. Other causes of ridgy cops may be found in a loose backing-off sector; in a carriage not being firmly fixed in the square; and occasionally, through not gearing the quadrant pinion deep enough, ridges have been produced.

(2) Cops may be longer at the outer ends of the carriage than those nearer the headstock. Weakness in the faller shafts is the chief cause of this defect; and in some cases, if the weights are too heavy and placed too near the outer end of the carriage, the same fault arises. Faller shafts must be strong enough to resist torsion, and the weighting must be arranged to obtain a uniform strain throughout.

(3) Cops may be soft throughout the mule. It is quite possible that cops should be made soft, especially if cotton of a poor quality is being used; such cotton cannot resist breakage so well as better cotton, and accordingly the weighting of the under faller must be much less in order to prevent breakages when backing-off takes place or winding. Cops become soft, however, when such a condition is neither necessary nor desirable, and it may arise from the following causes:—

Winding may be badly performed, that is, the quadrant may turn the spindles too slowly. To remedy this the quadrant must be put back a little. It is a frequent practice to put the quadrant forward so as to obtain easy winding and avoid breakages; but it results in a soft cop.

If the driving strap touches the fast pulley during the run-in, the result will be soft cops; see, therefore, that the strap when on the loose pulley is quite clear of the fast one.

Sometimes, after years of work, the highest point of the shaper rail, where the two inclines meet, becomes worn and flat; this makes the shoulder of the cop larger and softer throughout its length.

Faller rods sticking in the stands is an occasional cause for soft cops; and sometimes a difference in level between the slips on each side of the headstock has produced the same result.

If, during backing-off, the tin roller slips a little on the shaft we get soft cops.

When the softness appears in the cop after a certain length has been made all right, and the mule has previously made a good parallel cop, it shows that the nosing has not been carefully performed, or has even been neglected. To cover this kind of carelessness the winding is slackened a little, which causes larger shoulders to be made and correspondingly softer cops.

(4) Cops may be soft close by the headstock only. In such a case, the drawing-up scrolls may be too large for the length of stretch. Anything wrong with the scrolls, such as being loose on the shaft, wrongly set, or not in right position, will give to the carriage during the run-in an irregular movement, and so cause softer cops at one part than another. If the back shaft is too weak we get a similar result.

(5) When thick and soft noses are made, the following may be sought for as the cause:—

Delaying the use of the nose peg or nosing motion.

The under faller being depressed too soon before the copping faller has been unlocked.

When the backing-off chain is too slack we get a frequent cause of soft noses, and correspondingly, if it acts too quickly. A defect in the shaper rail which causes a hollow at the upper end of the chase produces slack yarn at the termination of winding, and in this way either snarls are made or soft noses.

We can now deal with badly-shaped cops whose faults may be traced directly to the shaper. A few typical cases will be examined, and remedies suggested whereby a correct form may be obtained.

In the first place, a thorough examination must be made to see that the shaper mechanism is in perfect working order; the studs firmly fixed, the bowls quite round and set without "winding"; the faller sickles, especially the faller sector, connected to the faller leg, must be securely fastened on the faller rods; no dirt or waste ought to lie about the copping motion to prevent the free working along the slides of the inclines; and the faller rods ought to be well supported in their stands, and perfectly free to turn.

Before flying to the shaper for a remedy, one must be quite sure that the fault does not arise from some other defect in the machine. For instance, the motions may not be acting in unison; a faller wire in its movement may catch some chain or bracket; or weights may be touching the floor or some fixing, and thus producing irregularities in tension. These and a number of other apparently small matters all have some influence in affecting the building of the cop, and no improvement can then be effected through the shaper.

In the accompanying sketch, Fig. 81, a few well-recognised faults in cops are shown. The one marked A may be taken as a standard by which the others may be compared. B represents a cop ridgy in its body part instead of being perfectly parallel. All the other cops show in their full lines variations from the dotted lines, which should give a cop similar to A; we can in this way see whether too much or too little yarn has been placed on the parts that are faulty. Of course it will be understood that a combination of the faults illustrated may be found in a cop at one time: for instance, a ridgy cop can have a hollow bottom and a soft nose.

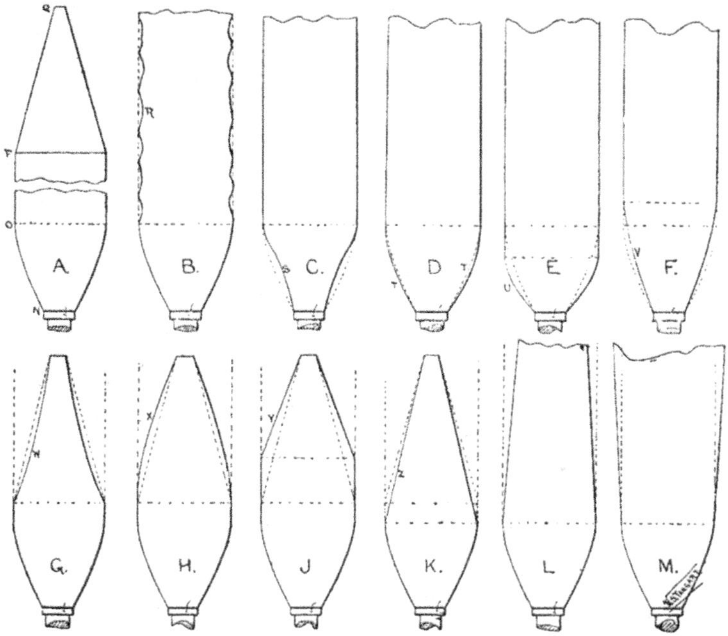

Fig. 81.

In suggesting alteration in the shaper for remedying the defects illustrated, a warning must be given that the utmost care possible is absolutely necessary in making the correction by filing the plates. It should always be done gradually, and in order to gauge the alteration accurately a template or exact copy of the plate should be made previous to the filing; and its position should be definitely noted, otherwise the labour of hours may be wasted and the work required to be begun again. A few words of general purport, but nevertheless important, will probably fix the general principles in mind and serve as a guide in working.

The bottom cone of the cop is formed by the earlier portions of the front and back inclines, namely, the upper short flat portion of the front incline and the short curved portion of the back incline; in most machines marks are put on the plates showing the starting point. Faults in the lower conical portion of the cop may be corrected by attention to these parts of the plates.

The long parallel body of the cop is formed by the straight portions of the front and back inclines; remedies for defects that appear in the body may be sought for at these points of the plates.

The upper cone, or chase of the cop, is obtained from the shaper rail, to which attention must be directed when faults appear in that part of the cop.

Desired alterations in length of chase, length of the body, or length of the bottom cone, can be obtained by raising or lowering the plates that support the rail.

A bulging or "lump" on any part of the cop is due to a "hollow" in the rail or plates, and its location is easy, by observing the position of the bowl on the rail as the faller wire is guiding the yarn on the lump. Correspondingly, a hollow place on the cop is due to the rail or plates being too high or lumpy at the point; its position can be found as above.

Let us now trace out the alterations required to correct the faults in the cops illustrated in Fig. 81.

(B) Ridgy Cop.—This fault is due (if not to the causes

already given) to a similar condition of the straight portions of the front or back inclines, or both. Examine their profile and see whether it is irregular; if so, file carefully until a good profile is obtained. When, instead of a number of ridges, there is only one or two, locate the spots as directed above, and file until a straight body is given. Be careful to note whether a series of irregularities on the body of the cop are ridges or hollows. This is important, for as in the sketch B the bottom is finished the correct diameter, so the irregularities at R are really hollows. By filing the plate to correct the hollows the coning portion would be slightly shortened, and consequently the next cop would be thinner in diameter. To keep the diameter correct, therefore, we must take a parallel filing off the coning part of the incline, so as to have the same length as before.

Again referring to Fig. 81: two common faults are represented at C and D. The narrow thin form S in C is due to a too quick fall of the rail on the coning parts of the inclines; while at T in D it is thicker than is desirable, and is caused by too little or slow a fall of the rail on the coning parts. A series of diagrams in Fig. 82 will enable us to point out the necessary changes to be made in order to correct these faults. To thicken out the bottom at S, Fig. 81, to fill up to the dotted line, the rail must fall slower, and to do this we must not start so high up; the beginning must be brought a little nearer to the finish (see No. 1, Fig. 82). If the full lines represented the starting points for the cop C, an alteration to the dotted lines would cause a slower descent of the rail (because the fall is not so steep), and produce a thickening of the bottom cone. At the same time as this is done we reduce the vertical height through which the rail falls while on the coning part, and consequently shorten the bottom cone. As a rule this is not an objection, as a hollow bottom is generally associated with a long one, and so both faults are corrected. In regard to D, the thinning of the bottom is brought about by adopting an opposite course to that suggested for the specimen C; the coning is therefore lengthened a little by starting a little higher up the plates.

In the specimens marked E and F we also find a state that is occasionally troublesome; E, it will be seen, is too short in the bottom cone, while F is equally too long. To a certain extent they are only a variation of the effects noticed in the cops C and D; this fact, however, will be treated of a little later.

In the first place, to deal with E: the shoulder requires to be raised, and to do this it is necessary to have a greater fall of the rail between the start and finish of the coning parts. This is effected by filing off a portion of the plate at the finishing point of the coning and going up to nothing at the starting point. (See No. 2, in Fig. 82, as marked by dotted lines.)

In the cop F the length from the bottom end of the cop to the shoulder requires to be shortened, consequently we adopt an opposite course to the above; for instance, the fall in the coning part must be lessened, and to do this the filing must start at nothing on the finishing point of the coning part and finish at the necessary depth at the starting point. No. 3 in Fig. 82 will illustrate this.

In the above remarks the suggestions made are probably the most delicate matters that can be found in the whole range of cotton spinning. The operation of filing a plate on the coning part is one that requires the utmost care, and generally another fault appears whilst correcting the first one. The few remarks already made will indicate to the thoughtful reader in what way this arises. It was shown in regard to the specimen C that thickening the bottom meant also shortening it; and in making D thinner we also lengthened it. Something of the kind occurs when

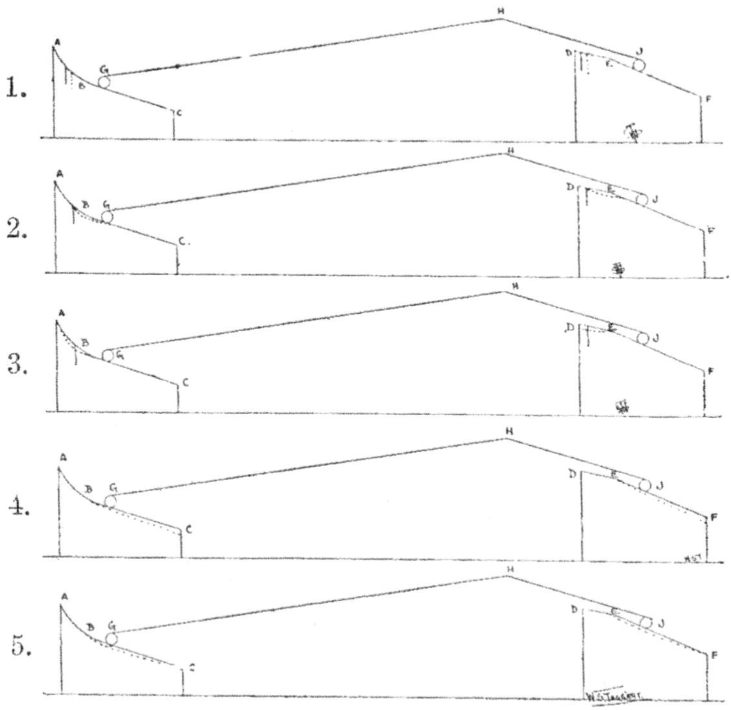

Fig. 82.

dealing with the faults at E and D. To lengthen the cone part at E by filing the plates as suggested, we naturally lessen the vertical distance between the start and finish; but at the same time we thin the bottom, and care is required that by doing so we do not exceed the proportions wanted. The sketch shows how necessary it is to thin as well as to lengthen it; but great care is absolutely necessary in order to maintain a good shape even when the correct length is obtained. To shorten the cone at F we adopt a course of filing that is not nearly so difficult as at E, but the thickening effect on the cop is inevitable, although not relatively so great as at E. The chief difficulty lies in the connection of the coning parts with the straight inclined parts of the plates; and it is sometimes necessary, to keep the coning parts their original lengths (although the vertical fall has been reduced, or *vice versâ*), to file a parallel strip off the full length of the straight part.

While the diagrams may help in conveying an idea of the parts to be filed under certain circumstances, it is only by practical experience that the amount, and the form the filing ought to take, can be decided.

To continue the specimens, we will consider faults that may arise in connection with the nose of the cop. Four faulty cops are shown in G, H, J, and K. In G a hollow cone is made, and this at once suggests that the long rail is too high in the middle; a correction can be made by filing the rail flatter. In respect to H, where the chase is round, a remedy is almost invariably found by making the rail with less fall between the highest point and the lowest; or the highest point may be lowered by means of the adjusting screw which is connected with the bowl on the front plate; then file a little off the outer end of the rail, and continue to nothing about the middle of the rail, or even further if it is found necessary. If the hollow is not a general one, as shown in the sketch, but only local, then locate the spot on the rail which produces it, and file on either side of it.

Specimens J and K are easily corrected by simply altering the vertical height between the highest and lowest points of the rail.

When the cops are not formed parallel, they may either get smaller in diameter as the cop builds, as at L; or larger in diameter, as at M. In either case carefully examine the faller sector; as a rule the centre of the stud connecting the sector to the faller leg is in a line with or opposite to the centre of the faller rod when the faller wire is in the centre of the spindle blade. If the wire is higher than the centre of spindle blade the cop will become smaller in diameter as it builds; and if lower, the cop will be formed with an enlarging diameter. Granted that the sector is in the correct position, a remedy may be suggested by filing the plates for L as shown in No. 5, and for M as at No. 4 in Fig. 82.

In all these diagrams the amount of filing is exaggerated in order that the "direction" might be distinctly shown; in almost all cases very little filing will be required, but whatever is done must be done carefully.

A brief description, with illustrations, of the faller wires during the inward and outward run of the carriage was given in an earlier portion of these pages, and in the preceding notes on the shaper, the position and movement of the winding faller wire have been fully discussed; it therefore remains to direct a little attention to the other action and methods of controlling the counter, or under faller wire.

Weighting the Fallers.—In different districts various names are given to the same features of the mule, so care must be taken in reading descriptions to follow out the references to the illustrations. In the accompanying drawings, Figs. 83, 84, and 85, the counter or under faller rod is shown at A, while B represents the copping or winding faller rod. The wires carried by these two rods are marked H and J respectively. As the carriage comes out, and the spinning process is going on, these two wires are inoperative, both occupying positions, as already illustrated, close to the spindle point, but perfectly clear of the yarn being twisted. They are practically locked during the whole of the run-out. When the carriage comes to the end of the stretch, backing-off takes place, and, as the yarn is unwound from the bare part of the spindle, the winding faller wire J comes down in position ready for winding, while the counter faller wire H rises in order to take up the surplus yarn that has been unwound. In doing this it makes the yarn taut between the rollers and the cop. So far we have simply indicated that the wire is carried by a series of sickles G from the counter faller rod A, and they being all on one side, the tendency is rather for the wire H to fall rather than rise. In order, therefore, to cause H to rise and put tension in the yarn, the sickles and wire must be balanced on the opposite side of the rod A. To do this a sector C is keyed to the faller rod, and to it is attached a chain, which is hooked at its lower end to a weighted lever E centred at F. There are several of these chains and levers in the length of the mule, and their direct effect is to lift up the wire H as backing-off proceeds.

Now it will readily be seen that this arrangement is a very important feature; on the careful balancing effect of the weighted lever E depends the amount of tension in the yarn during the winding, for since A is free to oscillate in its bearings, it is quite possible that H might be forced too high and severely strain or even break the whole of the "ends," as the yarn is sometimes termed; on the other hand, the weight may be insufficient, and so produce a slackness that would result in the yarn running into snarls, and, in addition to the bad yarn so produced, making a soft and misshaped cop. To carefully adjust the balance, arrangements are made on E for applying additional weights W until the required tension is obtained. The skill of the minder is shown in his ability to gauge the tension, and while for the stronger yarns great care is necessary, it becomes a very delicate operation when the finer qualities are being spun. The character of the cop, so far as its density is concerned, is regulated by the tension,

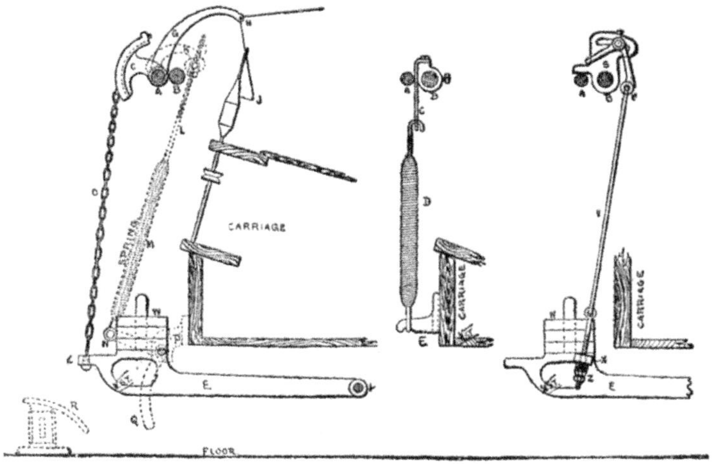

FIG. 83. FIG. 84. FIG. 85.

and it is an easy matter to make it too hard or too soft by carelessness. We have already pointed out how essential it is that the levers must be perfectly free from interference whilst they are in action, and also how it may be necessary, in order to neutralise tension, to weight the levers more, nearer the headstock than at the outer ends of the carriage.

Directly associated with the arrangement just described is the device illustrated in Figs. 84 and 85. They are shown in separate drawings in order to avoid complication. As the carriage runs in, the copping faller is guiding yarn on the spindle by virtue of its connection to the shaper. It is doing this in opposition to a strong spring D, Fig. 84, which is attached to the winding faller B by a strip of leather C and at its lower end to a fixed bracket E. The only effect of the spring at this point is to keep the shaper bowl pressed against the rail; when, however, the run-in is complete and the faller leg is unlocked, the tension in the spring D instantly causes the winding faller wire to move upwards a little above the spindle point. To prevent the wire being pulled up too far, a bracket S in the rod, Fig. 85, has a projection which, when the winding faller reaches its correct position, comes on the top of the counter faller rod and so prevents any further movement. The action of the spring D must be of a very definite character and strong enough to overcome, during the outward run, the weighting of the counter faller.

This last remark will be understood on reference to Fig. 85. It was remarked a short time ago that the two faller wires occupied certain positions during the outward run of the carriage. These positions are regulated by an indirect connection of the faller rods to each other. On the copping faller rod the bracket S has attached to it at T a link U, which is connected by a rod V to a projection on the weighted lever E. During the inward run, the rod V is adjusted to be free from the influence of the weight by so arranging it that it is able to pass freely through the hole in the projection X; when, however, the spring D pulls the copping faller wire upwards, on the completion of the run-in, it also pulls the rod V in the same direction, and in doing so, the nuts Z on the lower end of the rod come into contact with the projection on the lever, and so lift the lever also upwards, and consequently relieve the counter faller of their weight. This lifting of the lever E is a definite amount, because the spring D can only pull E upward until the projection on S rests on the counter faller A. The counter faller wire, being free from the balancing effect of the levers E, naturally falls by gravity until the chain D, Fig. 83, is again taut, and when this occurs the two wires occupy their correct position for the outward run, and the spring D must be strong enough to maintain them in this position so long as spinning is in progress.

Easing Motions.—The dotted portion of the drawing (Fig. 83) belongs to a class of mechanism called easing motions. The present illustration is given here for the purpose of explaining the necessity and effect of such a device.

It has been explained how the weighted levers E are partially supported from the copping faller rod B during the outward run of the carriage. When, however, the traverse is finished, a "change" takes place for the purpose of backing-off, and at this moment of changing, the full weight —or, rather, effect—of the weighted levers E would be thrown suddenly on to the counter faller A. Such an action is not necessary nor desirable; rather the reverse, for it is easy to understand that the free yarn unwound from the bare spindle during backing-off requires to be taken up gradually and gently. To bring this about, an "easing" device is applied, whereby the weighted levers are partially freed from the influence of the weighted levers E. On the opposite side of the faller rod A to that in which the sector C is fixed, is fastened a lever K, to which is attached a spring M, by means of an adjusting screw L; the other end of the spring is hooked to one arm of an L lever carried by the bracket P. The L lever has its other arm Q in such a position that just before the carriage arrives out, it comes into contact with an inclined floor bracket R, and this, preventing Q from going further forward, depresses the other arm N, and so puts tension into the spring M. This tension in M, acting in the opposite direction to the pull of the weighted lever E, neutralises a great part of the weight, and prevents the shock that would otherwise come upon the counter faller wire; in other words, it eases the faller rod considerably, and allows it to move upward in a much more gentle manner as backing-off proceeds. When backing-off is complete, and the carriage runs in, the arm Q of the lever moves gradually out of contact with the inclined bracket R, and so, by destroying the tension in the spring M, permits the full effect of the weighted lever E to fall upon the counter faller wire, and so maintains the tension in the yarn.

In passing, it may be observed that several of the weighted levers are placed in the full length of the mule, but the easing motion is only applied to two or three of them, a little discrimination being necessary in setting and disposing them in order to obtain the best results. Means are supplied for the necessary adjustments both for tension and position, the screw L supplying the former, and a regulation of the bracket R the latter.

The Effect of a Tapered Spindle on the Winding.—Mention has been made of the necessity for tapering the spindle blade towards its point, and that this fact had an important bearing upon the problem of winding after the cop bottom was finished. We will now examine this question as fully as possible, so that, in considering the mechanical methods adopted for compensating for the taper, we shall be able the better to judge of their adaptability for the purpose. It will be advisable to recapitulate a little, so we will begin by showing the necessary winding effect required for building the "chase" of the cop at the point when the cop bottom is finished. Figs. 86 and 87 will illustrate the explanation. In Fig. 86 the chase of the cop at A D is shown soon after the cop bottom is complete. At this time we will assume that the full diameter of the cop is $1\frac{1}{4}$ inch and the diameter of the small end of the cop on the bare spindle at D is $\frac{5}{16}$ inch. To wind correctly this conical surface A D, the speed of the spindle must increase in the ratio denoted by the full-lined hyperbolic

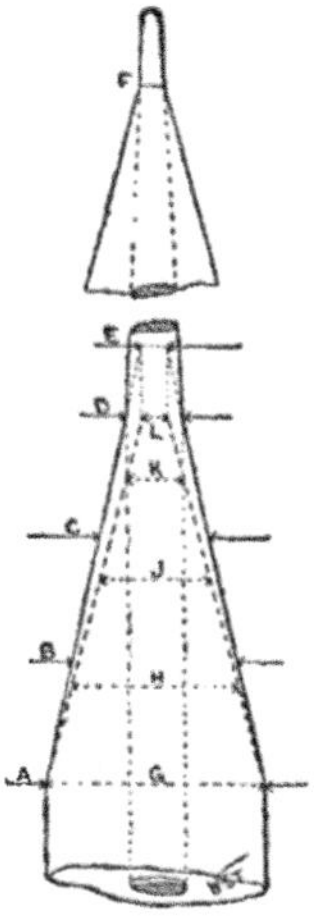

Fig. 86.

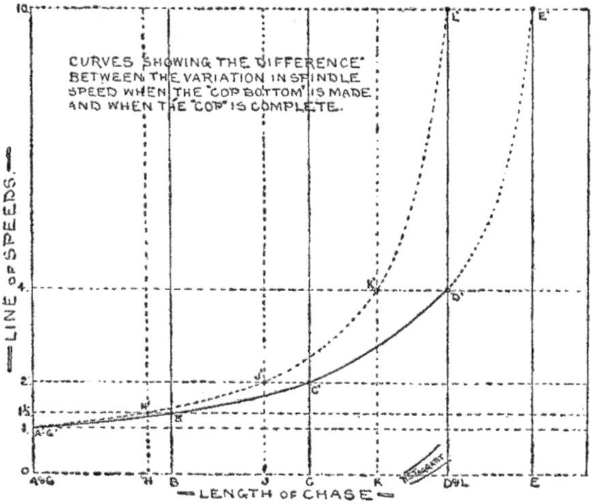

Fig. 87.

curve, as shown in the diagram, Fig. 87. For instance, suppose one revolution winds on a certain length of yarn at A, then the "rate" of spindle speed must be such that the same length could be wound on at B, C, and D, and in order to do this the proportionate speed at these points must be increased to $1\frac{1}{3}$, 2, and 4 times respectively the speed at A. It is a very simple matter to obtain these numbers, for, knowing the diameters at the points A, B, C, and D, the proportion each of them bears to the diameter of A will give us the proportionate speed.

Suppose A is $1\frac{1}{4}$ inch diameter and its speed 1,

then B is $\frac{15}{16}$ inch diameter and its speed is $\frac{1\frac{1}{4}}{\frac{15}{16}} = \frac{5}{4} \times \frac{16}{15} = 1\frac{1}{3}$,

and C is $\frac{5}{8}$ inch ,, ,, $\frac{1\frac{1}{4}}{\frac{5}{8}} = \frac{5}{4} \times \frac{8}{5} = 2$,

and D is $\frac{5}{16}$ inch ,, ,, $\frac{1\frac{1}{4}}{\frac{5}{16}} = \frac{5}{4} \times \frac{16}{5} = 4$.

These numbers, measured by any convenient scale along the speed line A 10 and horizontal lines drawn through them to meet perpendicular lines erected upon the chase line A E, as at B, C, and D, will give points through which the curve A^1 D^1 can be drawn. The line A D is drawn any length to represent the chase, and the points B and C are marked off equi-distant from each other in the same way as B and C on the chase line in Fig. 86. As the cop builds from the cop bottom upwards, the body remains the same diameter, namely, $1\frac{1}{4}$ inch, but the nose of the cop gradually becomes smaller until near the end of the spindle, when it is wound on a diameter of probably $\frac{1}{8}$ inch. In consequence of this reduction in the terminal diameter of the conical chase, a new set of conditions are introduced, which have an important bearing on the problem of winding. For the moment, we will assume that the chase A D lengthens until it meets the smaller diameter, as at A E; if this occurred, all that would be required would be a continuation of the accelerated speed which formed the portion A D, and in the curve in Fig. 87 the speed curve would be lengthened as shown by the dotted line D^1 E^1. In such a case one might reasonably say that a "nosing" motion would be a device for accelerating the speed of the spindle as the yarn was being wound on the nose of the cop; indeed, the word "nosing" is derived from such an idea, and it is perpetrated partly because of the prevailing idea that such a method of reasoning is fairly correct, and partly because the irregularities of winding show themselves more at the nose than in other parts of the cop. But we know that instead of the chase lengthening towards the spindle end, it is generally kept either the same length throughout, or is made a little shorter. Let us see what effect is produced when the chase is kept the same length. At F in Fig. 86 is shown a part of the chase; for convenience and comparison, the diameter at F is brought down to L, and we then see and are able to compare the two chases A D and A L, each being the same length. Starting from the same diameter at A, the initial speed in each will be the same, but there is a great difference between the diameters at D and L in the proportion of $\frac{5}{16}$ to $\frac{1}{8}$, and this reduction is one that, starting at A, works down to L. Such a reduction in the conicity of the chase means that there *ought* to be a proportional increase in the speed of the spindle *throughout* the chase, and not, as in the last assumption, only at the nose. By drawing out the speed curve for the new chase A L, and plotting it as in the dotted curve in Fig 87, we can form a very clear idea of the difference that should exist between the speed of the spindle when the *cop bottom* is completed and when the *cop* is completed. The speed when winding at D is only four times what it is at A, whilst when winding at F or L it is ten times greater than at A, and for proper winding this speed must be attained through a gradual acceleration starting from the bottom of the cop chase at A until L is reached. To show the great difference this makes, the corresponding divisions on the chase at B, C, and D are marked on the chase A L at H, J, and K; this means that, suppose the spindle is making twice as many revolutions when winding the yarn on at C as at A, it must make on the new chase twice as many revolutions at J as at A. Each of the other divisions can be interpreted in the same way, and it will help greatly in comprehending the vital importance of effecting a change in the rate of winding as the cop lengthens, and that this change should commence at the bottom of the chase.

In this analysis the two extremes have been taken as illustrations, but during each added layer the variation throughout the chase should take place corresponding to the smaller diameter on which the chase finishes. The fact that the quadrant is imperfect in not giving the correct curve of speeds, as in Fig. 87, does not in the least interfere with the reasoning employed, for the difference between the two chases remains proportionately the same.

Nosing Motions.—Various means are adopted to obtain the desired change of speed to compensate for the taper of the spindle, all of which depend more or less upon two principles of action. In one case, the effect is produced by moving the winding chain out of a straight line during the run-in of the carriage, thereby unwinding more chain from the winding drum. In the other method, the winding chain is shortened, and at the same time it is arranged that the shortening effect causes a scroll portion of the winding drum to come into action; and the act of the chain working on a smaller diameter produces an increased speed in the spindles.

Both systems will be briefly examined in diagram before describing the actual mechanism used. In Fig. 88 the quadrant and its connections with the spindle are shown; the chain is represented as perfectly straight between the screw H and the winding drum B, and it will work in this position while the cop bottom is being formed. To obtain an increased speed of spindle as the cop lengthens, the chain is, by suitable means, gradually depressed during the run-in, its position under these circumstances being shown by the dotted lines. The fact of moving the chain from D to C unwinds a portion of it from the drum B, and as it is done gradually the spindles are in the same degree increased in speed. Fig. 89 will give a better idea of the action. When the quadrant occupies the position at A B, the chain passes to the drum from B to F in a straight line. When the winding reaches the second position, the quadrant is at A G, and the chain during the interval has been depressed

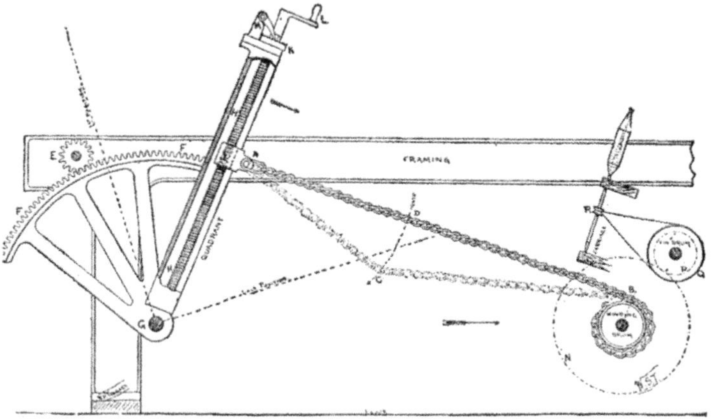

FIG. 88.

slightly from the straight line G H to the extent shown at N G. In the third position the depression of the chain has been increased to J P from the straight line D K; and in the last position it has been still further increased, as indicated at L Q. The depression of the chain has therefore been gradual, and in the right direction. It is unnecessary at this point to inquire into the question of the exact amount of depression required, it being sufficient to show that for practical purposes a near approximation is obtained. When the cop has commenced to form up the tapered part of the spindle, the depression of the chain is only of the slightest character; but it increases gradually as each layer is added.

In the second method of increasing the speed of the spindle, the chain passes over a small bowl B on the quadrant arm, Figs. 90 and 91, and on to another bowl C, which is actuated so that as the cop builds it can wind on a little of the chain after each draw. This action, of course, shortens the winding chain; but so long as the chain was wound on a cylindrical winding drum, the shortening would have no effect whatever on the speed of the spindles. In combination with the shortening, therefore, the winding drum F is made with a scroll end, so that while the chain is wound round C on the quadrant, it pulls over the winding drum and changes the finishing point from D to E. When the cop bottom is completed, the winding chain finishes winding at D; by winding the chain on at C each draw, the drum F will be pulled round, and when the cop is complete, the winding of the chase finishes at E. Consequently, during a portion of the winding, the chain has been unwound from a gradually reduced diameter, and this chain unwound from the smaller diameter results in an increasing speed of spindles as compared with the unwinding of the same length of chain from a larger diameter on the winding drum.

Nosing motions assume a variety of forms, the majority of which depend upon an action which depresses the chain. As a rule, each machine-maker has some automatic motion which is applied specially to the machine he makes; but a type common to all mules, and which is still extensively used, is that known as the "nose peg." As it is not automatic in its action, a certain amount of skill is required on the part of the person in charge of the mule. Fig. 92 illustrates its application to the quadrant. Its essential feature consists of a slotted bracket projecting from the upper end of the quadrant arm; it may be curved, as shown, or straight out, and in some cases it is disposed angularly to the arm, partly with the idea of gaining

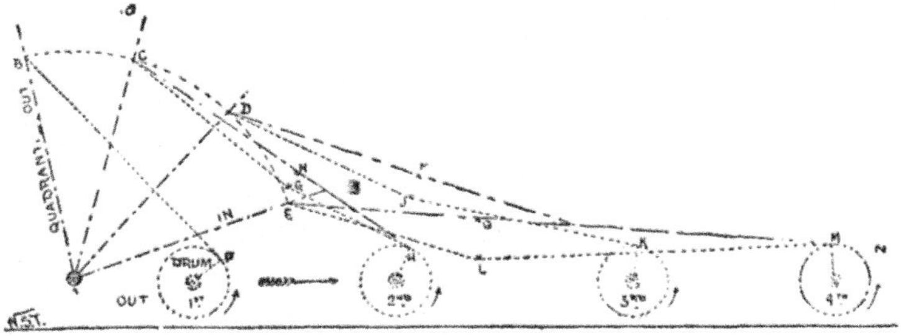

FIG. 89.

FIG. 90.

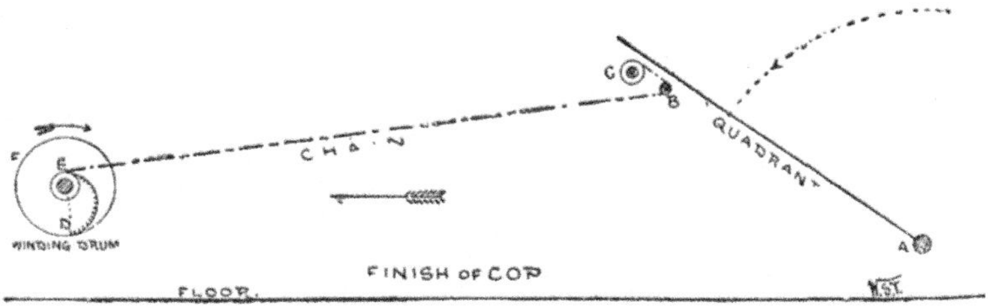

FIG 91.

strength and partly to suit the requirements of winding. In the slot of the bracket is placed a stud C, having a winged nut, which enables it to be readily adjusted and fastened in the position suitable for the increasing length of the cop. As shown in the diagram, the carriage has commenced its inward run and the yarn is being laid during the downward movement of the faller wire. The carriage continues to move in, and as the quadrant follows, a point is reached when the stud or nose peg C comes into contact with the winding chain; further movement of the quadrant then causes C to press on the chain and move it out of a straight line, so that when the carriage arrives in, the chain and quadrant occupy the position shown in the dotted portion of the sketch. The extra chain unwound from the winding drum by this depression of the chain produces an acceleration in the speed of the spindles.

As already pointed out, it is not necessary to use the nose peg until the cop bottom is complete. When this occurs the peg C must be set in such a position that it only slightly depresses the chain during the few succeeding draws. During the lengthening of the cop the nosing peg must be moved a little further away from the quadrant arm every few draws, and in this way the depression of the chain is increased in the ratio considered correct by the minder, the necessary movement being one entirely dependent on his judgment. It is probably on this account that the nose peg is still so generally employed, its convenience and the absence of complication being its great features.

It has, however, some very serious faults, its chief one being obvious if the previous explanation has been carefully followed. We have shown that to be theoretically correct a nosing motion ought to commence to act directly the chase is begun, its action being extremely gentle and gradual at first. In the nose peg this is a condition practically impossible. In the first place, to refer to the sketch, the peg C would not come into action until the quadrant arm occupied the position as indicated by the dotted line H G, so that the acceleration of the spindles would only take place as the quadrant moved from H G to

H D ; in other words, only the nose of the cop would be affected by the extra speed. This is apparently contradictory to what reason would lead one to require of the motion ; but an important consideration, when pointed out, will explain why such an anomalous action gives good results. The yarn when wound on the thicker parts of the chase is put on a comparatively soft and yielding foundation, but as it nears the nose the foundation becomes more solid, and therefore the tension put in in consequence of all the acceleration being thrown into that part, simply causes

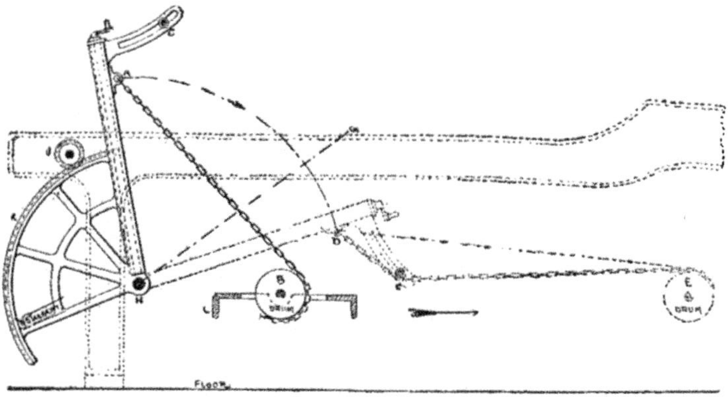

FIG. 92.

the yarn to wind tighter without enabling it to influence the shape of the chase. This difference in the character of the foundation even renders it in many cases advisable from a practical point of view not to cause acceleration to commence until past the middle of the chase. The fault of the nose peg arises from the fact that when it begins to depress the chain it depresses it too quickly at first, and not in the correct ratio. This often leads to spoiled yarn, partly through being strained and partly in snarls, and carelessness in moving the peg only after long intervals increases its inherent fault. To get the best results the peg C must be moved regularly and often during a set, and this means that a slight movement often repeated prevents any great increase of speed being given to the spindles, and the minder is able to better gauge the tension of the yarn as the nose is wound, and so prevent its being strained, or, on the other hand, soft noses being made.

The "nosing peg" system, it will be noticed, depends for its success entirely upon the skill of the minder. Many attempts have been made to eliminate this factor, so as to obtain an automatic motion, but the problem is one that contains several conditions which are so variable and uncertain that the success which some of them attain can only be described as comparative or local, and as due rather to the additional skill of the minders, who, after a reasonable experience of them, can adapt them to the special character of their machine and the work they perform.

To make a nosing motion automatic, it must be actuated directly or indirectly from the copping faller rod. The cop lengthens as the faller wire rises and places the yarn on a smaller diameter of the spindle; it is to compensate for this that the nosing motion is necessary. Some motions are therefore worked directly from the faller rod; but since the faller rod receives its movement from the shaper, it amounts to the same thing to use this feature of the mule for operating the automatic mechanism, and in one or two cases the oscillation of the quadrant has been taken advantage of to obtain a regulation. This latter method is, however, bad in principle, though occasionally it works well; more will be said on this point when dealing with governor motions.

The first illustration is taken from a well-known source, and, as will be seen from the sketch Fig. 93, it is actuated from the faller rod. A bracket N containing the mechanism is fixed to the quadrant screw box. An arm K is centred on a stud L, and its other end carries two catches, or pawls, which a spring presses into the teeth of a portion of a circular rack struck from L as a centre. A projection J of the arm K carries a hook to which is attached a chain M; this chain first passes over a bowl H and from this point goes forward and is fixed to the winding chain at G. We can now see that, according to the position of the arm K, the chain M can be so arranged that it will have no pulling effect on the winding chain, but by raising K we

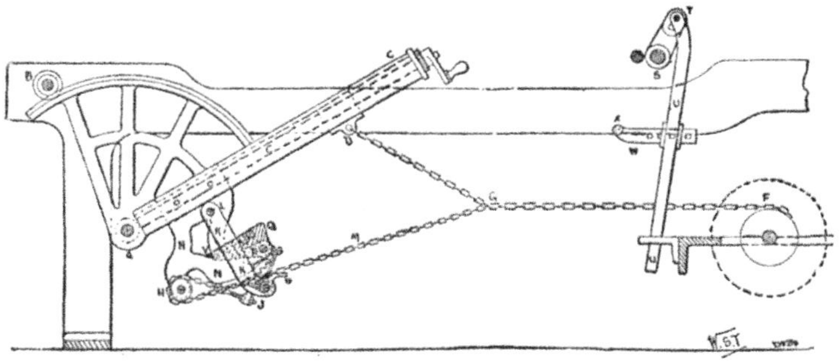

FIG. 93.

practically shorten the nosing motion chain and consequently the forward oscillation of the quadrant causes the winding chain to be pulled out of its course, as shown in the drawing. The higher the position of the arm K on the toothed portion of the bracket N, the more is the winding chain depressed. It is therefore an easy matter to use this motion just as one would use the nose peg without using its automatic features. In many cases this is an advantage, because yarn is a material that requires considerable humouring in some of its conditions, and means ought always to be provided for helping or retarding an intended automatic action such as a nosing motion.

The following means are provided for the automatic working. A projection on the arm K carries a stud P, on which is swivelled an incline Q V; this incline is in contact with the arm K through an adjusting screw, so that if the incline is lifted up it will also raise the arm K. On the copping faller rod S is a lever carrying one end of a rod U, which is guided in a bracket fixed to the square. The rod carries a finger W, which can be readily adjusted in position. As the carriage moves out, the finger naturally occupies a high position, and so comes into contact with the lower portion of the incline V, which is arranged to swivel out of the way; but directly the backing-off is finished, the finger has fallen much lower, so that as the carriage moves in, the projection X of the finger W comes into contact with the back of the incline and lifts up the finger a little. This lifting up of the arm K through the action of the finger W upon the incline Q V each draw, gradually shortens the chain M and gives the necessary increased acceleration to the winding drum F. Practically it is almost impossible to raise the arm K each draw, because owing to the large number of layers in the cop it would be necessary to have in the short portion of the circular rack such a number of fine teeth as to make the motion unworkable; a reasonable number of teeth are cut and a double effect is produced by having two catches, and by this means a permanent lift is produced when the faller wire has been raised high enough to cause a pawl to catch in every half tooth. To render the motion more efficient, the bowl H is made of a cam shape to suit the conditions, as near as possible, as laid down in a previous description. After doffing, the catches are released from the rack and the arm lowered to its starting point. It may be remarked, that if the arm is purposely or inadvertently lowered during the building of the cop the next draw simply lifts it into its correct position again.

A nose peg in very common use is of the form shown in Fig. 94; by comparing this with the drawing, Fig. 92, it will be seen to vary from it in the direction of the slot along which the peg D is moved as the cop builds. Much more care is required in moving the peg along a slot which

FIG. 94.

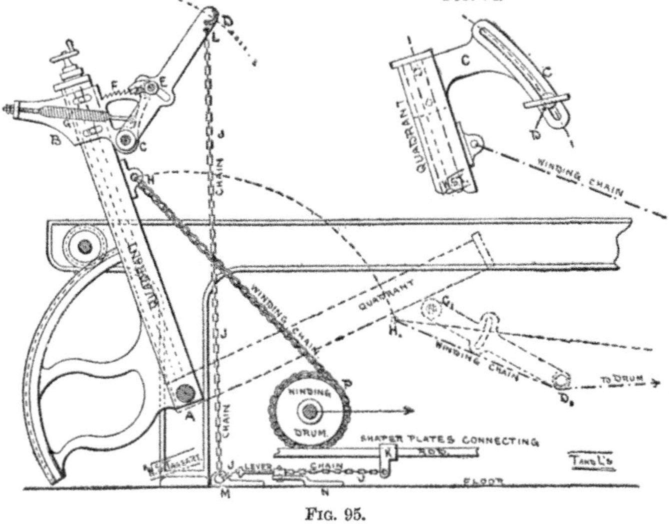

FIG. 95.

lies parallel or is only slightly inclined to the quadrant arm than when the slot is more nearly at right angles to it. The regulation is obtained in a much shorter length of slot, and consequently each movement requires to be very little, and well judged, otherwise too much nosing will be obtained. The slot is frequently made in the form of a curve drawn out empirically, rather than upon any fundamental reason, and so long as the movement of the peg depends upon the judgment of the minder this is a matter of little moment, and in any case variation from what might be considered a correct form of curve will not interfere greatly with the efficacy of the motion.

Fig. 95 is a motion used by a well-known firm, who employ it extensively, and who find that it gives very good results in practice. It might be termed an automatic method of moving the peg along the slot in Fig. 94. Instead of a slotted bracket, a lever C is used, centred on a bracket B, fastened to the upper end of the quadrant arm. This bracket contains a portion of a circular rack F, into which engage two pawls or catches E, carried by the lever C; a spring G keeps the catches engaged. The lever C is set so as to begin to depress the winding chain directly the tapered portion of the spindle is reached. Afterwards it is automatically lowered by a chain connection to the shaper. It has already been observed that the regular movement of the shaper is sometimes taken advantage of in actuating the nosing motion, and in this case a bracket K is fixed on the rod which connects the front and back shaper-plates. To this is attached a chain J, and in order to pass it to the other side of the headstock a bell-crank lever is used centred on a floor fixing N; the chain, after being guided through the eye of the bracket M, is taken upwards and attached to the nosing lever at L. By this means, each movement of the shaper draws the lever C downwards, and causes the end of it D to come into contact with the winding chain sooner each draw. The dotted lines show one position of the motion when it is acting upon the winding chain.

The next example of a nosing motion is a practical illustration of the principle explained in connection with Fig. 90. It is of a much more complicated character than any of the motions previously given, but apart from this disadvantage it works sufficiently well for the purpose. It belongs to that class of motion which is actuated from the shaper mechanism, so that, like all automatic nosing motions, it works in an uncompromising manner and entirely independent of any peculiar characteristics of the yarn or formation of cop. The following description will disclose the salient features of its action.

The shaper-plate M, Fig. 96, is moved forward by the screw L; attached to this part of the shaper is a bracket J, to which is connected a chain F. The chain is guided over the back surface of a hanging lever centred at its upper end on a stud G; from here it passes round the lower end of the quadrant screw box and on in an upward direction to a curved lever E. This lever, as well as a ratchet wheel D, is fixed on the end of a stud, which carries a small boss or scroll, to which the winding chain is fastened. The winding chain before passing to the winding drum R is taken over a guide bowl C, so that it is at this point that the winding chain must be considered to be attached. The bracket B, which carries the whole of the arrangement at C, D, and F, corresponds to the quadrant nut of the ordinary motions, and up to the point when the cop bottom is finished, B is caused to travel up the screw in the usual way for forming it. When the nut B occupies its lowest position on the screw, the chain F is slack, but it is drawn tighter as the nut is worked upwards; during this period there is very little effect produced in shortening the winding chain, but when the cop bottom is finished, every additional movement of the shaper-plate M takes up the chain F and pulls over the lever E, so that the winding chain is wound up at a quicker rate on the scroll at D. It will be noticed, however, that the mere movement of the screw is not depended upon for pulling over the lever E. The extremely small amount of motion given to the bracket J by the screw L is only capable of moving the lever E after several draws, because E can only take up another position after sufficient movement of the bracket J enables the catch to escape half a tooth in the ratchet wheel D. Until this occurs there would be a great strain on the chain F if the hanging lever G was fixed. To relieve this strain the lever is made pendent from G, so that during the inward run of the carriage the chain F is free from strain. During the outward run, however, an inclined bracket P on one of the arms of the quadrant comes in contact with a bowl Q carried by the hanging lever and presses it backwards, thus tightening the chain F and pulling over the lever E. In Fig. 97 the other extreme position is shown for the shaper-plate M and the lever E, and from it we see the shortening effect produced on the winding chain.

As before observed, this shortening of the winding chain would have no effect on the problem if the winding action depended on the ordinary straight winding drum. To obtain the necessary variation, therefore, the drum R is made as a scroll for a portion of its length, and as the chain is shortened the drum is pulled round so that it is brought into action sooner, and the direct effect is to cause the chain to finish unwinding from a smaller portion of this scroll part after each movement of E.

Sufficient examples have now been given to convey a good general idea of the various methods adopted for compensating for the taper of the spindle, and those who have carefully followed the descriptions will scarcely have failed to notice that in no case does a motion follow out the conclusions arrived at when the theory of the action was described. In the first place, the arrangements start

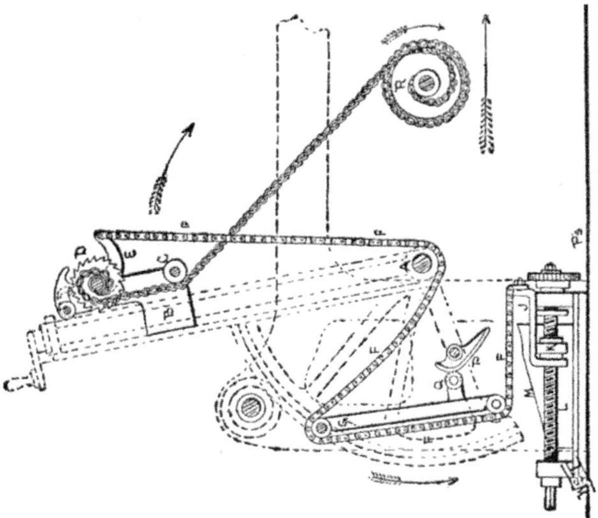

Fig. 97.

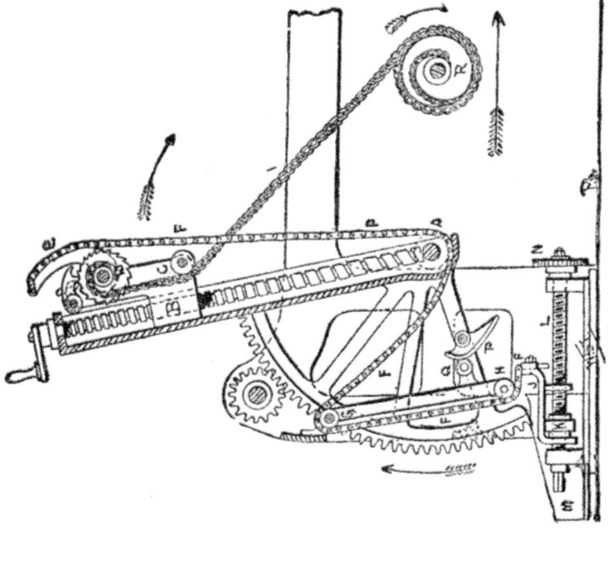

Fig. 96.

too late; secondly, their starting point varies for each

layer instead of remaining constant throughout the set; thirdly, no attempt is made to vary the action of the motion in conformity with the actual conditions of winding; and, lastly, there is a serious disadvantage in the absence of means for the motions to adapt themselves to the inherent irregularities and characteristics of the yarn being spun.

Note.—Attention is called to a slight error in the drawings, Figs. 96 and 97. The ratchet wheel D is shown with its teeth in the wrong direction.

Governor Motions.—Governor or strapping motions are the names usually given to the appliances which automatically regulate the position of the nut upon the quadrant during the building of the cop bottom. A variety of means have been employed for performing the operation, but very few have been found to stand the test of practical experience, and on English machines at least the arrangements are confined within very narrow limits, the methods varying only in details of construction and the time of action.

The subject is not without interest, and affords opportunities for diverse opinions as to the correct mode of action, so there will be some advantage in considering the matter in detail. It will be assumed that the nut on the quadrant is in the correct position for causing the bare spindle to turn the correct number of times, during the inward traverse of the carriage, for winding on the length of yarn in the stretch. So far as we are at present concerned, this first layer is supposed to be wound on with a regular tension, and there is nothing to suggest that the position of the nut on the quadrant should be altered from the start to the finish of the layer—that is, during the whole of the inward run the nut must remain in the same position. Now before the next layer can be put on the cop bottom, it will be necessary to raise the quadrant nut in order to compensate for the increased diameter of spindle, and we can see clearly that the time to do this must be during the outward run of the carriage, so that the nut is in the right position for "starting" the next layer at the correct speed.

From this reasoning it might be concluded that an automatic motion should be so arranged that *after* each layer is added it moves the nut up the quadrant. Practical considerations, which will be mentioned subsequently, prevent this conclusion from being accepted as fundamental, though in essentials it ought to be the foundation of a governor motion. As it is, we find that the subject is treated more as a question of opinion, and naturally there is an absence of unanimity in regard to it. When dealing with the quadrant it was pointed out that the rate at which the nut must be moved up the screw was a varying one, gradually decreasing from a quick movement at the beginning. In a governor motion this must be taken into account, and if we depended on pure reasoning we should expect that every layer would require its share of movement. In addition to these considerations, it must be remembered that the yarn itself is an ever-varying factor, and that there are inherent peculiarities in the cops, while the faulty or imperfect character of the connections to the faller rods must be taken into account, for they modify to a large extent the presumed ideal of a governor motion. A perfect governor motion might be summed up as possessing the following points:—

(1) To give a movement to the quadrant nut for each layer added to the cop bottom.

(2) To give a "correct" decreasing movement each draw.

(3) To compensate for peculiarities of cotton, yarn, cops, or connecting motions.

(4) To actuate the quadrant screw after one layer and before the commencement of the next one.

To the practical reader it need scarcely be pointed out that these conditions are never fulfilled, and it might almost be added that the difficulties in the way of fulfilment have hitherto prevented any success being attained when the attempt has been made.

Instead, therefore, of the governor motions working under ideal conditions, we generally find them entirely under the control of the yarn itself, and actuated either before or after the run-in of the carriage. Simplicity and convenience are the deciding factors in the case, and while good average results are obtained, the erratic and faulty character of many motions leaves much to be desired in the direction of a governor motion founded on correct principles.

By permitting the yarn to actuate its own winding we practically combine the first, second, and third conditions enumerated above, and by so doing take advantage of the tension-regulating action of the counter faller. The yarn unwound during backing-off is taken up by the counter-faller wire, and as winding proceeds a certain tension is maintained by means of the weighting arrangement already described. It is easy to see that if through any cause the tension is lessened or increased, the wire will yield, and we can also understand that one of the chief causes of any variation in the tension of the yarn will be irregularity in the winding. On this effect the action of a governor motion is generally based. For instance, suppose the first layer has been put on the spindle correctly, the next layer will naturally require a slower speed of spindle, and to do this the nut must be moved up the quadrant screw. Instead of doing this in anticipation, most mules *commence* to wind the next layer with the screw in the same position as for the first layer. The almost immediate effect is that the larger diameter winds on too much yarn, and naturally puts so much tension in the yarn that the faller wire is pulled down. This, as will be shown subsequently, brings about, through suitable mechanism, a change in the position of the nut, which gives the required speed to the spindle. In such an action as this we get the third condition incorporated with the first two, and the yarn, as it is being wound, is relied upon to do all the regulating required. It will be noticed, however, that a serious evil is introduced in the great increase of tension that is put into the yarn at the commencement of winding, and this is especially noticeable at the commencement of the cop and in low and medium numbers; the fact that it is a progressively decreasing one, helps to neutralise it considerably, and possibly on this account, together with the presence of some personally adjustable feature of the motion, maintains such a mode of action as a base of those arrangements which are most successful.

A great difference of opinion exists in regard to the time when the nut ought to be moved upwards. The writer's opinion has been expressed above so far as the principle of action is concerned, but actuating the governor motion during the run-in has its advantages; for instance, carelessness is more easily and quickly corrected by this system, and, moreover, insufficient governing during one draw will be corrected during winding in the next. Apart from the features common to both methods, it may fairly be taken for granted that the more uniform tension and preparedness for the next draw in the regulation during the outward run will equalise the practical advantages of the regulation during the inward run. In either case the class of cotton and quality of yarn must decide the question from a practical point of view, but it cannot be too strongly impressed upon the reader that the best results can only be attained by keeping as closely as possible to the conditions laid down for a perfect motion, and for the best quality and finer yarns it is almost necessary that the last condition should be followed.

The following examples of "band" governing motions may be taken as typical of the kind which find most favour. They are termed "band" motions because a band is used to give motion to the quadrant screw. On the lower end of the quadrant screw is fixed a bevel wheel, which gears into another bevel cast or fixed on a band pulley, which rides loose on the quadrant shaft A, Fig. 98. An endless band is passed round this pulley and guided over a series of guide pulleys B, C, D, E, and F. Three of the guide pulleys, B, C, and E, are carried on fixed studs, but the other two, D and F, are carried on studs fixed to the carriage, so that as the carriage moves the pulleys travel backwards and forwards. The only effect this disposition of the pulleys has is to set up a certain amount of friction in the band, but since D and F are free to revolve on their studs, the friction is relieved by their motion, and the band remains unaltered in position. In order to produce some effect of the pulley on A, it will be necessary to grip or hold the band in some way, so that the movement of the carriage will draw it along. A variety of methods are adopted for doing this, several of which will be shown. Referring to Fig. 98, it will be seen that a lever or arm is fixed on each faller rod; one end of a chain is connected to the arm K on the copping faller rod T, and after passing round a pulley H is attached

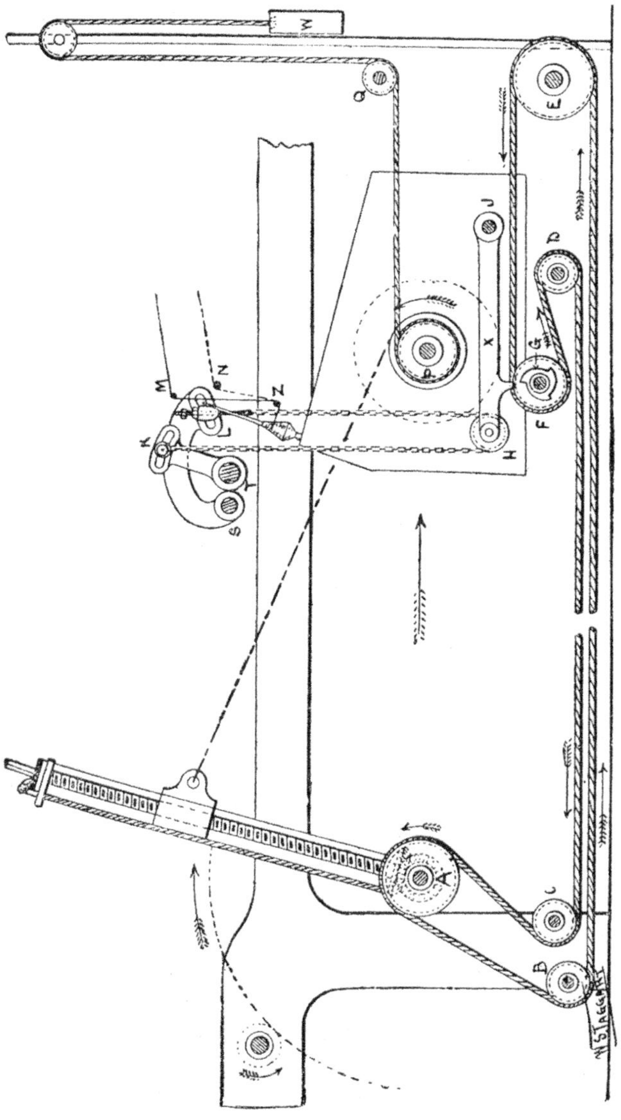

FIG. 98.

to an adjusting screw carried by the arm L. The pulley

H is supported by the end of a lever X centred on the carriage at J, and a projection on this lever is arranged so that it can be lowered into the path of a revolving toothed disc G fixed on the guide pulley F.

As the yarn is being wound on the cop bottom it passes over the faller wires L and M. As the wire L guides the yarn on the spindle, it of course moves, and naturally the arm K does the same, but this has very little effect on the chain, the lever being arranged in position so that it is passing along the upper part of the circle it describes; H, therefore, is affected very little by this movement of the faller wire. In the case of the counter-faller wire M it is different; the position of M depends upon the tension of the yarn as regulated by the faller weights. Therefore, directly a larger diameter of the cop or other circumstances cause the spindles to wind on too quickly, the tension is increased and the faller wire is pulled down, say, to N. The lowering of the wire M to N gives a similar movement to the arm L, and this immediately causes the end of the lever X to drop, and the projection falling into the path of the disc G prevents the rotation of the pulley F; this sets up sufficient friction in the rope to hold it so that the carriage takes the band forward and produces a movement in the pulley A in the opposite direction to that shown by the arrows. The revolution of A will continue to move the nut up the screw so long as the pulley F is held by the lever X; but since the nut in its higher position on the screw will revolve the spindles more slowly, the tension will be quickly relieved and the wire M will rise to its normal position and lift the lever X out of contact with the disc G and so permit F to revolve freely. This action takes place just as often as the yarn becomes sufficiently tight to draw down the wire M low enough to let the projection fall upon G. This occurs very frequently during the early part of the cop bottom, but at much longer intervals towards the finish.

It sometimes happens that the nut has not been moved high enough for a certain layer, and in such a case we should find that the tension at the commencement of the following run-in would cause the lever X to at once fall into contact with the disc G, and so complete the raising of the nut.

According to the quality of cotton or yarn, it is absolutely necessary to arrange for some means of adjustment either for modifying or increasing its sensitiveness; a regulating screw is therefore provided on L, which enables this to be done, and it is also used to lift the lever X out of position, so that, after the cop bottom is finished, any incidental irregularity of the yarn will not give a higher permanent position to the quadrant nut; this action must be left to the judgment of the minder.

No arrangement is made for any reduction in the tension of the yarn, because such a condition is scarcely possible, and indeed everything is done to prevent any lowering of the nut, a catch wheel being generally provided on the top of the screw box. The too easy movement of the screw is also prevented by means of a strong friction brake either on the top or on the pulley on the shaft A. Carelessness in allowing these brakes to become inoperative has frequently led to bad work and breakdown of ends.

The illustration, Fig. 98, also shows an arrangement for winding back the winding chain during the run-out of the carriage. A band is fastened to one end of the winding drum, and its other end is attached to a weight W after passing over the guide pulleys Q and R, carried by an upright rod. As the carriage makes its inward run the weight W is lifted up to near the top of the rod, so that during the outward run it falls, and in so doing turns the drum and winds on the chain.

Fig. 99 presents us with another arrangement of band governing motions. It differs from the previous motion only in the method of holding the band in order to give motion to the quadrant screw. On the faller rods M and L are fixed the levers K and J, and to these are attached an endless chain which passes over a solid loose bowl H carried by one end of a lever centred on a stud at G. The other end T of the lever is so arranged that when the tension of the yarn pulls the faller wire down, say, from P to Q, the weighted end will press T against a projection F on a bracket bolted to the front of the "square." The governor band passes through slots which keep it always in front of the projection F, so that when the chain permits H to fall, the band is forced against F, and the pressure is sufficient to hold it fast while the carriage carries it forward in the direction of the arrows, to give motion to the screw. As in Fig. 98, the action of the lever K on the copping faller gradually lifts H a little higher, and when the cop bottom is finished it will have been raised high enough to prevent it coming into action again unless an unusual amount of tension depresses the faller wire. Fig. 100 gives sufficient of a side view to enable the motion to be readily understood.

The front and side elevation of an interesting motion are given in Figs. 101 and 102, and although it now belongs to a numerous class of movements which have been found wanting, it has features which give it importance from a mechanical point of view.

On reference to the drawings, there is a small pinion B

cast to the back of the usual bevel C. A rack A is

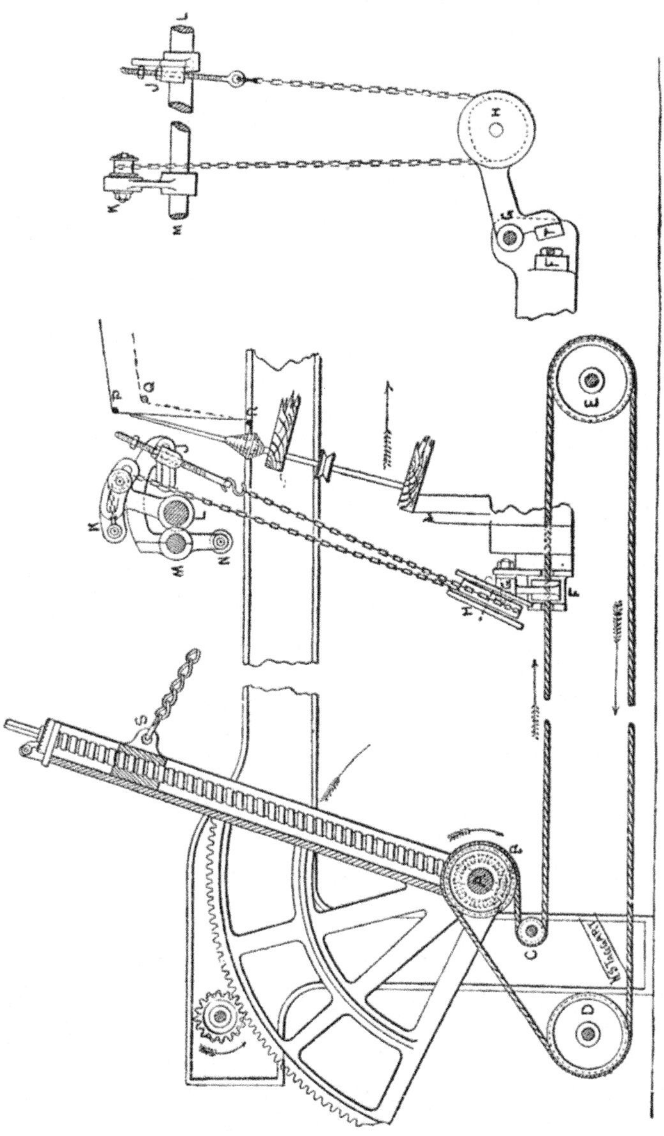

Fig. 99. Fig. 100.

arranged to gear into B, and the interesting feature of the

motion lies in the method adopted for regulating the number of teeth in the rack A to gear with B for each layer. It will be noticed that the rack A rests upon a sliding plate D by means of inclined projections, and that D rests upon a slide Q, one end of which is fastened to a rod E, while the other end slides upon a fixed box-like bracket which is firmly fastened to the headstock or floor. The rod E is supported by this bracket and floor fixings as shown. On it is also fastened a swivel catch R, which can be acted upon by a drop pendant N connected by chains to the usual connections K and L on the faller rods. Variations of tension in the yarn during the outward run will cause N to drop; in doing so it comes against the catch R and carries the rod E forward, and naturally also the slide Q with its small slide D and the rack A. As Q is pulled along a projection on the end of the slide, D comes into contact with a projection on the box bracket, and its further movement is stopped, but the rack continues its forward movement with Q. Previous to this the inclined projections on the under side of A have been in corresponding slots in the slide D, so that when D stops moving A is compelled, by means of its inclined projections, to slide up and occupy the position shown in the drawing; it will thus be seen that the rack A has been out of gear with the small wheel B during the inward run, and moreover the increased tension in the yarn has simply raised the rack up in such a position ready for the carriage, during the outward run, to push the slide Q and cause A to gear with B, and so turn the screw.

It will easily be seen that the full length of A would be employed each time the motion worked, unless an arrangement was made for regulating the number of teeth to be used to suit the size of the cop bottom. This is done by

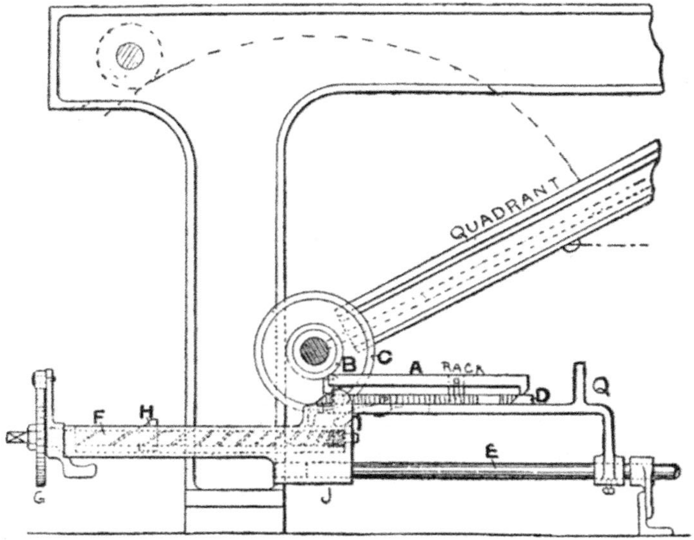

Fig. 101

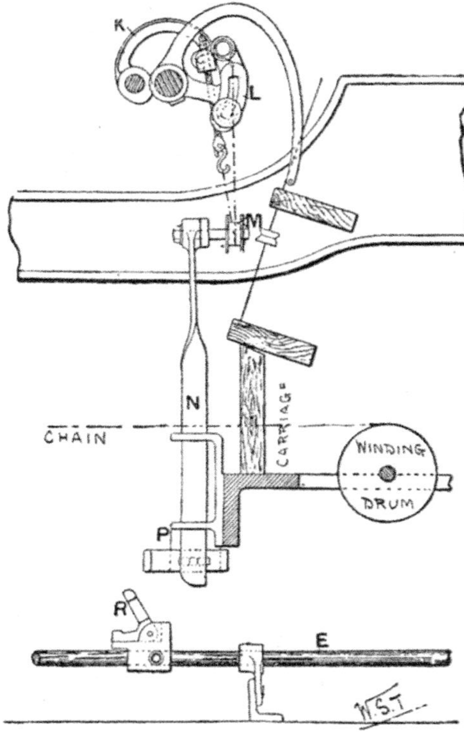

Fig. 102.

means of a screw F, on which is threaded a stop-washer H. On the slide D is swivelled a catch I which comes against H when the carriage pushes Q forward, and thus stops D from further movement. However, A continues with Q to move forward, and directly its projections come to the slots in D it falls down out of gear with D and finishes its forward movement out of gear with B. A ratchet wheel G on the end of the screw F is actuated by the end of the rod E, and H is moved along the screw so that the rack may, at the commencement of the cop bottom, use its full length in driving B; but as H moves forward, the stop on D, coming in contact with it sooner, causes A to drop out of gear with B earlier, and thus reduces gradually the number of teeth capable of driving the spur wheel. This continues until, when the cop bottom is complete, the rack will fall down before any of its teeth can touch B. The screw F is variable in its pitch in the proportion necessary for the shape of the cop bottom, and the whole motion is of a character to fulfil almost all the conditions required of a successful motion. The great objection lies in the fact that the slightest carelessness will result in a derangement or even a breakdown of some part of the mechanism, and it requires such a careful adjustment that it has at last been discarded in favour of motions with less mechanical difficulties in their application.

Our next example, Fig. 103, is very similar to the one given in Fig. 100; its main point of difference consists in so arranging the faller connections that instead of a lever being allowed to fall when the faller wire is pulled down, it is drawn up and an extension of it made to bear against the governor band. Reference to the drawings will make this clear; the faller levers D and C are both placed on the opposite side of the faller rods A and B to

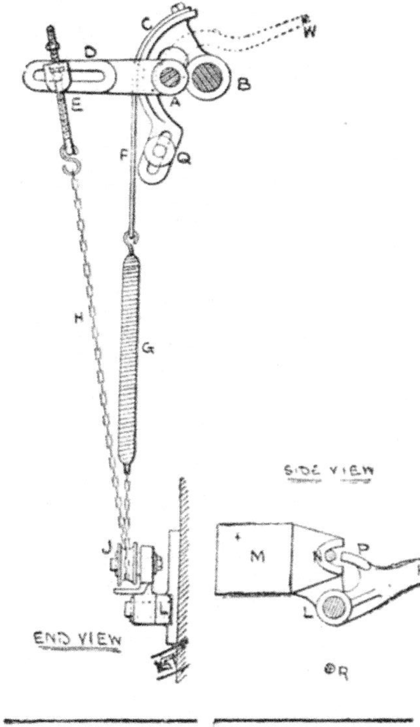

Fig. 103.

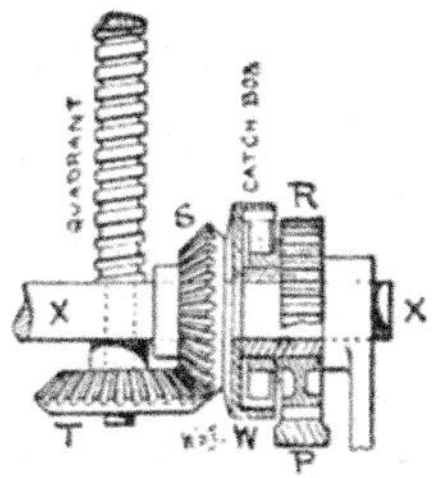

Fig. 105.

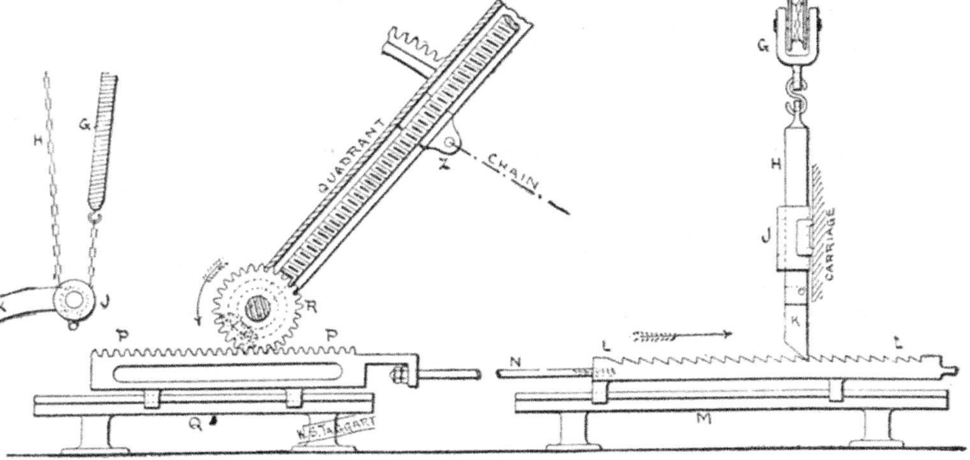

Fig. 104.

the previous example, so that their movements lift the lever K instead of permitting it to drop. When the bowl J is lifted, a projection P on the lever K is brought against the governor band N, which passes through a recessed portion of the bracket M; in this way the pressure is sufficient to hold the band, and as it is connected to the quadrant pulley in the same way as in Fig. 103, it naturally gives motion to the screw.

In order to regulate the pressure put on the band at N, a spring G is connected to the leather band F and the chain H; any excessive movement of the faller wire W, therefore, simply stretches the spring G.

As the cop bottom enlarges, the sector C is lowered, and this lowers the lever K further away from the band N, until at last it is low enough to remain out of action during the building of the body of the cop. For the same reason it is sometimes found advisable to have the projection P as near as possible to the band N when the cop is commencing, and to effect this a bowl Q is arranged on the sector C, which the leather band F passes over; a sensitive action is thus obtained for the first layers, but afterwards such a degree of sensitiveness is not so necessary, and Q therefore works clear of F. The usual adjusting screw E is provided, and in addition slots at D and at Q enable a high degree of exactness to be obtained in setting the motion. In practice, this motion has been found to be unusually successful.

Figs. 104 and 105 present us with another form of rack governor motion which has been found particularly suitable for fine spinning mules. To the usual faller connections D and C is connected a chain F which passes over a carrier bowl carried by a small frame G, which in its turn is hooked on to a drop pendant H; this slides in a bracket

J fixed to the carriage square, and at its lower end is a swivel piece K. When the faller wire is pulled down, the lever D is lowered, and the drop pendant H K falls into contact with a rack L, and the movement of the carriage takes the rack forward on its slide M in the direction of the arrow. To L is connected a rod N, whose other end is screwed to a rack P, which gears with a small pinion R mounted loose upon the shaft X (see Fig. 105). The wheel R is arranged to drive the bevel S, and therefore the quadrant screw through the catch-box W; consequently the forward movement of the rack P can be made to give motion to the nut Z. The return or outward movement of the carriage, by means of a finger bolted to the square, pushes the rack P back without operating the screw. In place of this, the catch-box may be arranged to be inoperative when the rack P is moved by the inward run of the carriage, and during the outward run to act upon the quadrant. The rack L is made long enough to enable the full length of the rack P to be used when such is necessary, as in the earlier layers, and also to use small portions when the cop bottom is getting finished or when the tension is only slightly altered.

Sufficient examples of governor motions have now been given to show how near to self-acting the winding operation has been brought. At the same time the observant reader will have noticed the disadvantages associated with the various automatic arrangements described, and to any one with a practical knowledge of the subject, such disadvantages are almost considered inherent, and in most cases prove a source of difficulty when a motion is first applied.

It has already been explained how necessary it was to move the nut up the quadrant screw at a gradually diminished rate. When the action is performed by the minder he finds it necessary to turn the screw several times for the earlier layers and only occasionally for the last layers. This calls into play a certain amount of care and judgment, which can be modified by making the screw with a varying pitch, as at B, Fig. 106. By this means a single revolution of the screw will move the nut a good distance upwards when the cop bottom is started, while a similar turn when the cop bottom is complete only moves the nut a short distance. The example given at B is taken from an actual screw as used by a well-known firm of machinists. The rate at which the nut would travel up the screw is shown in full lines in the diagram, and from it we get a clear idea of the diminishing rate of its upward movement. Such a screw as this has been in use almost from the first introduction of the quadrant, and, probably from practical motives, it has remained until the present time; in order, however, to prevent misconception, it is as well to point out that the screw B only goes a little way towards giving the nut its correct but varying movement for a "uniform" turning of the screw. A quadrant screw to do this ought to be made as shown at A, yielding a curve as dotted on the diagram. This curve represents the true movement of the nut up the quadrant, and its variation from the curve of B is considerable. The vertical lines may be taken as representing complete turns of the screw, so that in the case of B the first turn would lift the nut from I to E, while for screw A the nut would be lifted from I to F, more than twice the distance. A similar difference, but in the opposite direction, is noticeable at the upper end of the screw, B having a much quicker movement than A for each turn. As a comparison, a uniform screw is shown at C, having an equal number of threads, as A and B. Its rate is naturally represented as a straight line in the diagram, and we see very clearly its difference from the variable pitch.

In the application of a governor motion, some firms have discarded the variable screw, and rely upon the motion giving to the nut its correct variable movement; but we see that although the tension of the yarn is likely to be increased above the required amount early in the inward run when the cop bottom starts, and late in the run-in as

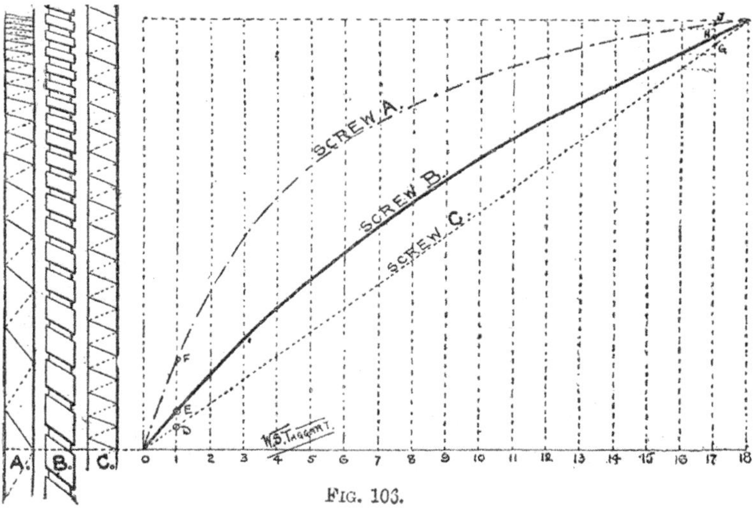

FIG. 103.

it finishes, it may easily happen that in the former case it acts on the band or rack too late to move the nut high enough, and in the latter case too early, and so gives the nut too much movement; we therefore get the action spread over two draws for the first case, and there is no remedy, outside the minder, for the second case. A correction for this is found by some makers in still retaining the variable screw, and its difference from the correct form may be looked on as a convenient compromise between the two extremes at A and C.

Long-lever Mule.—The descriptions of the general actions of the mule have so far been confined to the cam-shaft principle of working the changes, but, as already stated, this is not the only method. The "long-lever" mule, as it is aptly termed, to distinguish it from the "cam-shaft" mule, is one that receives a very extensive application, and its range of work from the lowest to the highest numbers gives to its mechanism an unusual and important interest. Because the long-lever mule is made by firms who have a very high reputation for machinery adapted for good quality and high numbers, it is sometimes thought that it is only suitable for such purposes; this is a mistake, and it is desirable to emphasise the fact that its working is equally satisfactory on the lowest counts.

A general outline description of the long lever and its action will now be given, and reference will be made to the drawing, Fig. 107. This sketch embodies the principal features of the mechanism, but it must be looked on rather as a key diagram than as an illustration of detail. An attempt will be made, as when dealing with the cam shaft, to describe and illustrate all features essential to a clear understanding of various actions.

On reference to Fig. 107, the long lever A B is centred on a stud C fixed in the framing of the headstock. By giving movement to the ends of this lever we can bring about changes in the working of the mule which permit spinning, backing-off, winding, and drawing-up to be performed. In spinning, the strap is on the fast pulley W, and this both turns the spindles and takes the carriage out. While this is going on the strap must be kept on W, and the backing-off wheel V must be kept out of contact with the leather cone on W. This latter effect is produced by a stud E on the long lever coming against the lever,

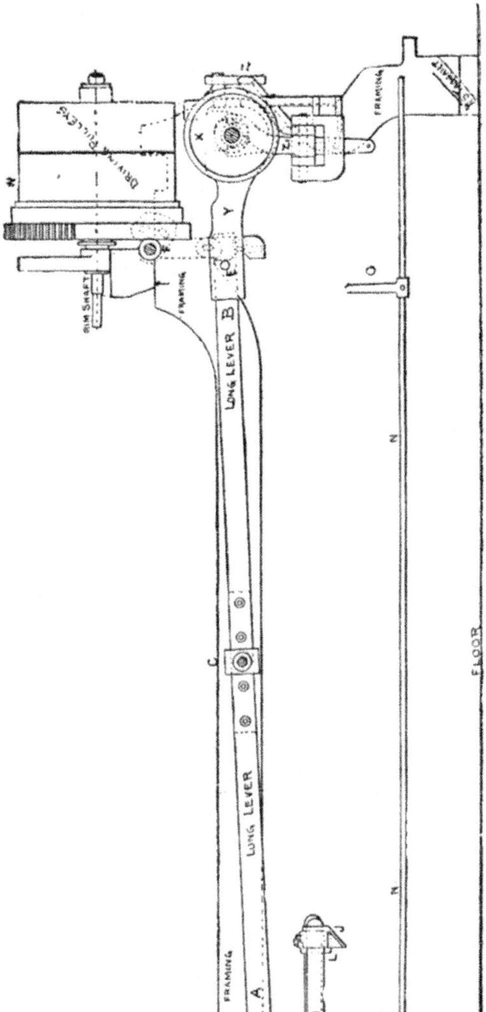
DRIVING PULLEYS
RIM SHAFT
FRAMING
LONG LEVER
B
LONG LEVER
A
C
N
O
FRAMING
FLOOR
WEIGHT
FRAMING

FIG. 107.

which puts V in and out of gear with W; in the position shown it is impossible for V to be moved. On the other hand, the long lever must be locked in this position, so we find that at the outer end of the long lever a stud D is held by a catch G on a bell-cranked lever centred at H. As the carriage comes out, the quadrant drum shaft S is giving motion to a wheel R, upon whose face is a special cam groove. A slide U, the lower end of which carries a heavy weight, is raised by the cam groove, and the end A of the long lever is free so far as this weight is concerned, but the upward movement of the slide U puts a spring into tension, which is attached to a lever T, upon which the long lever rests. There is, therefore, a force pulling A upwards, which is resisted by the catch at G. When the carriage arrives out, a bowl on the square comes into contact with the incline J and releases the catch G, the end A instantly moving upwards, and the other end B falling. The stud E, coming opposite a recess on the backing-off lever F, permits a spring to pull F forward and so puts the wheel V into contact with W.

The backing-off action completed, a bowl on the square is moved forward and lifts an incline L, carried by another catch lever K centred on H. The projecting catch on K has prevented the stud E falling further than the recess in F, but now that K is released the stud E is forced further downwards and in this movement takes the lever F out of its way and consequently V out of contact with W.

Matters remain in this condition during the drawing-up or run-in, but as this nears completion the carriage comes into contact with a finger O on a rod N, which is attached to a lower portion M of the lever K L; by moving O forward, the catch at K is taken from under the stud D and the end of the long lever suddenly falls under the influence of the weight and again assumes the position shown in the drawing.

The other end of the lever is bolted to a piece Y, which carries a stud acting upon a lever Z, whose fulcrum is at Z^1. Special forms of incline slots in Z permit the stud on Y to be inoperative until drawing-up commences; the fall of the weight, therefore, puts a catch box into gear by means of the lever Z, and enables the drawing-up to be effected. The movement of E in an upward direction takes the catch box out of gear and leaves the back shaft stopped during backing-off.

A detailed description and enlarged drawings of the long-lever mule will now be given. The principal operations performed are spinning, backing-off, and drawing-up and winding: all these actions are directly worked from the long lever, and, as in the cam shaft, the cycle of movements for producing them is termed the "changes"; discussing them in their order, we will take spinning for first consideration.

Spinning.—On reference to the drawings, Figs. 108, 109, and 110, it may be pointed out that they represent the position of the mechanism during the operation of spinning. In Fig. 109 duplex driving is shown, and the strap is supposed to be on the two fast pulleys. Under this condition the backing-off cone wheel G^1 is out of contact with the backing-off cone on the pulley H^1, so that the continuous and independent driving by band of the drawing-up pulley U^1 has no effect on the rim shaft. The rim shaft being driven, we have the motion transmitted through the rim pulley to the spindles. At the same time the rim shaft drives the front roller (see Fig. 16), and the front roller drives the back shaft through the wheels T, O, E, P, and

Q; from the back shaft, motion is given to the carriage during its outward run.

While these actions are going on, the scroll shaft T^1 is rendered inoperative by keeping the drawing-up cone

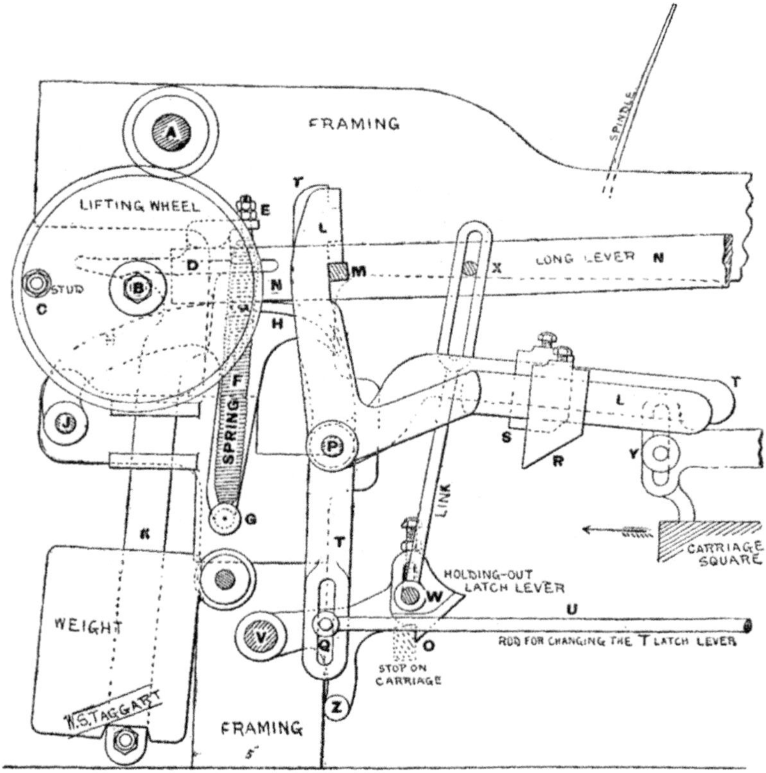

Fig. 108.

clutch Q^1 out of gear with the cone R^1, so that although the drawing-up pulley U^1 is being driven, it has no effect on the scroll shaft. The end of the long lever in Fig. 109 has therefore three very important functions to perform while spinning is taking place. First, it must keep the backing-off cone wheel G¹ out of gear with the cone on H¹; secondly, it must keep the drawing-up cone clutch Q¹ out of gear; and, thirdly, it must keep the clutch wheel on the back shaft (Fig. 16) in gear.

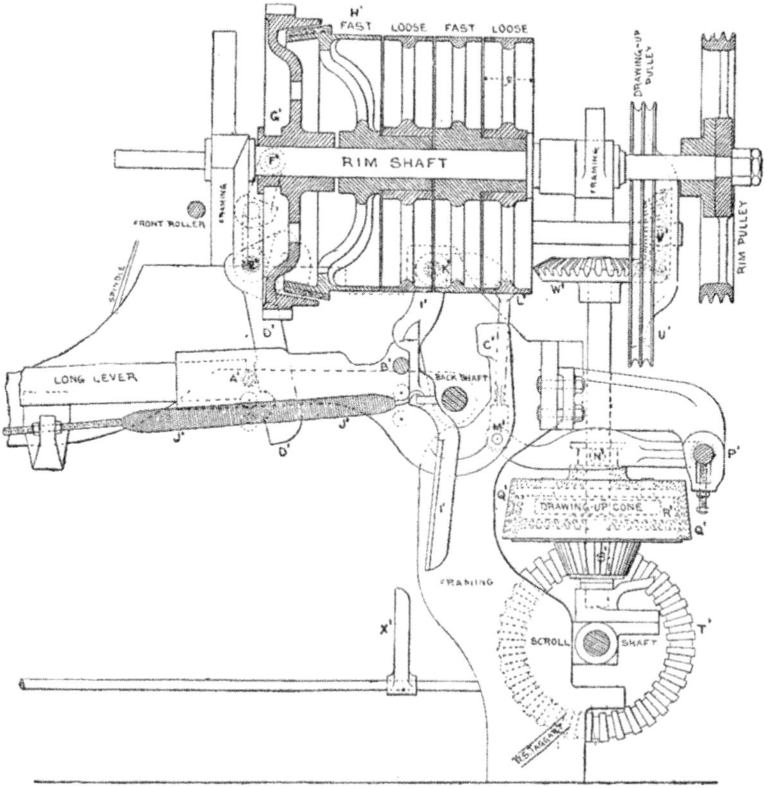

FIG. 109.

The first object is performed in the following manner :—The backing-off lever D¹ (see also Fig. 110) is pivoted on a shaft E¹; its upper end F¹ is forked to fit into a grooved boss on the backing-off wheel G¹, while its lower end bears against a stud A^1 carried by the long lever; in this position the lever D^1 is locked, so that it is impossible for the wheel G^1 to go into contact with the pulley H^1. The second object is effected as follows:—A lever N^1 is carried from a stud at P^1, and the lever is forked and fits a groove on the upper part of the cone dish Q^1. The other end of the lever N^1 is connected by a link at M^1 to the lever I^1 centred on a stud K^1 (see also Fig. 32, but the reference letters are not the same). A stud B^1 on the long lever bears against a prepared part of the lever I^1 as shown, and so long as the stud occupies this position the two halves of the cone clutch Q are prevented from going into contact, and the drawing-up pulley U^1 cannot drive the scroll bevel T^1. The third effect is produced by a stud on the end of the long lever at C working in a groove on one extremity of a lever, the other extremity of which is forked and fits the groove on the clutch wheel on the back shaft, as shown in Fig. 16.

The corresponding position of the outer end of the long lever during spinning is shown in Fig. 108. It occupies its lowest point, and in this position it is held by a stud M, coming under the catch of the L lever centred at P; the long lever is therefore locked during the whole of the outward run of the carriage. As already explained, this movement of the carriage gives rotation to the shaft A for the purpose of working the quadrant. Its motion is taken advantage of to drive by means of a pinion the lifting wheel; a stud C is carried round by this wheel, and in the course of its revolution it comes against the underside of a projection D on the drop weight lever K, and so raises it. A projection on K at E carries one end of a spring F, the other end of which is connected to a projection G of a specially-formed lever H, whose centre is at J.

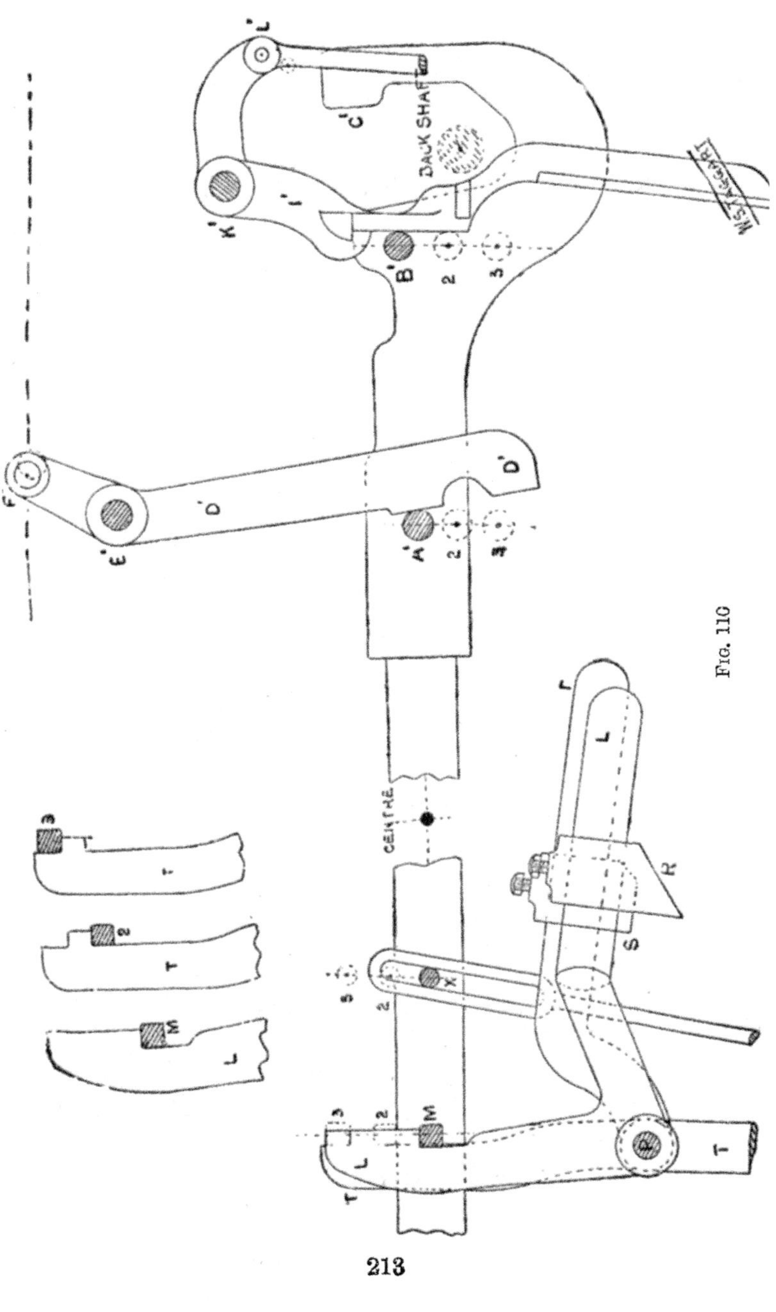

FIG. 110

The upper surface of H bears against the long lever, and the tension put into the spring F, as a consequence of E D K being lifted by the stud C, tends to force the long lever upwards; so long, however, as the stud M is held by the L lever the tension in the spring has no effect.

Backing-off.—The carriage at last reaches its outermost position; at this moment a stud Y, on the end of a lever centred on the square, comes against the inclined tappet R on the L lever and lifts it up, thus freeing the stud M from its catch. Directly this occurs, the tension in the spring F instantly forces the long lever upwards, but it can only ascend a short distance, because, although L has been moved out of the way, the lever T has not been touched, so that the projection on T acts as a stop to the further upward movement of M (see Fig. 110). By referring now to Fig. 109, we shall see what effect this movement of the long lever has upon the levers D^1 and I^1. We already thoroughly understand that when the carriage has completed its outward run, the spinning process is over, and so the spindles must first be stopped and then immediately reversed for backing-off. The ascent of M in Fig. 108 means the descent of A^1 in Fig. 109, and it will occupy the position of the middle dotted circle, or as shown at 2 in Fig. 110. A strong spring in tension (not shown in the sketch) immediately pulls the lever D^1 forward, a recess cut in the face of the lever permitting this to be done. This at once puts the wheel G^1 in gear with the fast pulley H^1, and as the straps have been moved on to the loose pulleys, the drawing-up pulley U^1 is enabled to drive the rim shaft through the pinion which gears into G^1 (this is shown clearly in Fig. 57). The direction in which it is driven is also in the opposite direction to that in which the straps drive when they are on the fast pulleys; the spindles are therefore reversed.

It will be noticed that the drawing-up cone Q must still be kept out of gear during this backing-off action; for this purpose the lever I^1 is so arranged that on the descent of the stud B^1 it simply comes on to a lower portion of the straight surface of I^1 and produces no effect on the lever itself.

The act of reversing the rotation of the rim shaft effects, through special mechanism which will be described subsequently, a movement in the fallers, and one effect of this movement is to lift up the end of the lever which carries the bowl Y in Fig. 108. The lifting of Y brings it against the tappet carried by the T catch lever, and of course this moves T on one side, freeing the stud M, so that the tension in the spring F forces the long lever still further upwards, and as it moves upward it passes from the lower side of the projection on T to the upper side, where it rests (Fig. 110).

Drawing-up.—This second change of the long lever causes the end in Fig. 109 to fall to its lowest point (see also 3 in Fig. 110). Its effect on the lever D^1 is to force it backward as the stud A^1 moves out of the recess, and this necessarily takes the cone wheel G^1 out of contact with H^1, and so stops the spindles. At the same time, the stud B^1 falls clear to the lever I^1, and a strong spring J immediately pulls the lever forward; this action, through the link and the forked lever N^1, forces the cone dish Q^1 into gear with the cone clutch, and so permits the drawing-up pulley U^1 to drive the scroll shaft and cause the inward run of the carriage.

A locking arrangement is provided for the carriage on completing its outward run, in the form of a holding-out catch W. A stop O on the square comes against an incline W, lifts it, and passes under, so that the incline falls back and locks the carriage in position; this latch must be lifted before the inward run can take place. Connected to the incline is a link, the slotted upper end of which fits a pin X on the long lever. This pin X moves clear in the slot of the link during the change from spinning to backing-off, but when backing-off is complete and the long lever makes its second upward movement, the pin X comes against the top of the slot and lifts the link, which raises the catch W out of the way of the stop O, and sets the carriage free to make its inward run.

During the run-in the straps are on the loose pulleys; winding is taking place, and the long lever is locked in position by the T latch lever (as shown in Fig. 110); the stud M occupies its highest position, and the studs A^1 and B^1 occupy their lowest positions; the stud C on the lifting wheel is clear of the projection D, and therefore the full effect of the heavy weight on the drop lever K comes on the end of the long lever. So long, however, as the stud M is supported by the T lever, the weight is inoperative.

As the carriage completes its inward run, it comes against the stop X^1, fixed on a rod which is connected to an extension of the T latch lever at Q. The forward movement given to the rod pulls the lever T on one side and permits the full effect of the weight to come on the long lever and to pull it down in one movement to its lowest position. The studs A^1 and B^1 in Fig. 109 move up to their highest points, and in doing so assume the positions shown in the drawings ready for spinning. The upward movement of the stud B^1 is not allowed to move the lever I^1 on one side; this is effected by a stud on the carriage (see Fig. 32), which comes against the lower end of the lever and lifts the drawing-up cone clutch completely out of gear.

The drawings have been made as complete as possible to enable the descriptions to be clearly understood, but with this object in view several details have been kept out, such as the backing-off motion, the chain-tightening motion,

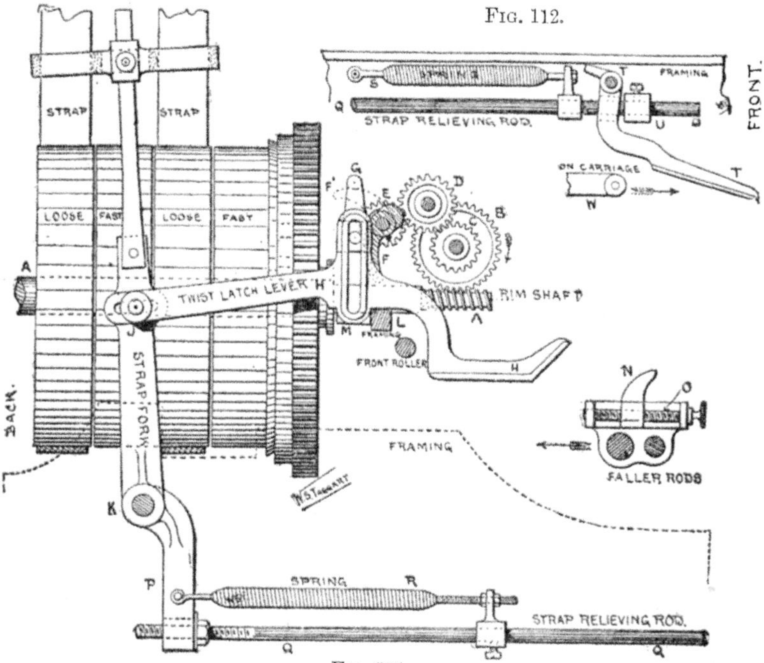

FIG. 112.

FIG. 111.

the strap-fork arrangement, etc. These, however, will be fully dealt with.

Changing Strap from Fast to Loose Pulley. Strap-relieving Motion. Hastening Motion.—There are several methods of changing the strap from the loose to the fast pulleys, and *vice versâ.* One of these is illustrated in Figs. 111 and 112. The duplex system of driving is shown. As the carriage moves out, the strap is on the fast pulley; as it arrives within a few inches of the finish of the stretch, a stud W (Fig. 112) on the carriage comes into contact with a pendant lever T, centred on the framing; this lever is moved forward, and A, a projection thereon, presses against a stop-washer fastened on the rod Q and moves the rod also forward. Attached to the rod at the outer end is a spring S, whose other end is fixed to the framing; the other end of the rod Q (Fig. 111) passes through a slot in the lower part of a lever P, which is fastened on the strap-fork shaft K. The forward movement of the rod causes Q simply to move freely in the slot of P; but a spring R, attached to the rod and to the lever P, is put into tension; and with this tension existing in the spring R there is a strong force tending to move P forward and put the strap from the fast to the loose pulley. This action would of course directly occur under some circumstances, but frequently an arrangement is provided whereby spinning continues until the necessary amount of twist has been put in the yarn. Until this occurs the strap-fork is locked by means of the twist latch lever H, which is attached to the strap-fork at J and a projection at M fitting over a portion of the framing at L, where it is locked. This lever is set free in the following manner:—A screw is formed on the end of the rim shaft A, into which gears a worm wheel called the "twist wheel" B. Through the gearing C, D, and E a short shaft is driven, whose end carries a tumbler F. This tumbler, though free on the shaft of E, is, through a pin, capable of being carried round. As it revolves, it comes against a projection G fastened on the upper part of the twist latch lever, and lifts it until the projection M rises clear of the catch L; directly this happens, the tension in the spring R pulls the strap-fork over, and changes the strap from the fast to the loose pulley. Backing-off then takes place, and afterwards the carriage is drawn in by the drawing-up band.

When the lever H is freed from the catch L on one side, the spring R pulls the strap-fork over, and with it the twist latch lever, so that this lever passes over the top of L, falls down on the other side, and again becomes locked; the strap-fork therefore cannot be moved from the loose to the fast pulleys until H is again set free. Now it will be noticed that the tension put into the spring S by the carriage moving T forward is not affected when the spring R acts on the strap-fork; Q makes no movement at the moment the strap changes; P is simply pulled over, and now abuts against the nut on rod Q, the tension in the spring S remaining. Although this tension has a tendency to move the strap back to the fast pulley, it cannot do so, because it causes the twist latch lever to press against the projection L on the side opposite to the position it occupied when the carriage was going out. The illustration, Fig. 111 shows the position during the run-in of the carriage.

On the faller rods a small bracket is loosely fitted, carrying a screw O on which is fitted a tumbler N. The use of the screw O enables the position of N to be carefully regulated according to the circumstances of the case, and moreover N is so arranged that it can easily be turned over so as to avoid coming into contact with H. In the position shown, the carriage is moving in, and naturally N will come into contact with H and lift it; this frees the twist latch lever from L, and permits the tension in S to pull the rod Q backwards; the nut on Q being against P, forces P backwards, and so removes the strap from the loose to the fast pulley. By making N so that it can be moved on one side, the mule is enabled to be stopped when it completes its inward run, because it prevents the strap from being put on to the fast pulley so long as the twist lever H is locked on the catch; by turning N over, the lever H is untouched when the carriage gets in, and as the strap is not changed the mule stops.

Adjustment is provided in every possible direction in order to obtain perfect harmony in the working of the several actions; while the inclined under-surface of the lever T permits a gradual movement of the strap from fast to loose pulley to be effected.

The special arrangement shown in Fig. 112 is generally called a "strap-relieving motion," and the arrangement on the fallers at N may be designated a "hastening motion."

Backing-off Chain and Faller Sector.—The effect of the backing-off action on the copping faller has already been thoroughly explained; it is therefore only necessary to describe and illustrate the method adopted for this purpose in the mule under discussion.

When the backing-off cone wheel is put into contact with the fast pulley (while the strap is still on the loose pulley) the rim shaft is driven in the opposite direction by the small pinion on the drawing-up shaft (Fig. 57). This reverses the direction of the revolution of the tin roller and of the spindles. A chain M, Fig. 113, is attached to a snail or small scroll on the tin-drum shaft, and passes on to a pulley K carried by a slide bar J. The pulley K may be either single or compound, according to the movement required; in this case a compound one is shown, to one of which pair of pulleys another chain is attached, its other end being hooked on to the faller sector fixed on the counter faller A.

Through this faller sector on A, the movement of the faller wire, as it lays the yarn on the spindle, is regulated. (For a full description of this, see p. 93 *et seq.*) When the carriage has almost finished the run-in, the lower projecting end G of the faller leg C comes against a floor bracket H, and the slight further movement of the carriage forces the faller leg at D from the position it occupies on the slide E, as the winding proceeds. Directly D is free from E it falls down into the position shown in the draw-

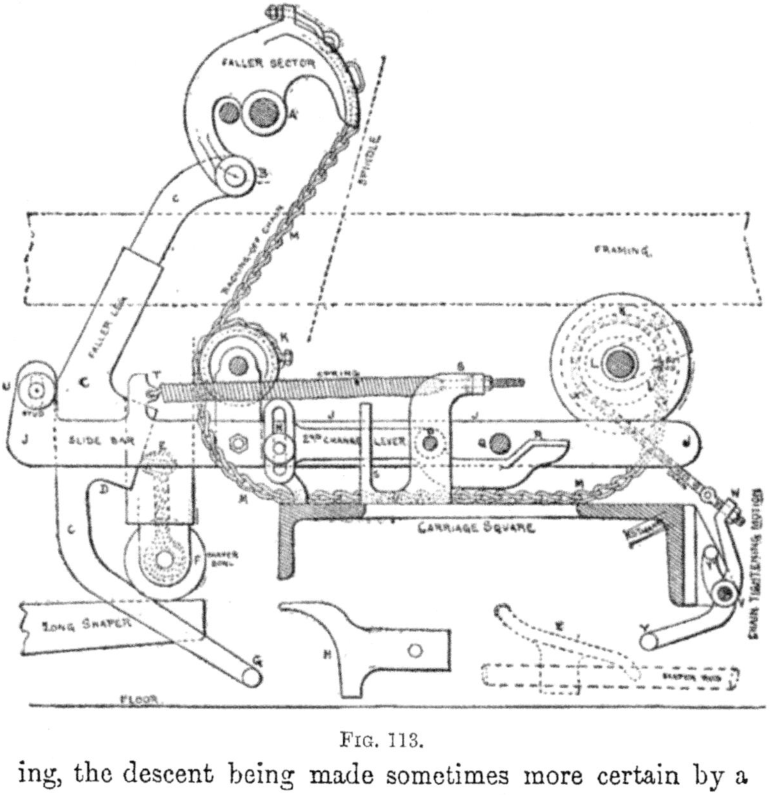

FIG. 113.

ing, the descent being made sometimes more certain by a spring (not shown in the drawing) attached to C and the carriage end. As C is forced on one side by G coming into contact with H, it pushes forward a slide bar J, by virtue of a projecting stud on J being in contact with C. The forward movement of J puts tension into the spring attached to J at T and to a fixed bracket at S. This state of things, with the positions shown in Fig. 317, continues during the whole of the outward run. While backing-off, the reversal of the tin drum pulls down the faller sector through the chain M, and this also tends to pull backwards the slide bar J because the pulleys at K are carried by it. The pulling down of the sector raises the faller leg, and at last it is lifted sufficiently high to allow the ledge at D to slip over the projection at E. The tension in the spring and the pull of the chain M cause the faller leg to shoot instantly over E, whereupon the faller becomes locked and ready to be actuated from the shaper through the bowl F. As the slide bar J shoots backwards, a stud Q thereon comes against an inclined part R of a lever centred on a bracket at P. Its other end carries a bowl N, so that, directly backing-off is completed by the locking of the fallers, the almost simultaneous raising of the bowl N forces upwards the latch lever T, which releases the long lever and brings about the change for drawing-up (see Fig. 108).

Backing-off Chain-tightening Motion.—The arrangement for tightening the backing-off chain is also shown in Fig. 113. A lever Y, centred at V, has a chain X attached to one end W. The other end of the chain is fixed to a small pulley on the tin-roller shaft, mounted in such a way that any pull on the chain X will give a movement to the snail round which the chain M is wound. A little tightening movement of the chain M is required at first, so the lever Y is arranged to just come into contact with the incline Z, carried by the shaper rod.

Note.—The incline Z is really at the outer end of the headstock; it is placed in the position shown in the drawing simply for clearness.

As the cop builds, the necessity arises for having the backing-off chain M tight, so that since Z moves forward with the shaper, Y is brought into contact with the incline earlier each draw, and in this way a little more of the chain M is wound on the scroll previous to backing-off, so that at last we get a practically tight chain, which is capable of acting immediately on the faller sector.

Backing-off Motion.—There are one or two very important variations of the mechanism shown in Fig. 108, the improvements primarily consisting of methods intended to quicken the backing-off action and render it more certain. One of these variations is shown in the accompanying drawing, Fig. 114.

The carriage is moving outwards; the straps are on the fast pulleys; the backing-off and the drawing-up cone frictions are out of gear. As the carriage is completing its run-out, a stud or bowl at M, carried by a lever centred at N, comes in contact with a bracket L on a long rectangular backing-off rod; the rod is moved forward, and as a consequence the studs B and C carried by it are moved, so that B comes under the end of the lever E, and C is moved out of contact with the backing-off lever D; a spring K attached to the rod and to D is also put into tension. The force exerted by the spring K cannot, however, pull the backing-off cone clutch into gear (which is its intention), because a projecting arm J on the strap-fork has a stop *h* which prevents the lever D from moving.

The result of the stud B coming under the end of the lever E is to prevent the drawing-up cone clutch from going into gear until such time as is necessary. Both cone clutches are therefore locked during the time the stud M is moving forward the bracket L and its rod. It will also be noticed that the long lever, by means of its stud J, is keeping the lever H from permitting the drawing-up clutch to go into gear. While the stud M is still moving L forward, a stud on the carriage comes against the incline on the T lever, and lifts it; this at once releases the long lever from the catch V on the T lever, and sets the lever H free from the stud J, so that now the spring *g*, which is in tension, exerts its full pressure to pull the lever E downward; as long, however, as the stud B is under the end of E, the cone clutch remains out of gear. After a short interval (depending on the number of twists put in at the end of the stretch when the carriage is stopped) the twist latch lever is released (as already described), and the strap-fork is moved on to the loose pulley, and its projecting arm J being raised, sets free the backing-off lever D, and permits the spring K to pull D forward and so put the cone friction into gear. The actual backing-off action now commences; the reversal of the tin drum winds on the chain, pulls down the faller sector, and lifts the faller leg. The chain passes over a pulley S carried by the lever centred at N, so that its pull tends to draw the faller leg forward through the connecting link P. A spring R, in tension, also tends to pull forward the lever N M and consequently the faller leg. The gradual rising of the faller leg, as the chain is wound on the tin drum, at last brings the recess U opposite the slide Q, which rests on the shaper. Immediately this occurs the combined pull of the spring R and the chain causes the faller leg to shoot forward over Q, and the lever N M is drawn backwards. Four actions simultaneously occur in consequence of this movement of the lever N M. First: M is taken out of contact with L, which permits the spring K to pull the backing-off rod backwards. Second: a lever O, working on the same centre N as the lever N M, is lifted, and,

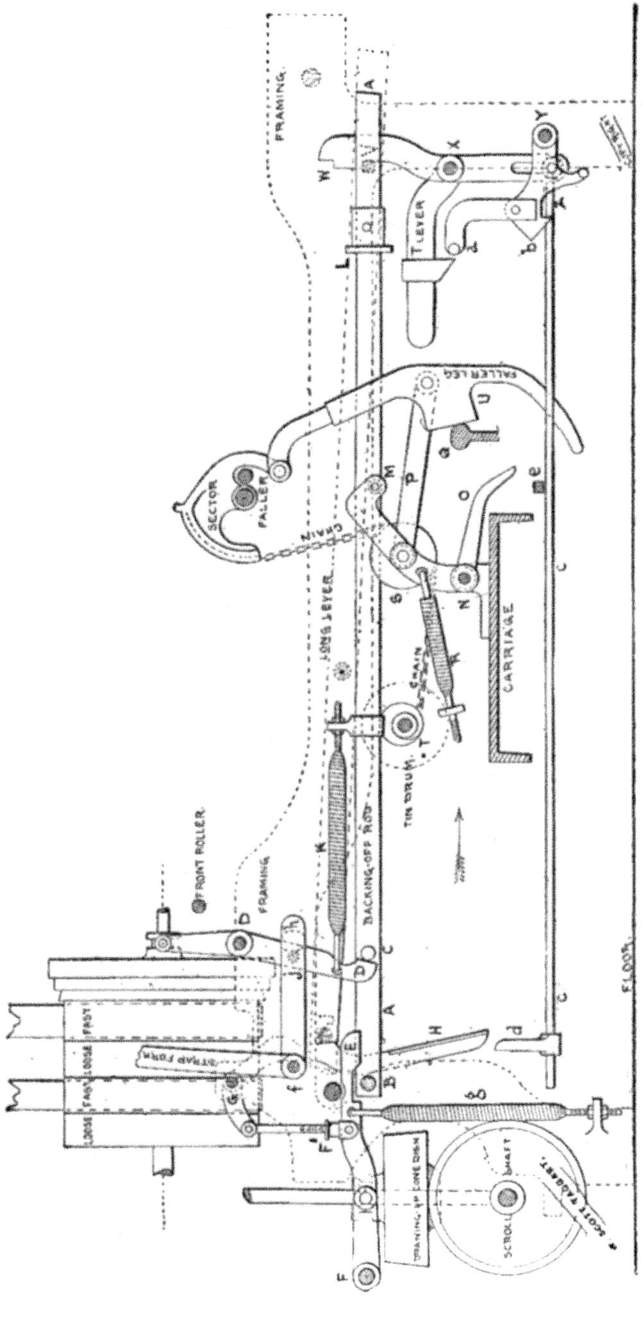
FRAMING
LONG LEVER
BACKING-OFF ROD
CARRIAGE
TIN DRUM
CHAIN
SECTOR
FALLER
FALLER LEG
T LEVER
FRONT ROLLER
STRAP FORK
FAST
LOOSE
FLOOR
SCROLL
SHAFT
SCOTT TAGGART

FIG. 114.

coming into contact with the incline *b*, lifts it, and so frees the carriage which has been locked by the recess at Z fitting over the projection *e*. Third: the backing-off cone clutch is taken out of gear by the stud C coming against D and moving it backwards. Fourth: the faller leg, through being pulled over the slide Q, puts the copping faller in direct connection with the shaper.

The first action, in moving the stud B out of contact with the end E of the lever F E, at once permits the spring *g* to pull it downward, and so puts the drawing-up cone clutch into gear, which action causes the carriage to be drawn in. At the same time the stud C, coming against D at the moment M releases L, moves D backwards and takes the backing-off cone clutch out of gear.

The carriage now makes its inward run; the stud on the carriage comes against H and lifts the drawing-up cone clutch out of gear, and so stops the carriage. At the same time the finger *d* is moved forward, and this releases, through the rod *c*, the long lever from the catch at W. This brings the stud J into the position shown in the drawing, and prevents the cone clutch from falling into gear again; it also puts the catch box on the back shaft into gear, and so permits the front rollers to bring the carriage out. Simultaneously the incline on the faller rod releases the twist latch lever, and so changes the straps from the loose to the fast pulleys. When these actions are all finished their respective mechanisms occupy the positions shown in the illustration, Fig. 114.

Fine Spinning Details.—A number of important details of the self-actor are only used when the machine is employed in spinning fine numbers. This discrimination between fine and coarse counts of yarn arises from causes that are not entirely obvious; indeed, as we shall see, the

Note.—See Appendix for further details of Fine Spinning Mules.

reasons generally advanced for the use of some of the additional movements are as applicable to the spinning of very good coarse numbers as they are to fine numbers. To give an illustration of this we may point to the fact that several motions that were formerly only found on fine spinning mules are now to be seen on almost any mule from which good work is produced. Moreover, high numbers are produced in a far less degree than formerly, and the skill that used to be displayed on counts such as 150's to 300's is now turned to account in producing lower numbers of a superior quality; and where 100's was necessary to give double 50's, we now find 50's by itself equalling the previous practice. It is no uncommon thing to see mules equipped for spinning high counts used for much lower numbers. The following may be taken as suggesting the difference of treatment between fine and coarse numbers:—

Fine numbers are spun from longer and better cotton than coarse numbers. Long cottons are weaker than short cottons. More draft can be used when spinning fine numbers than in coarse numbers, because of the length of fibres. Fine numbers are twisted more than low numbers; and fine numbers, owing to the delicate fibres, are strained through this extra twist, so that some means must be found to relieve them; while for a similar reason the operation of spinning must be performed very slowly compared with the speed for low numbers.

Double-Speed Driving.—Some of the actions already described operate so promptly that the suddenness of action so produced tends to strain the yarn. To overcome this difficulty, a more gradual stopping and starting is adopted, and, moreover, friction is reduced to the smallest possible degree. Some of the arrangements of mechanism for dealing with the points mentioned above will now be given, and the first example will illustrate what is generally termed "double-speed" driving.

We have seen that the counter shaft controls the whole mechanism of the mule. It is at this point that a change is usually made if it is desired to alter the relative speeds of the various actions that are performed. Now in fine spinning it is absolutely necessary to perform the spinning process very slowly, but there is no necessity to work slowly while the other actions are in operation; a form of driving is therefore adopted which is alternately slow and fast. Fig. 115 shows the usual method adopted. On the counter shaft, instead of a pair of pulleys, fast and loose, driven from the line shaft, there are arranged two sets of pulleys as at A and B, each set consisting of three pulleys, two loose and one fast.

When spinning is taking place and the carriage is travelling outwards the counter shaft is driven from the line shaft through the fast middle pulley at B; at the same time the other strap from the line shaft is running on the middle loose pulley at A. This driving continues until the carriage gets out, and, as backing-off can be performed quickly without danger to the yarn, a quicker speed is obtained by moving the strap-forks P so that the straps are moved to the right from the fast to the loose pulley at B, and from the loose to the fast pulley at A. The set of pulleys at A being smaller than at B, we get by this means a quicker speed for the backing-off, and this extra speed is maintained during the run-in of the carriage.

Fig. 115 fully illustrates the arrangement for moving the strap. A bar N, upon which the strap-forks P are mounted, abuts against a lever V. It is connected directly by levers M, A and B to the upright rod L and through the lever at K to the setting-on rod J. The bar

N is also connected indirectly by an arrangement of lever and wheels to the back shaft, from which some of its movement is controlled. A lever R, centred on the back shaft, rests upon a cam Q, which is driven by a train of wheels at a certain fixed speed. The revolution of this

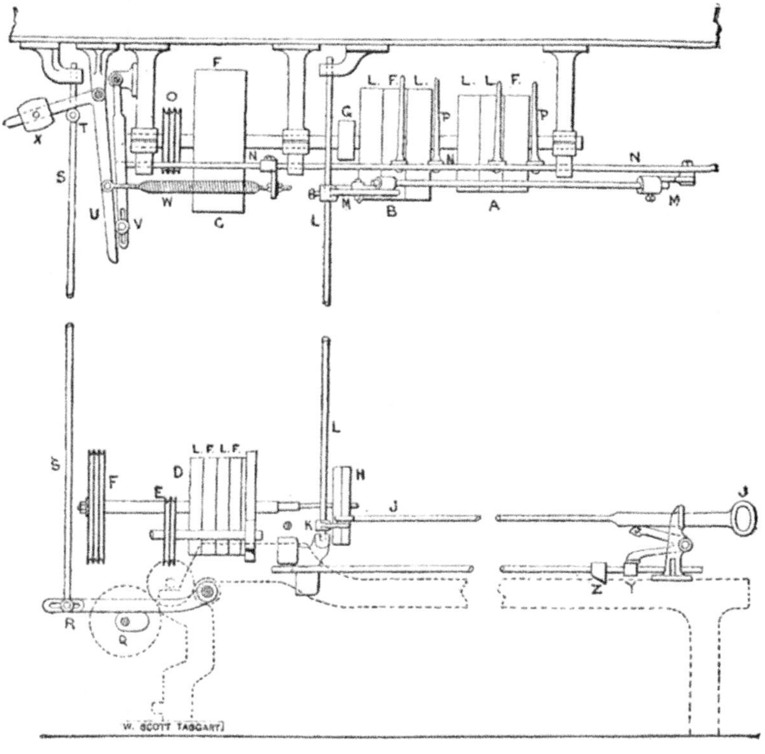

FIG. 115.

cam lifts the lever R, and with it an upright rod S to which it is attached; the rod S carries at its upper end a bowl T, which comes against one arm of the bell-cranked lever, whose other arm U bears against the lever V. The lifting of the lever R takes U out of contact with V, and puts tension in the spring W, which tends to pull the strap-forks from the double-speed fast pulley at A and put the single-speed strap on the fast pulley at B. This cannot be done, however, until the carriage arrives fully in, when the setting-on rod J is unlocked, which releases the rod N and permits the spring W to pull the strap-forks P forward and allows the single-speed fast pulley to be driven. As the carriage moves out, the cam Q allows the rods S to fall and leaves the weight X pressing U against V and tending to force N back again. This pressure is exerted during the run-out, but the strap-forks are not moved until the carriage, coming against Z, frees the setting-on rod and permits the strap-forks to be pushed back by the weight X.

When it is necessary to stop the mule completely the straps from the line shaft can readily be moved on the end loose pulleys of each set at A and B. The drawing-up pulley E is driven from the pulley O on the counter shaft, and through O the mule receives the change of speed. The pulley at H, driven from G on the counter shaft, is a special winding motion, another example of which will now be given. When dealing with another maker's type of mule further on in the book, a second example is illustrated of double-speed driving obtained directly from the rim shaft. See Fig. 135.

Winding Motion.—In spinning very fine counts, the change of the fallers when the carriage gets in, and winding, as performed by the quadrant, is completed, results in a momentary freeing of a certain length of yarn while the faller wires move into their new position. The fineness of the yarn and the twists it contains at once tend to form snarls and even cut yarn. Therefore a method is adopted to take up this length of yarn by giving the spindles a few extra turns, independently of the quadrant, just as the carriage is finishing the run-in and the fallers are about to change.

Fig. 116 represents an arrangement for performing this operation. The carriage is coming in, winding by the quadrant is in progress, and the strap is on the loose pulley C. On the rim shaft are placed two narrow pulleys, fast and loose, as at A and B. A strap from the counter shaft is on the loose pulley B, so that the rim shaft is

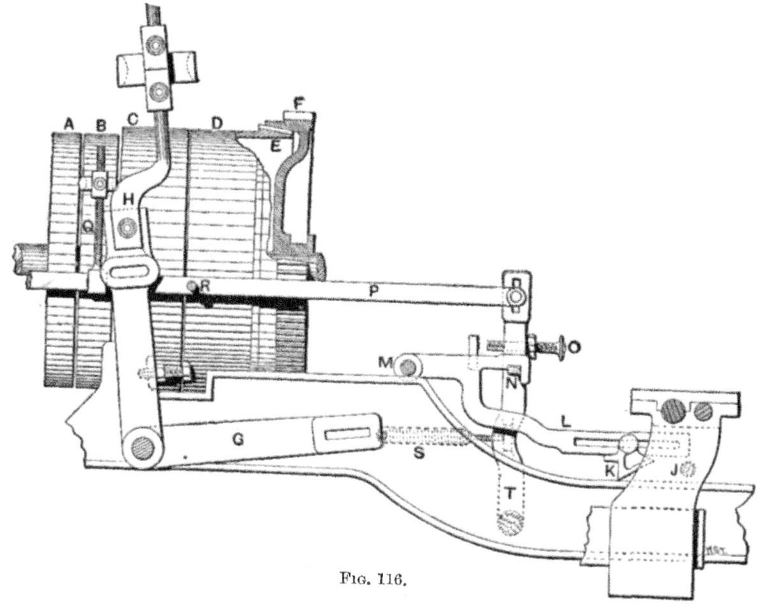

FIG. 116.

stationary. On the carriage is fixed a stud and bowl J, which, as the carriage nears the finish of the inward run, comes into contact with an incline K carried by a lever L fulcrumed at M. The stud J lifts K upwards, and in doing so sets free a projection N on the upright lever T, which L has previously held locked in the position shown in the drawing. Immediately T is free, a strong spring S attached to it pulls it over, and by means of a bar link P, connected to the upper part of T, the movement takes the strap-fork Q, which is attached to P, from the loose pulley B to the fast pulley A. Directly this happens the rim shaft begins to revolve, and consequently the spindles—which has the effect of taking up the yarn so that no snarls can be formed. At the same time a pin R on the link P is set so that it just comes into contact with the strap-fork H. The change of the main driving strap from C to D now takes place for the outward run and spinning, and as the strap-fork H changes, it moves back the link P by means of the stud R. This changes the winding strap from A to the loose pulley B. For regulating purposes K can be adjusted so that the extra winding can be made to commence up to 8 inches before the carriage gets in, and in addition the adjusting screw O enables the amount of strap that is considered necessary for driving A to be very delicately regulated. See also Fig. 117.

Drawing-up by Belt.—For fine spinning, as already explained, the drawing-up cone clutch is dispensed with, and, in its place, drawing-up is performed by a strap, the "change" taking place by moving the strap from a fast to a loose pulley, as shown in the drawing, Fig. 33; see also Fig. 117. The arrangement is frequently employed on mules spinning counts 120's to 300's, and its object is to avoid the sudden change resulting when the cone clutch is put suddenly into gear; by the method shown a gradual movement is obtained, and all shock or suddenness of action is avoided.

Gain and Ratch.—In spinning fine numbers, it is a frequent practice, in fact almost an unavoidable one, to cause the carriage to run at a slightly quicker rate than the surface speed of the front roller, which results in what is termed "gain." Further, this gain is augmented by sometimes stopping the rollers before the carriage has

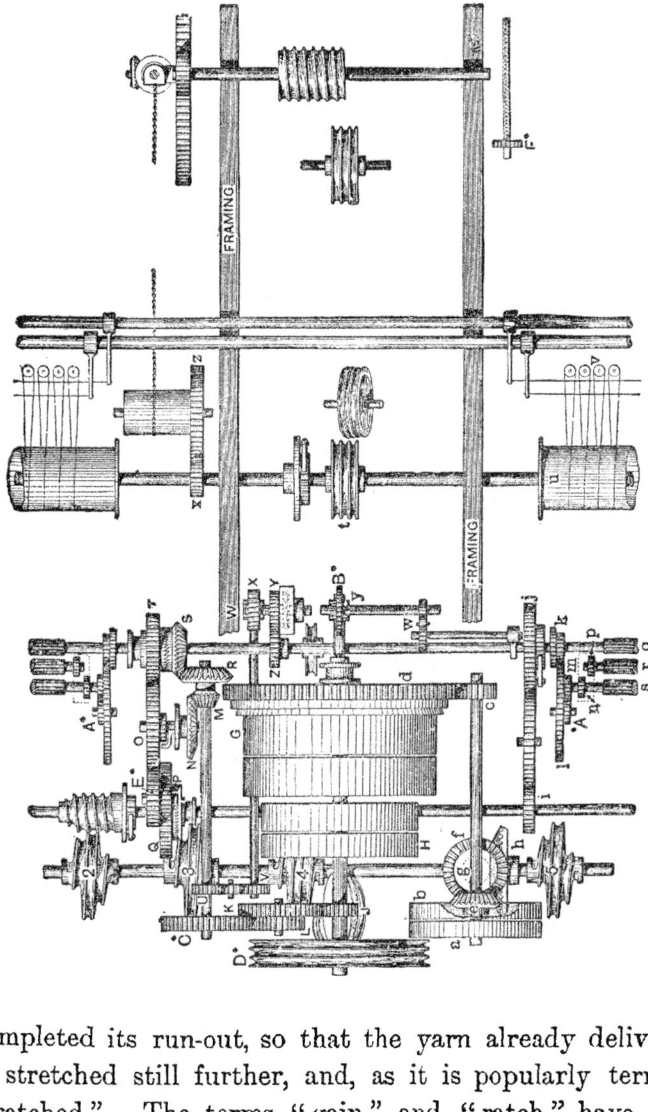

GEARING OF S.A. MULE FOR FINE NUMBERS

FIG. 117.

completed its run-out, so that the yarn already delivered is stretched still further, and, as it is popularly termed, "ratched." The terms "gain" and "ratch" have thus become almost standard expressions for these two operations, though the latter is frequently described as an "after-stretch motion." The effect of the "after-stretch" is naturally to draw out the thick and thin places in the yarn and make it more uniform in thickness.

Gain in the carriage is not confined to high numbers, though for ordinary medium numbers of twist it is seldom that gain is necessary. It is chiefly used for such numbers when weft yarn is made, and then only to a slight extent.

For yarn containing an unusual number of twists the opposite effect is often produced, and the carriage travels slower than the surface speed of the front roller; the extra yarn thus delivered is taken up by the extra twist put into it, and in this way the yarn is relieved of the strain to which it would otherwise be subject. Fig. 117 illustrates the gearing through which the relative speeds of the carriage and front roller can be altered. A change of the wheel L, or if necessary the wheels L and K, will regulate the speed of carriage and front roller in relation to the speed of spindle, but it will not alter the relative speeds of the carriage and roller; this is brought about by changing the pinion P through which the back shaft is driven from the front roller. This wheel is often called the "gain pinion," because of its function; see also Fig. 118.

Jacking Motion.—In Fig. 16 a sketch was given showing how the front roller is driven from the rim shaft, and the back shaft from the front roller. Fig. 57 also showed a similar arrangement. In the drawing, Fig. 117, another full gearing plan of the mule is given, which exhibits the gearing employed on a fine spinning mule; it therefore differs in a few details from the one illustrated in Fig. 57. The pulleys H are the extra winding pulleys, whose function was described in connection with Fig. 116.

On reference to Fig. 117 it will be noticed that the wheel work, between the rim shaft, the front roller, and the back shaft, is different from that given in Fig. 16, inasmuch as we find an extra pair of bevels M N driving the carrier wheel O. To explain the function of this arrangement, and also to describe other features of the gearing, an illustration

Fig. 119.

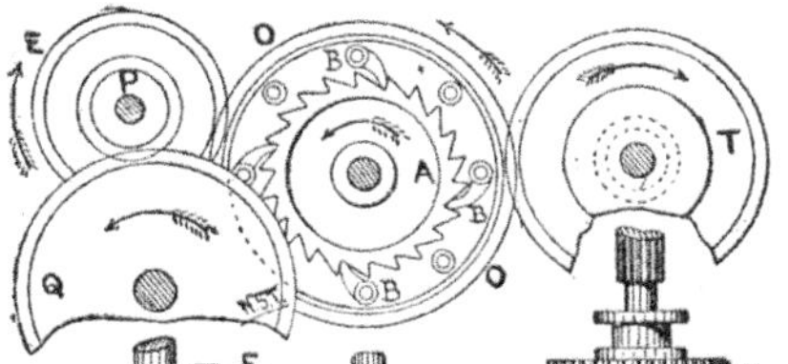

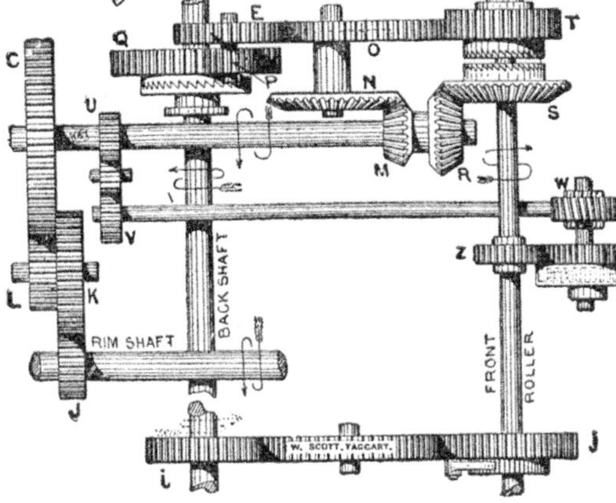

Fig. 120.

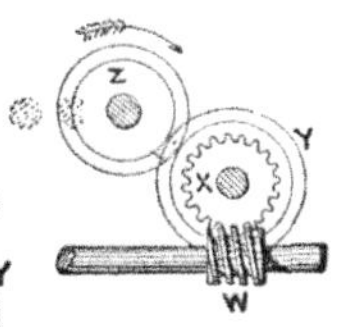

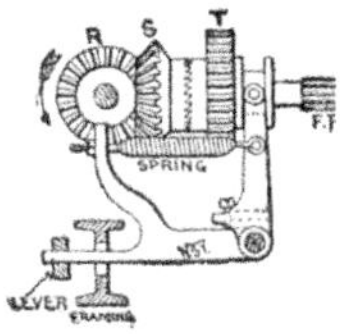

Fig. 118. Fig. 121

is given in Fig. 118. As the rim shaft drives the front roller through J, K, L, C, R and S, we have the yarn delivered consistently with the requirements of twist, gain, and ratch; and the back shaft, driven through T, O, E, P and Q, drives the carriage out, in harmony with these factors. At the same time, by introducing the bevel wheels M N, a connection is made between the rim shaft and the back shaft which is quite independent of the front roller.

This independence is obtained by attaching to the boss on N a ratchet wheel A (see Fig. 119), which revolves within the carrier wheel O. When the carriage is going out, and is being driven from the front roller, all the wheels are revolving in the direction shown in Fig. 119; and since O is receiving a greater speed from T than the ratchet wheel A is receiving from the bevels M N, a number of catches or pawls B carried by O simply slip over the teeth of A, and so the two wheels A and O revolve independently of each other; and the bevels M N, so far as this part of their work is concerned, are useless. When, however, the front roller is stopped, by separating the clutch catch box between T and S (Fig. 121 is almost self-explanatory of how this is performed) the wheel O will receive no motion from T; but since M N continue to be driven from the rim shaft, the teeth of A will engage with the clutches carried by O, and cause it to revolve and so drive the back shaft. In this way we continue the movement of the carriage when the rollers are stopped, and thus obtain what has been previously described as the "after-stretch" or "ratch." The wheels M, N and O are frequently spoken of as being the "jacking motion." Before leaving this feature it is as well to point out that this motion is not a necessary adjunct to the gearing through which we can drive the carriage at a quicker or slower rate than the surface speed of the front roller, and thereby obtain a drag or a gain.

Roller-turning Motion whilst Twisting at the Head.—Previous allusions have been made to what is termed "twisting at the head." By this we understand that, after the mule has completed its outward run, the front rollers are stopped, but the spindles continue to revolve and so put an extra number of twists into the yarn. These extra twists naturally put tension in the yarn because their tendency is to shorten it; the strain so occasioned would prove damaging by causing a good many breakages; to relieve the yarn, the rollers are therefore caused to deliver a very small amount of cotton at a much reduced rate as compared with that at which they revolve when the carriage is moving. The effect is obtained in the following manner:—When the carriage stops, the catch box, Fig. 118, between T and S is naturally thrown out of gear, so that although S is driven, it simply rides loose on the shaft. On the side shaft, which carries M and R, is a pinion U, which drives through V another side shaft, on the other end of which is a worm W, from which the front roller can be driven through the worm wheel X and the pinions Y and Z. On the back of Y is a catch box, which is inoperative when the front roller runs at its ordinary speed in the same way as A is in Fig. 119. But when the front roller is stopped the catches in Y permit the wheel to drive Z, and so we obtain from U a very slow movement of the front roller to compensate for the small amount of yarn taken up through the twisting action when the carriage is out and extra twist is being put in. "Jacking-delivery motion" is the name sometimes given to the arrangement, but it is much better to call it a "roller-turning motion whilst twisting at the head."

Roller-delivery Motion whilst Winding.—Another motion very often used, but upon the merits of which there is an amount of reasonable scepticism, is the one called the "roller-delivery motion whilst winding." As its name implies, its object is to turn the rollers while the carriage is coming in and winding is taking place. The reason for this action is generally sought for in the fact that an increased production is thereby obtained. This can readily be confirmed, for if the stretch is 64 inches and three more inches are delivered when the carriage comes in, the total length delivered each draw amounts to 67 inches. A better reason, however, than this of increased production can be deduced, namely, a strain-relieving effect on the yarn. We know that the yarn is made to assume a line something like the letter Z when the winding is taking place; this naturally puts some considerable strain on the yarn, and, indeed, everything is done to balance this strain as much as possible. Now it will clearly be recognised that this bending of the yarn can be safely done in a long length; but as the length gets shorter the strain will become greater, and to relieve it the rollers are made to deliver a little extra, and, of course, it comes in additionally as an advantage in the production.

In this connection there remains an important point which is the cause of a difference of opinion among spinners. Twisting has been completed, and winding commences; untwisted roving is now delivered, and a question arises as to whether the extra three inches delivered is as well twisted as the remaining 64 inches. There can be no doubt that the twists already in the yarn will run up to a considerable extent into the extra yarn, but it by no means follows that the three inches will receive an amount equal to any other three inches in the stretch. The probability is that it does not, except in well-twisted yarns and fine numbers — in both cases because of the combination of natural and artificial elasticity of the fibres. This doubt leaves room for the difference of opinion mentioned.

The gearing for giving the extra delivery is shown in Fig. 118, wherein *i* is a wheel on the back shaft, and from it the front roller is driven through *j*. A ratchet wheel by *j* is keyed on the front roller, and when the carriage is going out the wheel *j* runs in the opposite direction to the front roller, and so the ratchet wheel is not affected by the pawl catches. When the front roller is stopped, and the carriage runs in, the back shaft drives *j* and the catch or catches which *j* carries dip into the teeth of the ratchet wheel and turn it, and, consequently, the front roller. Another cause for the dissatisfaction as expressed by some for this motion will be understood from the fact that the extra material is delivered in a uniform manner from the beginning to the end of the run-in. This is not in accordance with reason: there ought to be (on condition that such a motion is practically necessary) an increasing delivery as the carriage approaches the beam; or, in other words, the front roller should deliver a little more in a given time towards the end of the run-in than what it delivers in the same time at the commencement. It is motions of this kind that now and again make the governor and nosing motions more difficult to work than they would otherwise be.

Backing-off Motion.—A further illustration of a "backing-off" motion is given in Fig. 122. It represents a well-known arrangement, and one that has been extensively applied to mules, especially to those of the "long-lever" system. Its action is as follows:—As the carriage moves out, and is on the point of completing the stretch, the end of the slide bar or gun lever F (this feature has already been fully described and illustrated, see Fig. 113) comes into contact with an adjusting screw A, carried by a hanging lever centred at B. To the lever at C is attached a link E, which carries one end of a long rod D; the other end of D abuts against a stop G on the lower portion of the backing-off lever whose fulcrum is at H. When the carriage moves the hanging lever forward, the rod D is moved out of contact with the stop G, and at the same time a spring M, connecting the rod and the backing-off lever, is put into tension, and consequently pulls the lever H forward; this action has the effect of moving the end J backwards, and so putting the backing-off cone wheel into gear with the cone clutch on the fast pulley. "Backing-off" now takes place, and when it is completed the faller leg locks; as this occurs the slide rod F shoots back and releases the hanging lever B. A spring S, which has been compressed by the previous forward movement of the rod, is also now relieved from constraint, and at once forces the rod D backwards, and, abutting as it does against the stop G, it moves back the backing-off lever H, and takes the backing-off clutch out of gear.

Fig. 123 shows a modification of the above arrangement. Instead of a hanging lever, a bell-crank lever is used fulcrumed at B; a bowl A is carried by one arm, while the other arm is connected to the rod D through the link E. The slide bar F is extended, and is formed with an incline, so that, as the carriage moves forward, it comes into contact with the bowl A, and depresses it, thus moving forward the rod D. When "backing-off" is finished, the shooting back of the slide bar F releases the bowl A, and, as before, the backing-off clutch is taken out of gear.

Roller Stand and Weighting.—The roller stand of the mule, Fig. 124, is very similar in most respects to the stand used on the fly frames. It consists of a principal bearing Q, bolted to the roller beam and carrying the front roller; a projecting arm R supports a slide S, which acts as the bearing for the middle and back rollers. These two rollers being generally set a fixed distance from one another, the slide S is made in one piece; but of course it

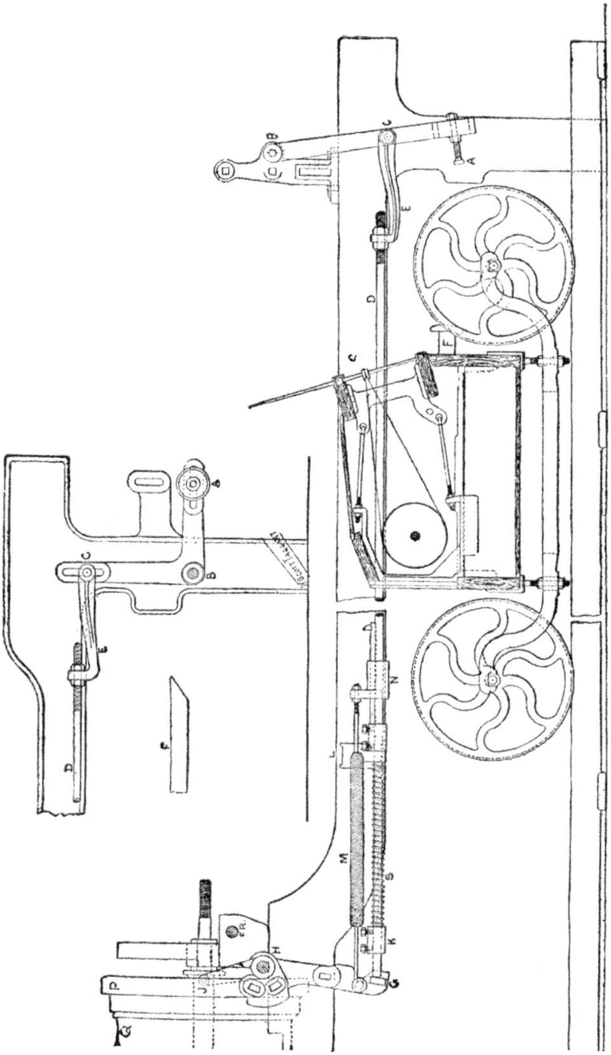

FIG. 122.

FIG. 123.

is necessary in many cases to make S in two parts, each carrying one of the rollers B and C, in a way similar to that shown in the fly-frame roller stand. The cap bar, for keeping the top rollers in position, is pivoted at J so that it can readily be moved over out of the way when the

Fig. 126.

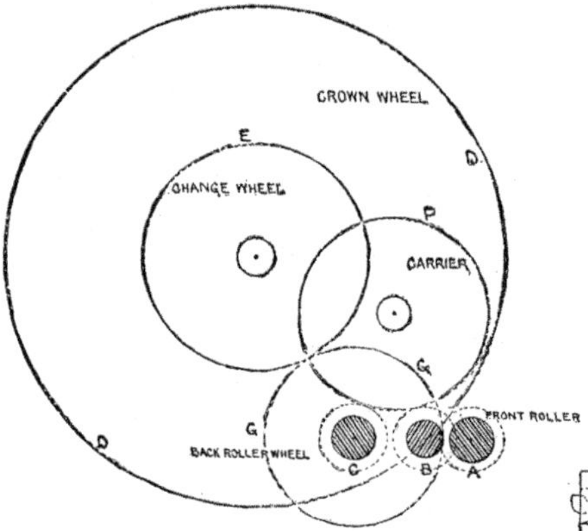

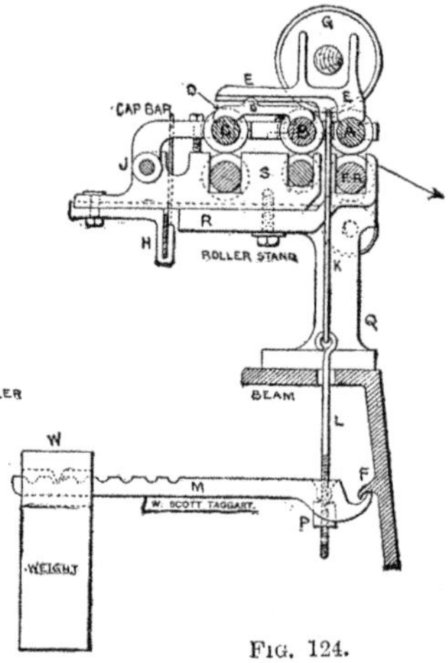

Fig. 124.

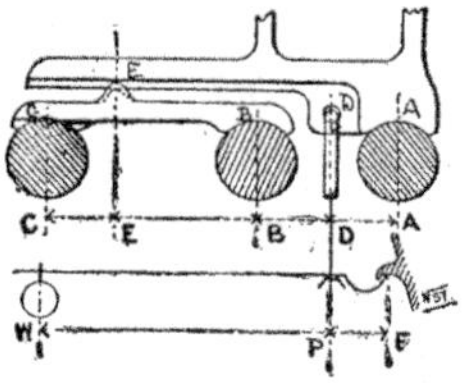

Fig. 125

rollers require attention. The traverse rod carrying the thread guides is shown at H, and is generally connected at the outer end of the roller beam to some cam arrangement that gives it a to-and-fro movement, and whose object is to cause an equal wear of the leather of the top rollers. The necessity for this traverse exists wherever leather-covered rollers are used, and a large number of special motions have been introduced during the past few years for obtaining the maximum amount of use of the leather covering. The best motions are undoubtedly those depending upon a uniform cam motion, arranged with a slightly accelerated movement at the change in the traverse. Motions that depend upon eccentrics or cranks, in whatsoever form, for the traverse, are as a rule wrong in principle, and are generally complicated and unnecessarily expensive.

The weighting of the rollers is an important matter. Two methods—dead weights and lever weighting—or their combination, may be adopted for obtaining the necessary pressure on the rollers. In Fig. 124 is shown a method frequently used in the mule. On the middle and back rollers B and C rests a lever D; a raised point on the upper part of D supports one end of a lever E whose other end rests upon the front roller A. To E is attached a wire link K, which in turn is connected to another wire link L, and this, passing through a hole in the roller beam, is supported by means of a nut P by a lever M whose fulcrum is at F; the lever M carries at its other end a weight W, the position of which can be varied for the purpose of obtaining a range of different pressures on the rollers. Fig. 125 will enable the effect of W to be thoroughly understood, and an example will be given showing how to calculate the pressure on each:—

The weight of	W = 4 lb.	The distance of	CE = $\frac{1}{2}$ in.
The distance of	WF = $7\frac{1}{2}$ in.	The ,,	CB = $1\frac{1}{2}$,,
The ,,	PF = $\frac{3}{4}$,,	The ,,	EB = 1 ,,
The ,,	AD = $\frac{5}{8}$,,	The ,,	ED = $1\frac{3}{8}$,,
The ,,	AE = 2 ,,		

The pull of the weight W at D will equal

$$\frac{\text{Weight} \times \text{WF}}{\text{PF}} = \frac{4 \times 7\frac{1}{2}}{\frac{3}{4}} = 40 \text{ lb.}$$

This 40 lb. will be distributed, part of it on A and the remainder on the point E.

The pressure on A will equal

$$\frac{\text{ED} \times 40}{\text{AE}} = \frac{1\frac{3}{8} \times 40}{2} = 27\tfrac{1}{2} \text{ lb.}$$

The pressure at $\text{E} = 40 - 27\frac{1}{2} = 12\frac{1}{2}$ lb., or the pressure at E will equal

$$\frac{\text{AD} \times 40}{\text{AE}} = \frac{\frac{5}{8} \times 40 \text{ lb.}}{2} = 12\tfrac{1}{2} \text{ lb.}$$

The pressure at B will equal

$$\frac{\text{CE} \times 12\frac{1}{2} \text{ lb.}}{\text{CB}} = \frac{\frac{1}{2} \times 12\frac{1}{2} \text{ lb.}}{1\frac{1}{2}} = 4{\cdot}166 \text{ lb.}$$

The pressure at C will equal $12\frac{1}{2} - 4{\cdot}16 = 8{\cdot}33$ lb., or the pressure at C will equal

$$\frac{\text{BE} \times 12\frac{1}{2} \text{ lb.}}{\text{CB}} = \frac{1 \times 12\frac{1}{2} \text{ lb.}}{1\frac{1}{2}} = 8{\cdot}33 \text{ lb.}$$

Direct weighting of the rollers is performed by placing a hook upon the roller and hanging a weight upon a link attached thereto.

The driving of the rollers is illustrated in Fig. 126. The front roller through A drives a large crown wheel D; on the axis of D is a wheel E, which drives the back roller. The back roller through C and the carrier E drives the middle roller wheel B. The necessary change (for draft) in the speed of C is obtained by changing the wheel E.

Figs. 127, 128, 129, 130, 131 and 132 represent the complete sets of rollers for working Japanese, Chinese, Indian, American, and Egyptian cottons. The particulars attached to them indicate the usual practice in the diameters and setting.

Another Example of Long-Lever Mule.—The machine now illustrated, where the changes are produced

Note.—This type of mule is fully illustrated in the Appendix.

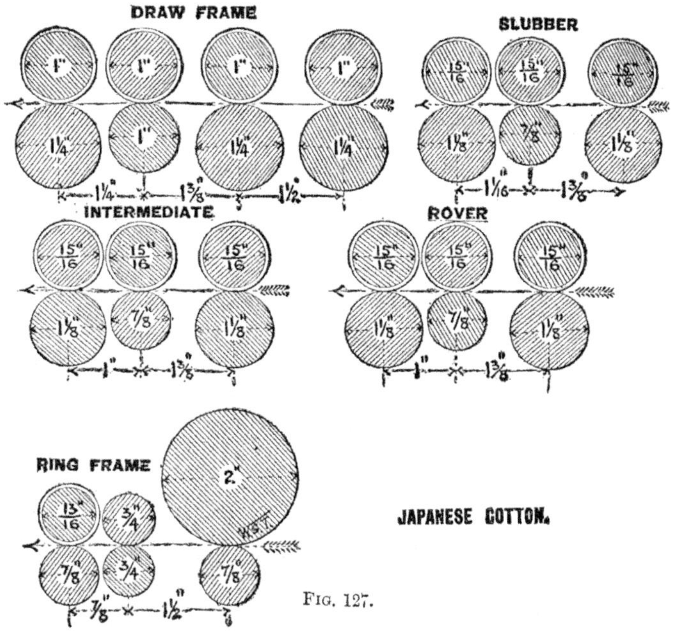

JAPANESE COTTON.

FIG. 127.

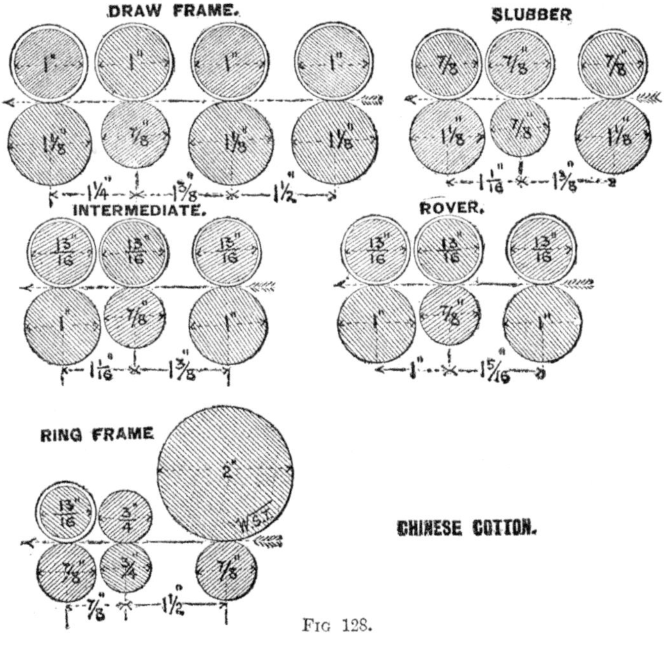

CHINESE COTTON.

FIG 128.

through the medium of a long lever, will be familiar to most of our readers, and its position in the production of the finer qualities of counts entitles it to some consideration in these notes. We therefore give a few details of its principal actions and the mechanism employed in producing them.

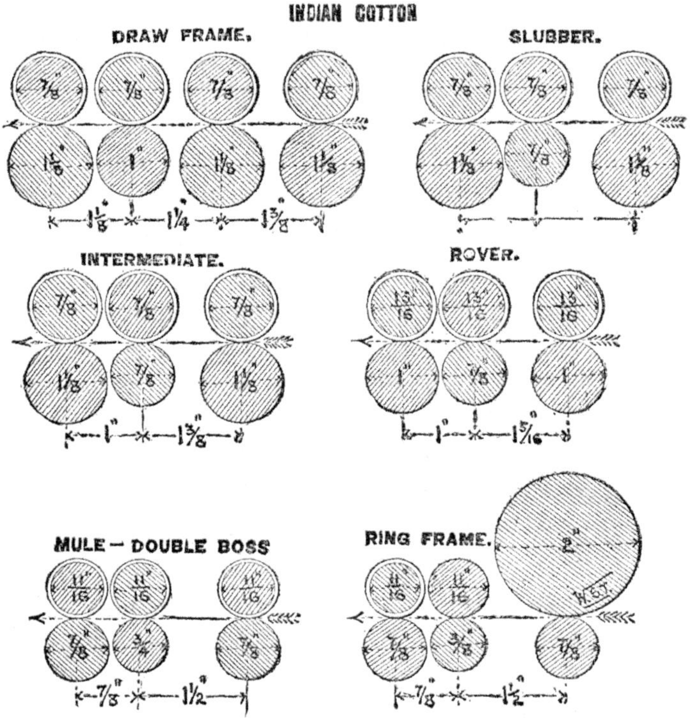

FIG. 129.

Figs. 133 and 134 illustrate the chief points of interest. In the former diagram the backing-off cone wheel and clutch are shown at Z. The bar or slide X is coupled up to the grooved boss of the backing-off cone wheel through a lever Y, so that any movement made by X will put the wheel in or out of gear with the cone clutch, which is fast on the rim shaft. The method of doing this is as follows:—A projection on the bar X carries a stud W, which locks itself into a notch cut in the under side of the connecting rod J; another projection on X carries one end of a spring O, whose other end is fixed to a portion of the

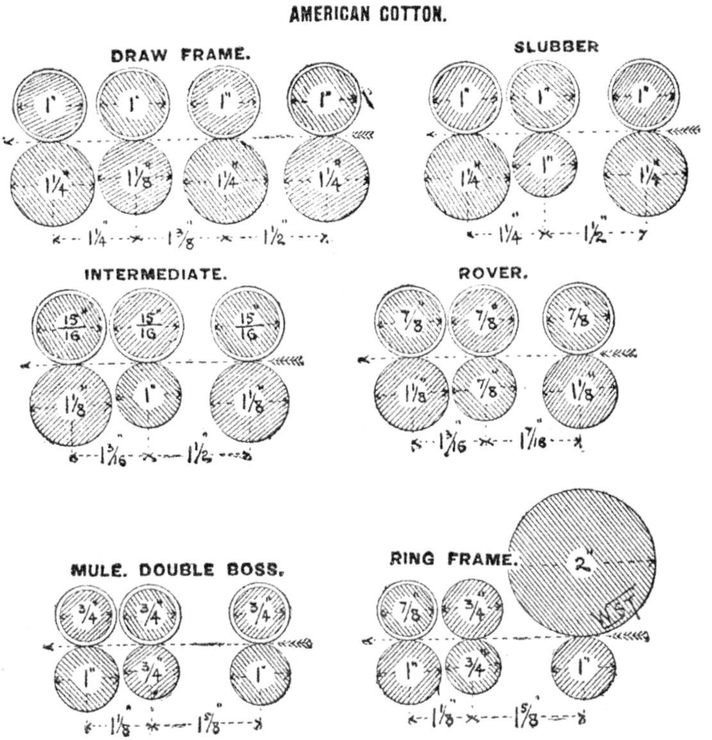

FIG. 130.

framing of the machine. A cam G, driven in the direction of the arrow from the rim shaft through the wheels A, B, C, D, E and F, comes into contact with an inclined swing or finger H, which hangs pendant from the stud on which the compound carrier B and C revolves. The revolution of G has the effect of pushing H forward, and in so doing the

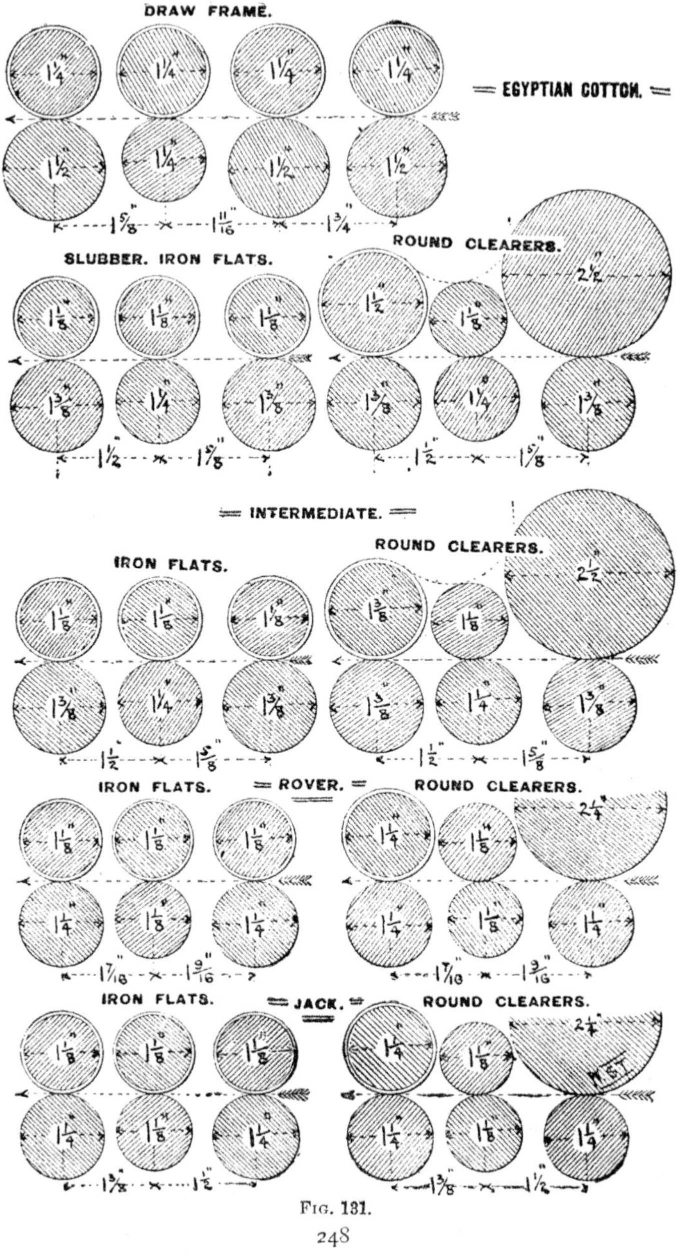
DRAW FRAME.
EGYPTIAN COTTON.
SLUBBER. IRON FLATS.
ROUND CLEARERS.
INTERMEDIATE.
IRON FLATS.
ROUND CLEARERS.
IRON FLATS.
ROVER.
ROUND CLEARERS.
IRON FLATS.
JACK.
ROUND CLEARERS.

Fig. 131.

connecting rod S, which is fastened to the swing H, is also moved forward, and consequently pulls the backing-off slide X in the same direction, thereby putting the backing-off cone wheel Z into gear with the cone clutch. Backing-off at once takes place, and of course this is arranged, through the gearing from the worm A on the rim shaft, to happen just as the carriage has arrived at the termination of the outward run, as shown in the diagram, Fig. 133. Immediately the backing-off is completed, the bar X is released in

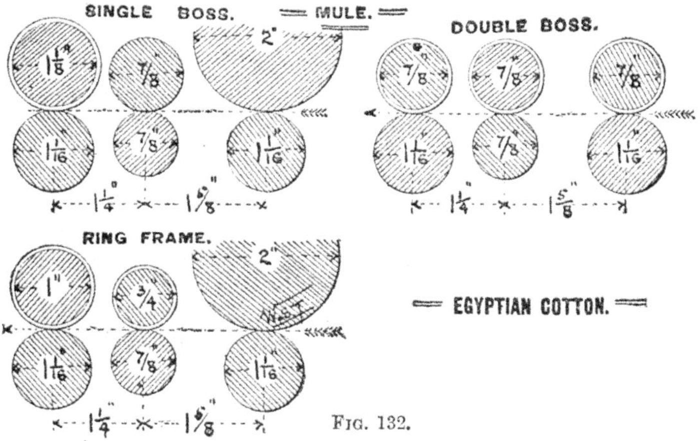

FIG. 132.

the following manner:—A long lever centred at M is connected to the rod J by a link K; its other end N carries an arm Q, whose lower end passes through the holding-out catch V, which is fulcrumed at U. The position that can be taken up by the holding-out catch is carefully adjusted through the nuts at Q, so that, as the carriage comes out, the snug at T, carried by the square, passes over the end of V and becomes locked by the catch. Backing-off is finished by the faller leg being locked; and, as this happens, the stud at S is oscillated and moves down the inclined finger at R, which presses against a projecting arm P of the down rod Q. This action at once forces the end N of the long lever in a downward direction, and correspondingly raises the end L, which at once releases the rod J from the stud W; K being set free is now pulled backwards by the spring O, and the cone wheel is taken out of gear with the

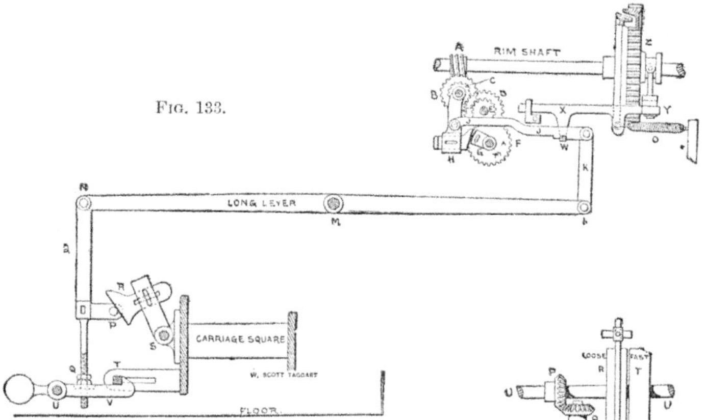

FIG. 133.

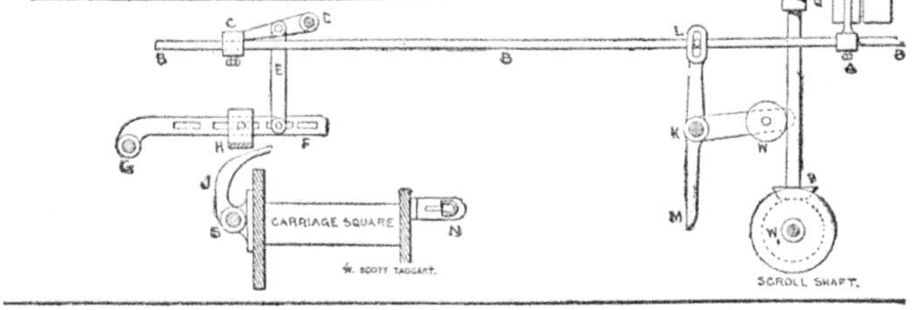

FIG. 134.

clutch, thus permitting the mule to run in. The same movement that lowers the long lever at N also presses down the holding-out catch V, and thereby unlocks the carriage.

Drawing-up.—The arrangement for drawing-up is illustrated in Fig. 134. The run-in of the carriage is effected through the pulleys R and T, fast and loose respectively, on the shaft U. During the run-out the strap is on the loose pulley, as shown in the diagram. The same movement of locking the faller leg acting through the stud S, as in the first sketch, also moves the arm J upward, and J comes into contact with a bracket H carried by a lever F centred at G. The upward pressure of J lifts the lever F, and through the link E releases a catch finger D, and takes it out of contact with a stop-washer C on the rod O; the weight W acting through L immediately pulls the rod O backwards, and transfers the strap to the fast pulley T. This action, it will be seen, takes place precisely as the backing-off is finished, so that no sooner is the rim shaft stopped than the strap on T commences to drive the scroll shaft W, through the bevels P and Q, and so draws up the carriage. Just as the carriage is arriving against the stops, a bowl N on the square comes against the lower end M of a lever fulcrumed at K, and presses it backwards. As a consequence, the upper end of the lever at L moves forward the rod B, and changes the strap again to the loose pulley, thereby stopping the mule. At the same time the strap is moved from the loose pulley on the rim shaft to the fast pulley, and spinning immediately commences. The rod B is locked in position during the outward run, by the finger at D.

Double-Speed Driving.—A recent improvement added to the mule is shown in Fig. 135, by which means a novel and satisfactory method of obtaining a double-speed effect is obtained. Briefly, it consists in the rim shaft being made in two parts, C and C_1. One rim shaft C carries a large rim pulley B, whilst the other rim shaft C_1 has a smaller rim pulley A keyed to it. The driving takes place through two fast pulleys D and F, the loose pulley being placed between them. When the strap is on the fast pulley F, the usual or slower speed of spindle is obtained through the rim pulley

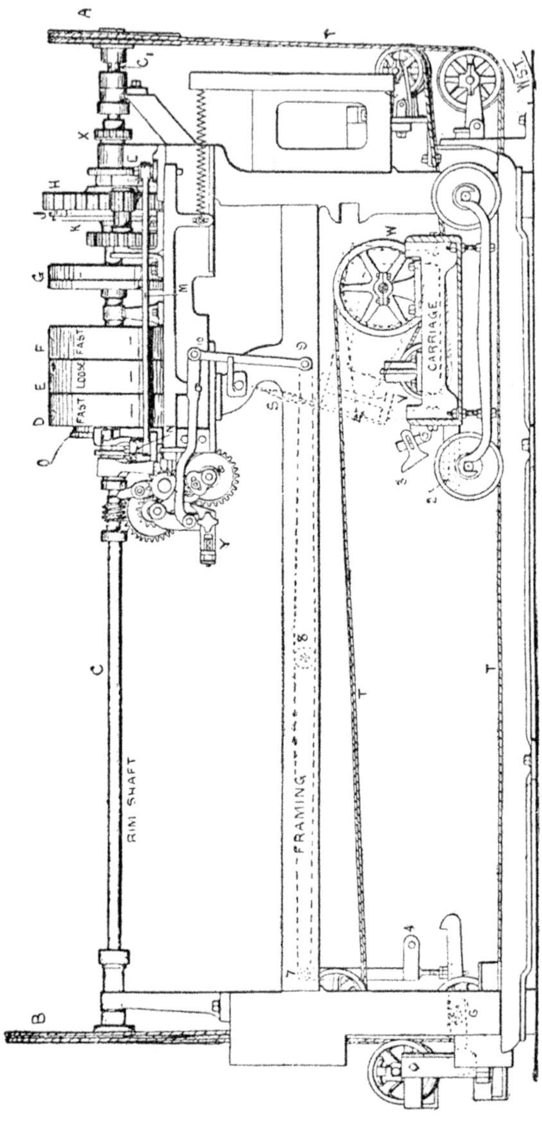

FIG. 135.

A; but a change is effected, as the carriage gets out, by

moving the strap on to the fast pulley D, whereupon the rim pulley B begins to drive the spindles at a greater speed than that obtained from A. The latter of course continues to revolve, but merely through its connection by band with B, and its movement does not affect the spindles in the least.

An additional and highly important improvement is effected in the arrangement by using two brake cone clutches at the points J and O for backing-off. By their means a double amount of friction is obtained for stopping the rim shaft ready for backing-off. Naturally this operation is performed very rapidly and effectively, and some time is saved in stopping and then reversing the spindles.

The illustration will serve the purpose of the relative positions of the details given in Figs. 133 and 134, the character in these two sketches being simply diagrammatic.

Snarls and Anti-snarling Motions.—We now touch on a subject which is always more or less a very troublesome feature in mules, and one that has been the occasion of innumerable devices being put on the market as remedies for "snarls." During the complete cycle of operations on the mule the yarn is supposed to be always slightly in tension; slack yarn must be avoided, and this is one great reason why governor, nosing, backing-off chain-tightening, etc., motions are employed, all having one object—that of keeping the yarn at a regular tension. If the yarn is permitted to become slack, it instantly doubles itself and forms into small curls or twisted loops, technically called "snarls"; and motions to prevent snarls forming are generally termed "anti-snarling motions." Two illustrations will be given of such motions; but first a few words as to why they are specially necessary will not be out of place.

Directly the carriage is on the point of coming against the stops, a change occurs, which moves the copping faller wire from the nose of the cop to a position just over the spindle point; at the same time the under-faller wire is lowered to a position just under the spindle point; and in these positions of the faller wires the yarn passes between them. The change of the faller wires to their new positions takes place very quickly, and, as it happens, a certain length of yarn is set free. The spindles are revolving during the change, so that the slack yarn resulting from the action just described is immediately wound on to the bare part of the spindle above the nose of the cop.

The action of winding the yarn on the bare part of the spindle blade is a very delicate one; moreover, it is ever varying; for as the cop gets longer the amount of yarn to be wound on becomes less, and very exact adjustments have to be made to enable the result to be attained at all satisfactorily. In spite of all that can be done in this direction there remains an amount of slack yarn, which runs into snarls and thus becomes deteriorated. Extra motions are therefore applied, and to bring about the desired result two methods are generally employed: either the carriage starts out slightly in advance of the rollers turning, or the carriage and rollers start simultaneously; but a little extra speed is given to the carriage for a few inches of the initial part of its outward run.

The first method is illustrated in Figs. 136, 137 and 138. The front roller A is driven in the usual manner through the bevels D and C; C rides loose on the shaft, and so does the other half of the catch box B. When the cam puts the catch box B and C into gear at the commencement of the outward run, no movement of the front roller will take place until the snugs L, cast on the catch box B, come against a disc K, which is fastened on the front

FIG. 136.

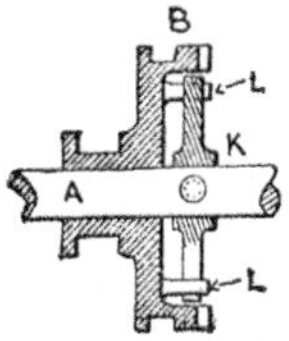

FIG. 137.

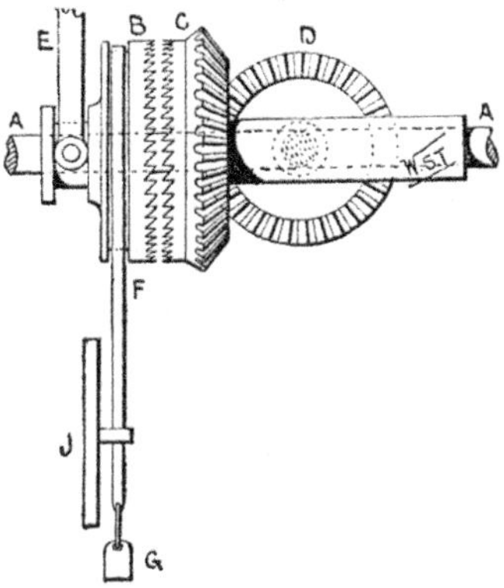

FIG. 138.

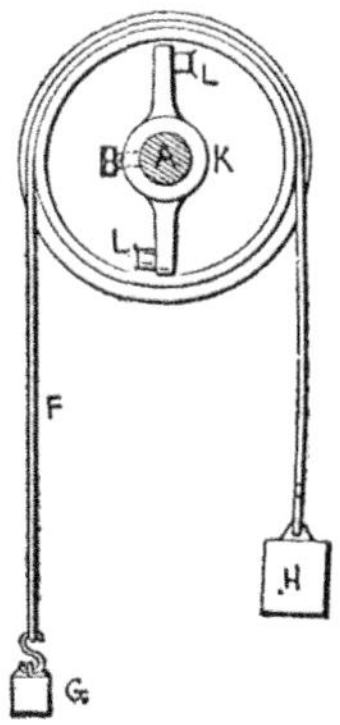

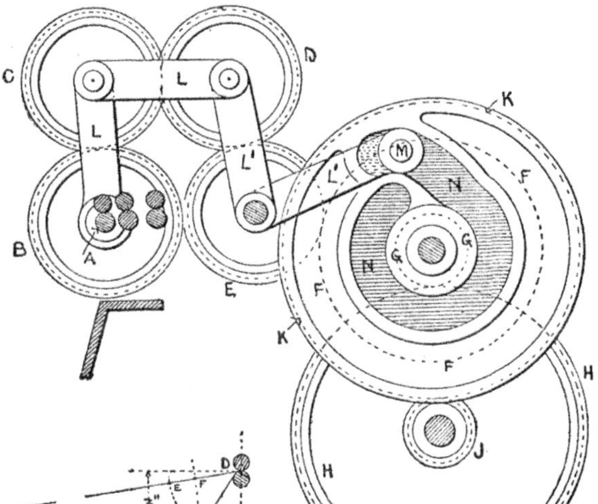

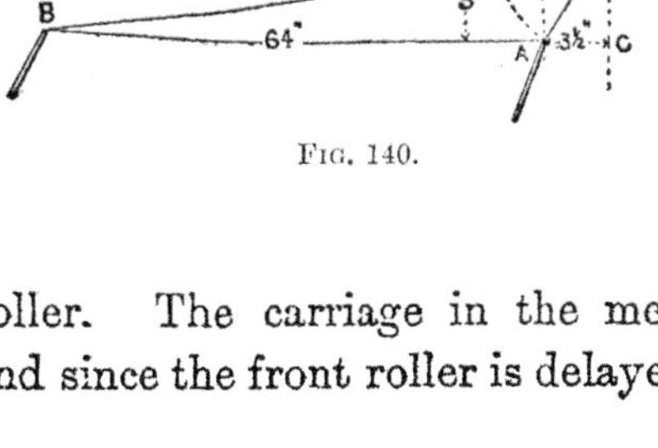

FIG. 140.

FIG. 139.

roller. The carriage in the meantime is travelling out, and since the front roller is delayed in its starting, the slack yarn is made tight, and any snarls that may be in are quickly taken out.

By referring to Fig. 138 it will easily be understood how the movement of the front roller, relative to the movement of the carriage, is regulated. When the cam takes the catch box B C out of gear preparatory to the running-in, B is free on the roller A; a leather band F passes over a groove on B, and each end of the band carries a weight; H is the heavier weight, and consequently it pulls over the part B and takes the snugs L with it out of contact with the disc K. The distance by which the snugs L can be moved away from the fingers of the disc is regulated by controlling the distance that H can fall, which is done by adjusting the vertical movement of the small weight at G; a stop on J easily effects this, so that by means of a wing nut the motion is entirely under the control of the minder. This is almost a necessary provision to make, for, from what has previously been said, it is clear that snarls will be larger and more frequent at the commencement of the building of the cop than at the finish. This arrangement permits the minder to regulate the motion to suit the varying conditions.

The second method is shown in Fig. 139. In this case the carriage is given a slightly additional speed over that of the front roller, and it is done in such an ingenious manner that we shall devote a few words to it.

Instead of driving the back shaft J from the front roller through the usual wheels B, E, F, G and H, the two wheels B and E are put out of gear and the back shaft is driven through the wheels B, C, D, E, F, G and H. The wheels C and D are on movable centres, and connected by links L, the last one L^1 being bell-cranked, with its centre on the stud-carrying wheel E and one end carrying a bowl M, which fits in a cam-shaped groove on the back of a wheel K driven from a wheel J on the back shaft. The action of the motion is as follows:—When the front roller commences to revolve, the back is driven by the gearing already mentioned; the wheel J then drives the wheel K, and turns the cam disc. This movement of the cam lowers the bowl M, and naturally pulls over the two upper wheels C and D. The direct effect of wheels moving over each other is to increase or decrease speed according to the direction in which they move. By observing the direction of the wheels C and D, it will be observed, first, that they will not effect any change in the speed of B, because B is driven from the rim shaft direct. A slightly increased movement is therefore given to the wheel E, which is transferred to the back shaft, and so we have the speed of the carriage accelerated. When the bowl M falls on to the circular portion of the groove no further movement of D and C takes place, and the carriage continues the remainder of its outward run at the usual speed. The amount of the excess speed given to the carriage is easily regulated by adjusting the cam so that a longer or shorter portion of the cam surface can be used. On the inward run of the carriage the revolution of the back shaft simply revolves the wheels, and the cam takes the wheels C and D back to their original position, ready for the next run-out. One advantage of this motion is that there is no loss in production, because the roller is not stopped, as in the first motion.

A feature of some interest to many people is illustrated in Fig. 140. The question occurs—Do the spindles wind on yarn equal to the length of the stretch? The sketch will settle the matter as far as actual measurements are concerned. When the spindle is close to the beam, the distances of the point A, horizontally and vertically, are shown; from these dimensions we can readily prove that the length of the yarn A D is 4·61 inches. As the carriage moves out, the angle of the yarn varies; and on the assumption that the spindle point travels 64 inches we shall get a length of yarn, between the spindle point at B and the front roller D, equal to 67·52 inches for productive purposes: therefore (and without taking into account the stretching of the yarn that may occur during the run-in) we clearly see that there is 62·91 inches to be wound on the spindle at each draw—which means that a stretch of 64 inches gives us a length of yarn $1\frac{1}{10}$ inch less, equal to a loss of about 1·7 per cent. The investigation opens up several interesting questions, but for the present purpose it is not necessary to go any deeper into the subject.

A variety of conditions arise to cause snarls, but these are usually remedied by attention to the following points: (1) Too great a movement of the nut up the quadrant for any given layer; when this occurs enough winding does not take place and slack yarn is the result. (2) Bad rovings, whether through poor piecings or irregular sliver. (3) Slack scroll bands. (4) Faulty nosing motions. (5) Insufficient weighting of the fallers. (6) Slipping of the winding catch. (7) Irregularities in the "changes." And (8) miscalculation in the amount of the drag.

Tubes and Starch for Cop Bottoms.—In commencing to build a cop bottom, we may either do it entirely on the bare spindle; or build it upon a short or long paper tube; or brush over the first two or three layers with starch. All these methods are adopted according to the class of work being done by the machine. The first one, however, is not often met with, so we will confine our attention to a few words on the use of tubes and starch. The object in using either of these methods is primarily to obtain a good foundation for the cop bottom so that in future use the cops can be passed on to a skewer without stabbing and spoiling the cop. The avoidance of waste in other directions is an important factor, for it is desirable to use if possible every inch of yarn wound on the spindle. From this point of view the use of a short tube pressed on the spindle where the cop bottom is formed ensures that a good opening is always left for a skewer to pass through, and another advantage is apparent, for in such a case all the yarn can be unwound without leaving waste. In some districts spinning finer counts, tubes are almost exclusively used, but they have their disadvantages, among which might be mentioned the following: extra labour is involved in putting them on the spindles, and this means a slightly increased cost for such labour; the tubes are pressed to their places, and sometimes in doffing they stick so fast that the cops are pulled out and of course waste is made; damaged tubes are a source of breakages whilst winding and unwinding, on account of rough edges; when ends have been allowed to remain down for a few draws it is not so convenient to push the cop up a little, so that the cop is nicked and spoiled yarn made; the few turns of yarn round the spindle, previous to putting on the tubes, accumulates so much that it becomes a little troublesome to occasionally clear the spindles. There are several appliances that dispense with a good deal of the labour involved in putting on the tubes; these are filled with the tubes during the working of the mule, so that when doffing is complete they are ready to be at once turned over on to the spindles without having to put on each tube separately.

Starching is performed by applying with a brush a little starch to each spindle before starting to build the cop bottom; when dry, it effectively prevents the hole closing under ordinary working conditions. If done properly and good starch is used very little waste is caused and very little labour is entailed, as a rapid movement of the brush (which is attached to a special box holding the starch which runs on to the brush) along the spindles enables the whole of the spindles to be starched in a minute or so. If a minder, in his desire to have a better cop bottom, starches twice and also puts on a layer or two before doing it, he of course sacrifices a little time and in addition causes more waste to be made at the loom, but this gives a much better cop and the fact induces the practice to continue. Bad starch, carelessness in starching, and the starch running down the spindle to the bolster bearing, are its great disadvantages, and a very frequent complaint results from the soft cop bottoms made. Longer tubes are generally used when a hard cop is desired and the yarn is spun rather soft.

Horse-Power required to drive the Self-Acting Mule.—It is now proposed to present, as briefly as possible in description and diagram, a digest of present knowledge as to the power required to drive the mule. Nothing will be said about the methods adopted to obtain the indications, beyond remarking that dynamometers of various kinds have been used, and careful observations taken of their results.

If one were to ask the question—What horse-power will a mule take to drive it? he would probably be answered, in a general way, that 110 to 120 spindles per I.H.P. for low cottons, and 130 spindles for finer cottons, are good averages. An answer of this kind is quite sufficient for ordinary purposes; and, as a rule, a result in such general terms can readily be obtained through the indications of the steam-engine. Like all general statements of facts, however, there is a tendency to overlook the circumstances and details which give to the statement its importance, and in consequence false ideas interfere with the true knowledge of the conditions that go to make up the average. Owing to the complicated and varying actions of the mule, it is by no means an easy matter to obtain exact results. When a dynamometer is used without an automatic recording apparatus, a large number of careful observations must be made so as to include as many complete draws as practicable; from the numbers thus obtained, as well as the intervals of time of their indications, a good average from each set of readings will be procured, from which it is possible to arrive at a comparatively accurate result. This result can be represented in a diagrammatic form similar to the indicator diagram of a steam-engine, and therefrom, in a similar manner, much of the inner working of the mule can be rendered intelligible.

In watching the actions of the mule for the purpose of indicating its power, three distinct actions stand out from the others, namely: the outward run of the carriage, during which the spindles run at their greatest speed when spinning; the pause or rest at the finish of the stretch, during which backing-off takes place; and the run-in of the carriage, during which the spun yarn is wound on the spindles. It will be apparent to any one who has watched the mule working that all these actions require different degrees of power to perform them. To a close observer another marked feature connected with the power absorbed will not be overlooked. When the carriage is at the roller beam the machine is practically stopped, so that the commencement of the run-out requires a very high power to overcome the inertia of such a large, heavy, and stationary mass as the carriage, especially as it is driven from the front roller, and to drive the spindles at the high rate of speed at which they run when spinning. This feature is properly included in the power absorbed during the outward run; but it is such a distinctive element in the power-diagram of the mule that it might almost be considered as quite separate from the power that really effects the drawing-out and twisting processes.

Apart from the extreme care required in making the observations of the readings of the dynamometer, equally careful attention should be paid to the speed of the counter shaft. It is upon this speed that the accuracy of the results depend, and therefore means must be adopted to denote the slightest variation of speed that takes place during the time the indications are made. One of the best ways of doing this is by means of the tachometer; when this speed indicator is applied to the counter shaft, variations are instantly shown. A noticeable feature in this connection will be observed when the carriage commences its outward run. The counter shaft is running at its full speed, with the driving belt on the loose pulley of the machine; immediately the belt is moved on to the fast pulley the whole carriage has to move and the spindles are driven at their full speed. It is almost impossible for this to occur instantaneously—the shock would be too great; so, in consequence of all the actions concerned being driven through belt or bands, we find, when this full power is thrown on them, a large percentage of slipping occurs, which allows the carriage and spindles to attain their maximum speed gradually. By the use of the tachometer it is easily seen that spindles do not attain their full speed until from 12 to 36 inches away from the roller beam, and in a few cases machines may be found in which the spindles only attain their maximum speed just as the outward run is finishing. This of course means that the power is considerably reduced from what it would be if the speed remained normal throughout. During the course of a large number of power tests on the mule, the writer has found invariably a large percentage of reduction of speed in the counter shaft at the moment when the carriage begins to move outward. In some mules this reduction is much greater than in others, ranging from 10 per cent up to as high as 35 per cent. The reader will therefore see the importance of observing very closely the variations in this important factor of the indications.

In the accompanying drawing (Fig. 141) three power diagrams of the mule are given. The one marked A is a machine with a normal spindle speed of 9100 revolutions per minute. The speed of counter shaft started at 410 and an interval of $2\frac{1}{2}$ seconds elapsed before it attained its normal speed of 460 revolutions per minute. The percentage of slippage is almost 11, which, in the opinion of the writer, may be considered a very low one. The fact that it is so low accounts for the very high power indicated as the carriage started out; and, although only a short mule of 636 spindles, the initial power required to move the machine was over 24·5 horse-power. All the belts were newly spliced and the bands renewed for the test, so that slipping was reduced to a minimum. Directly motion was imparted to the carriage and spindles the power rapidly fell, and the speed of the counter shaft rose until, after an interval of $2\frac{1}{2}$ seconds, normal conditions were attained. From this point onward, the power required to drive the spindles and carriage remained stationary until the outward run was completed and backing-off commenced.

The stoppage of the carriage and spindles naturally results in an almost instantaneous fall in the power absorbed, but there is sufficient movement going on in the mule to prevent its falling to zero. One of these movements is the action of backing-off, and as it occupies in Diagram A about $2\frac{1}{2}$ seconds, the power required in this interval is shown to be $1\frac{1}{2}$ H.P. This, it may be observed, is a comparatively high power to be absorbed during backing-off; it ought never to be above 1 I.H.P., even with the largest machine. It was noticed that the backing-off cone was extremely hot, and the cause of this may have given rise to the extra power.

When backing-off is finished, the drawing-up commences. At this point the moving of the carriage from a state of rest seems to give no occasion for much extra power—probably on account of the band working on the small diameter of the scroll, and so moving the carriage gradually in. The power during this action has risen to almost 3 H.P., and, naturally, after the carriage has passed its quickest speed halfway in, the force exerted naturally falls until the back stops are reached, when it is zero. The whole draw took 12 seconds to complete it.

The explanation of the other two diagrams, B and C, is similar to the above, but it is interesting to observe the difference in the power absorbed during the several actions. In B we have a mule of 1000 spindles of 8000 revolutions per minute. Its initial power is very low, in consequence of a large percentage of slippage (and this, it may be remarked, considerably reduces the average power of the machine, and gives to it an apparent advantage which it would not have if all its belts and bands were in perfect order). Its normal power in driving the spindles for twisting, bears comparison with diagram A, considering its

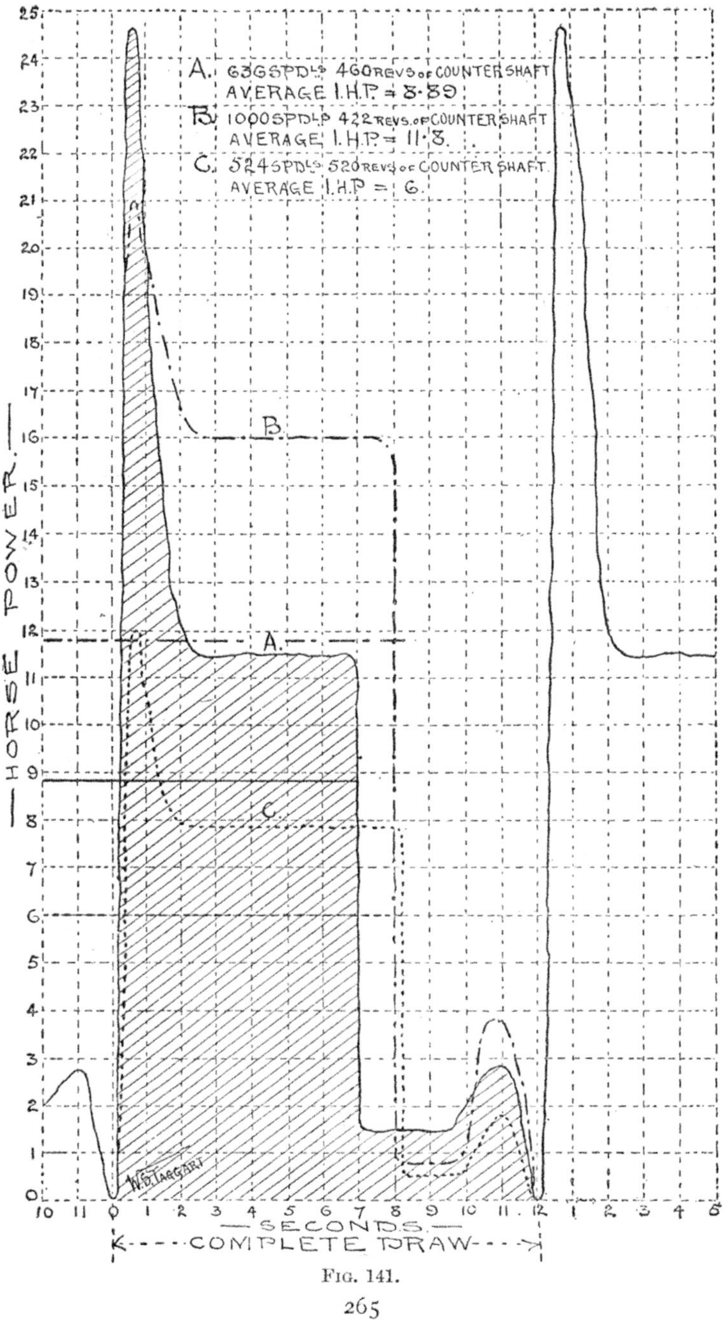

A. 636 SPDLS 460 REVS OF COUNTER SHAFT
AVERAGE I.H.P. = 8·89
B. 1000 SPDLS 422 REVS. OF COUNTER SHAFT
AVERAGE I.H.P. = 11·8.
C. 524 SPDLS 520 REVS OF COUNTER SHAFT
AVERAGE I.H.P. = 6.
B.
A.
C.
— HORSE POWER. —
— SECONDS —
COMPLETE DRAW
W.S. TAGGART

FIG. 141.

lower speed of spindle and the greater number of spindles. As one might expect, B absorbs more power than A during the drawing-up, and it is quite clear the backing-off process is much better arranged and requires less power to perform it. In Diagram C we have a machine a little over one-half of B, and whose motions almost correspond in respect to time. Its speed of spindle, however, is much greater, namely: 11,100 revolutions per minute, and this gives it a greater advantage when we compare their average results:—

In A the average number of spindles per horse-power	=71·5
In B ,, ,, ,, ,,	=84·7
In C ,, ,, ,, ,,	=87·3

The above averages, it must be remembered, include the driving of the counter shaft.

To show the difference between diagrams which are the result of a large number of individual readings, and one that is automatically recorded by the instrument itself, we give in Fig. 142 a drawing adapted from one issued by J. J. Rieter and Co., Winterthur. Its chief characteristics correspond very closely with those given in Fig. 141. The erratic motion indicated from C to D is probably owing as much to the dynamometer as to the mule; and the same may be said of the irregularity of the curve which indicates the backing-off and the drawing-up actions. A peculiar feature of the diagram is the line representing the backing-off from E to F. The writer has on several occasions observed a slight increase of power at the moment of putting the backing-off cone into gear, but it was so momentary and variable (in many cases it was entirely absent) that in the previous diagrams it is ignored. It represents the reversal of the spindles, and from this point the curve will be a gradual one to F, when drawing-up commences. Although the curve from E to F is automatically recorded, it does not follow that it is correct; the

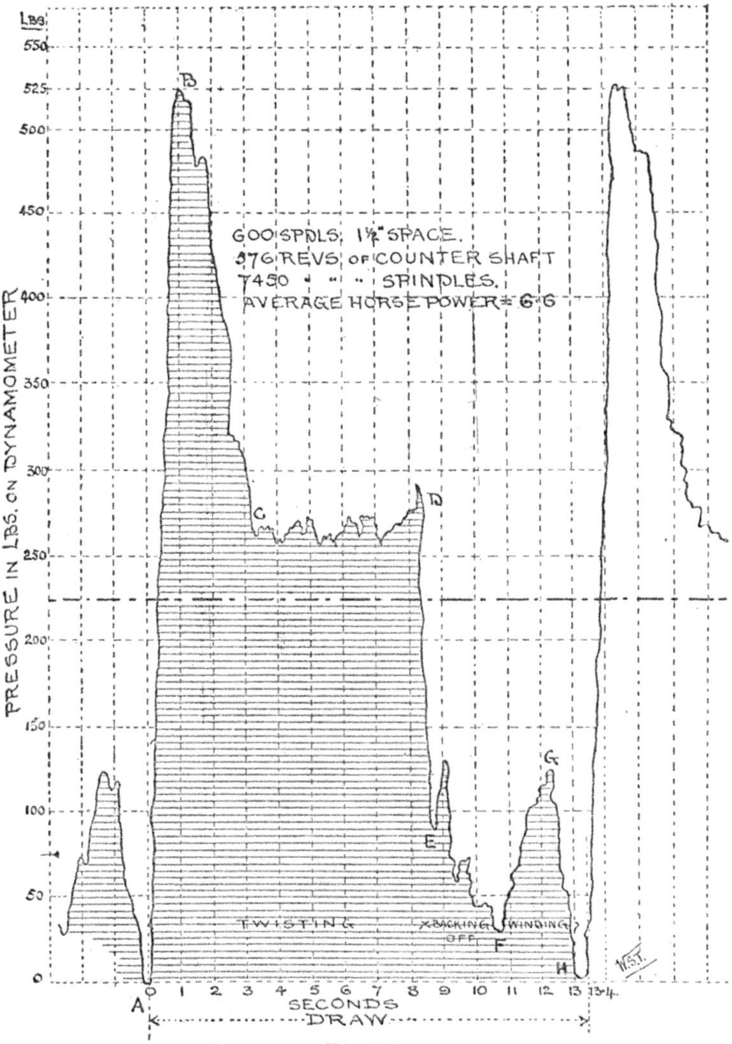

FIG. 142.

sudden changes which are brought about in the mule and the short intervals of time allowed for the actions do not allow the indicator to fall suddenly to the real pressure before another action comes into play and causes it to rise again.

In all dynamometrical indicators some means must be adopted to prevent the pointer from vibrating, on the same or a similar principle as the dashpot of a steam-engine governor. When the finger or pointer has been forced suddenly upwards, only a slow descent can be made, which depends on the character of the regulator used; and if, during the descent, another action comes into play, then we lose the real diagram that ought to be produced. An example of this is seen at E to F. The pressure falls suddenly at D, when the carriage stops; but before the figure has time to fall to the pressure that this stoppage represents, the backing-off takes place, and so the pointer must fall the remainder during the time backing-off takes place. The writer has experimented in this direction, and can speak from experience to the extent that such a curve as shown from E to F does not represent truly the actual conditions of power at that point.

On reference to the diagram again, it will be noticed that three seconds elapse before the normal speed of spindle is obtained. This is accounted for either by the slowness of the pointer in falling, or by the great power required to bring the spindles up to their maximum speed. In either case the average power is increased as a consequence, though, even so, the number of spindles per horse-power, namely 90·9, compares very favourably with any of those given in the first diagram. On the whole this diagram may be taken as representing very fairly good average results of a mule.

As showing to what extent good results can be obtained from a mule working under ordinary conditions, we reproduce in Fig. 143 an adapted diagram taken from one prepared by Sir Benjamin A. Dobson of Bolton; its

FIG 143.

average horse-power comes out very little lower than the normal power required to drive the spindles while twisting;

and the average for the number of spindles per horse-power, namely 105, is considerably higher than those obtained from the other diagrams illustrated.

Before just comparisons can be made on the power required to drive various types of mules, several important factors must be known: for instance, the diameter and shape of wharve, length and diameter of spindle, speed of spindle, and also the gauge. All these factors help to increase or decrease the power absorbed, according as they are greater or less. One-eighth of an inch increase in the diameter of wharve makes a considerable difference in the power; and when the length of spindle is increased even by only half an inch, the extra weight, revolved at 10,000 revolutions per minute, has some effect on the force required to drive it.

It may be pointed out that in none of the diagrams given does the number of spindles per horse-power approach those given in the earlier portion of these notes. It is fortunate that large margins are usually allowed for in the steam-engine, and that is why in large mills it is seldom that more than two or three mules at a time are working synchronically. In smaller mills, however, and especially old ones, it sometimes happens that a number of the mules get working in unison—to the considerable detriment of an already overloaded engine. It is to be hoped that as our knowledge extends of what power the mule really requires to drive it, a reduction in the extreme estimates at present in vogue will be made, so as to conform to the practical results of ordinary working conditions.

Calculations.—In connection with the calculations of the mule a remark made earlier in the book may be repeated,

Note.—See the author's book on *Cotton Spinning Calculations* for the gearing plans of other makers' mules.

namely—always get the speeds of quick running shafts, such as counter shafts, rim shafts, and spindles, by means of a speed indicator, which denotes the number of revolutions without necessitating the use of a watch. In this way there is no possibility of mistakes happening in expressing the actual revolutions per minute. No mill ought to be without one or two of such indicators, and no calculated speeds ought to be used when a speed indicator, such as the tachometer, is applicable. In the absence of an indicator, the following rules will be found serviceable:—

To Find the Speed of Rim Shaft per Minute.

$$\frac{\text{Revolutions of line shaft} \times \text{drum on line shaft} \times \text{drum on counter shaft}}{\text{Pulley on counter shaft} \times \text{pulley on rim shaft}}$$

$$\frac{250 \text{ revolutions} \times 30 \text{ in.} \times 24 \text{ in.}}{15 \text{ in.} \times 16 \text{ in.}} = 750 \text{ revolutions of rim shaft.}$$

From this point Fig. 144 will enable us to follow out the necessary calculations; reference may also be made to other sketches of gearing and driving which have appeared in these pages on the self-acting mule.

$$\text{Speed of spindles} = \frac{\text{Revolutions of rim shaft} \times D \times u}{t \times v}$$

$$= \frac{750 \times 18'' \times 6''}{10'' \times \frac{3}{4}''} = 10{,}800 \text{ revolutions of spindles.}$$

$$\text{Revolutions of spindle to one of rim} = \frac{\text{Revolutions of spindle per min.}}{\text{Revolutions of rim shaft per min.}}$$

$$\frac{10{,}800}{750} = 14{\cdot}4 \text{ calculated.}$$

It will be observed that the speeds so far have been "calculated," but it is almost unnecessary to point out that the use of belts and bands for driving occasion considerable slip. This must be taken notice of in all calculations, and for practical purposes 5 per cent is generally allowed. The above calculated speeds must therefore be reduced by this amount in order to obtain *actual* speeds. The speed of spindle, then, becomes 10,260 revolutions per minute.

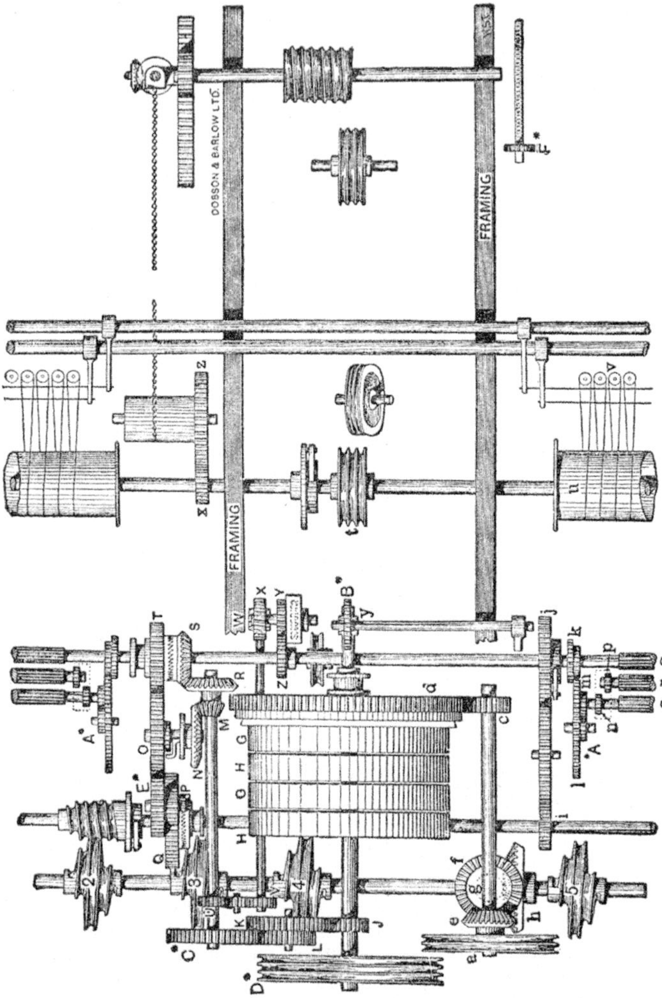

GEARING OF S.A. MULE

FIG. 144.

The Ratio of the Speeds of the rim and spindle is a very useful number to find, as it serves for a constant in finding the twist and twist wheel. The number so found as above, namely 14·4, simply means that the spindle revolves 14·4 times faster than the rim shaft.

Turns or Twist per Inch.

INDIAN AND AMERICAN COTTON

Mule Twist.	Multiply the square root of counts by	3·75
Mule Weft.	" " "	3·25

EGYPTIAN COTTON

Mule Twist.	Multiply the square root of counts by	3·606.
Mule Weft.	" " "	3·183.

Twist per Inch.

$$\text{Twist per inch} = \frac{\text{Length of yarn delivered or put up per draw}}{\text{Revolutions of the spindle per draw}}.$$

This is a well-nigh impossible rule to apply in actual practice, so in its place it is sometimes modified by assuming the rollers to run for one minute, and finding how much yarn would be delivered in that time. If the amount be then divided into the spindle speed per minute the result gives the twist per inch.

Sometimes the revolutions of the rim shaft per draw are first found by driving the mule very slowly; if this is then multiplied by the turns of spindle for one of rim, and the product divided by the length of stretch, we get the turns per inch. For instance—

$$\frac{\text{Revols. of rim per draw} \times \text{turns of spindle for one of rim}}{\text{Total travel of carriage}} = \text{Twist per inch.}$$

This cannot be relied upon for exact purposes, for there will clearly be far less slipping in driving the mule slowly than under ordinary speed conditions. If the twist wheel is used on the mule it is comparatively easy to adopt it as a basis for finding the twist per inch. For instance—

$$\frac{\text{Turns of spindle for one of rim} \times \text{twist wheel B}}{\text{Number of inches of yarn put up per draw}} = \text{Twists per inch.}$$

This is on the assumption that the twist wheel B moves

the strap on to the loose pulley after one revolution. If it revolves twice before changing the strap we should put the rule thus—

$$\frac{\text{Turns of spindle for one of rim} \times \text{twice the twist wheel B}}{\text{Number of inches of yarn put up per draw}} = \text{Twists per inch.}$$

$$\frac{14 \cdot 4 \times 2 \times 50}{65} = 22 \cdot 15.$$

Twist Wheel.—The foregoing rule also enables us to find the twist wheel for any given counts. For—

$$\frac{\text{Turns of spindle for one of rim}}{\text{Number of inches put up per draw}} = \text{Constant.}$$

$$\text{Constant} \times \text{twice the twist wheel} = \text{Twists per inch.}$$

Or,

$$\text{Constant} \times \text{twists per inch} = \text{Twice the number of teeth in the twist wheel.}$$

Example:—

$$\frac{14 \cdot 4}{65} = \cdot 221 \text{ Constant}$$

$$\cdot 221 \times 100 = 22 \cdot 1 \text{ twists per inch,}$$

$$\text{or } \cdot 221 \times 22 \cdot 1 = 100 \text{ (Twice the twist wheel).}$$

The twist wheel would therefore have 50 teeth.

Revolutions of Front Roller per draw—

$$\frac{\text{Length of the stretch} - \text{the gain}}{\text{Diameter of front roller} \times 3 \cdot 1416} = \text{Revolutions of front roller.}$$

If there is no gain in the mule, then this factor must be left out. Or,

$$\frac{\text{Revolutions of rim shaft per draw} \times J \times L \times R}{K \times S \times \text{diameter of F.R.} \times 3 \cdot 1416} = \text{Revs. of front roller.}$$

Revolutions of Front Roller per minute—

$$\frac{\text{Revolutions of rim shaft} \times J \times L \times R}{K \times U \times S} = \text{Revolutions of front roller.}$$

Back Change Wheel

$$= \frac{\text{Twice the twist wheel} \times \text{the rim spur}}{\text{Revolutions of front roller per draw}} = \text{wheel C}$$

$$= \frac{2 \times B \times J}{\text{Revolutions of front roller per draw}} = \text{wheel C.}$$

The Drafts in the rollers of the mule are worked out practically in the same way as in the previous machines, the arrangement of the gearing being the same. A is the draft wheel.

$$\text{Draft} = \frac{m \times l}{A \times k} \quad \text{(The front and back rollers are equal in diameter.)}$$

$$\text{Constant for draft} = \frac{m \times l \times \text{diameter of front roller}}{k \times \text{diameter of back roller}}.$$

$$\text{Draft} = \frac{\text{Constant}}{\text{Draft wheel}}.$$

$$\text{Draft wheel} = \frac{\text{Constant}}{\text{Draft}}.$$

$$\text{Draft wheel} = \frac{\text{Hank roving} \times \text{diameter of front roller} \times l \times m}{\text{Counts wanted} \times \text{diameter of back roller} \times k}.$$

$$\text{Total draft in mule} = \frac{m \times l \times \text{the stretch} + \text{roller motion}}{A \times k \times \text{length delivered by the rollers}}.$$

If the correct draft is known when spinning certain counts and it is wished to change to another count, using the same hank roving, it becomes a case of simple proportion of inverse order, for if count 40's has a change wheel of 26 teeth, then 50's will require not a larger wheel but a smaller one, so that the *Rule* is :—

$$\frac{\text{Draft wheel} \times \text{present counts}}{\text{required counts}} = \text{Draft wheel required.}$$

Under conditions of changing both the counts spun and also the hank roving the

$$\text{Draft wheel} = \frac{\text{Required hank roving} \times \text{present counts} \times \text{present draft wheel}}{\text{Present hank roving} \times \text{required counts}}.$$

When changing the twist wheel for a change in the counts, it is as well to use a foundation rule occasionally, such as the one already given; but for convenience, when once a correct twist wheel has been used, the practice is frequently adopted of using this as a standard from which to obtain the one required when the counts are changed. It is based on the fact that the twists per inch vary as the square root of the counts. From this we say that, if the square root of 60's counts requires an 80 twist wheel, the square root of 40's counts will require a proportionately less wheel.

Example:—60's counts are being spun, and the twist wheel has 80 teeth. What wheel is required for 40's counts?

The square root of 60 = 7·745.
The ,, ,, 40 = 6·324.
If 7·745 requires a wheel of 80 teeth,
Then 1 ,, ,, $\frac{80}{7{\cdot}745}$
And 6·324 ,, ,, $\frac{80 \times 6{\cdot}324}{7{\cdot}745}$

$$\therefore \frac{80 \times 6{\cdot}324}{7{\cdot}745} = 65 \text{ teeth.}$$

Readers will perhaps be unfamiliar with this rule, but it is based strictly upon reason. The following *Rule* is the one generally used:—

$$\text{Twist wheel} = \sqrt{\frac{\text{Twist wheel}^2 \times \text{required counts}}{\text{Present counts}}}.$$

The slightest acquaintance with equations will enable any one to prove that this rule is derived entirely from the first one. For the sake of a few who desire to know why such a form of rule is adopted, the explanation is given, as follows:—

If $\sqrt{60}$ requires a wheel of 80 teeth,
Then $\sqrt{1}$,, ,, $\frac{80}{\sqrt{60}}$
And $\sqrt{40}$,, ,, $\frac{80 \times \sqrt{40}}{\sqrt{60}}$

$$\frac{80 \times \sqrt{40}}{\sqrt{60}} = 65 \text{ teeth.}$$

Proof.—First square the expression, so—

$$\left(\frac{80 \times \sqrt{40}}{\sqrt{60}}\right)^2$$

This equals $\frac{80^2 \times 40}{60}$.

Then take the square root of this result, which may be expressed so—

$$\sqrt{\frac{80^2 \times 40}{60}}.$$

This gives us the familiar form of the rule; for 80 is the present twist wheel, 40 is the required number of the counts, and 60 is the number of the present count.

It is sometimes necessary to change the front and back roller wheels as well as the change wheels. If the reader is acquainted with simple equations there is no necessity to learn off a number of rules applicable to each case. From one rule he would obtain any change required.

Example :—Using the drawing for the letters on the wheels,

$$\text{Draft} = \frac{m \times l}{A \times k}.$$

$$\text{Front roller wheel } k = \frac{m \times l}{A \times \text{draft}}.$$

$$\text{Back roller wheel } m = \frac{A \times k \times \text{draft}}{l}.$$

$$\text{Top carrier wheel } l = \frac{A \times k \times \text{draft}}{m}.$$

$$\text{Change wheel } A = \frac{m \times l}{k \times \text{draft}}.$$

It is thus seen that, by using a simple form of rule, expressed as a formula, any one of the factors can be deduced, provided we know all the others.

CHAPTER III

THE RING SPINNING FRAME

General Description.—The ring frame, so far as its general mechanism is concerned, is probably one of the simplest and most easily understood machines in a cotton mill; and yet around the problem associated with its central action we find a peculiarly divided state of opinion, founded—as all opinions on ring spinning must be founded—on a mixture of theory and practice. The subject has a special interest of its own; partly because of the great and increasing rivalry of the ring frame with the mule, and partly because of the unsolved or only partly solved problems connected with its twisting action and its effect upon the yarn.

Before inquiring into the cause of this unusual interest, we shall first give a general description of the machine itself. The ring frame in its characteristics is practically the same machine as the flyer throstle, and in describing one we practically describe the mechanical arrangements of the other: the difference exists in the spindles and in the method of putting the twist into the yarn. In the flyer throstle the spindle is a plain rod of steel, surmounted by a flyer; this is driven from a tin roller, and its revolutions determine the amount of twist put into the yarn, exactly as in the fly frame. The differential speed between the bobbin and the flyer, for winding purposes, is obtained by letting the bobbin run loose on the spindle, and allowing the twisted yarn to drag it round. If the spindle runs say 7000 revolutions per minute, and yarn connects the flyer leg with the bobbin, the bobbin is dragged round at the same speed as the spindle; but the rollers deliver roving, and this when twisted decreases the tension between the flyer and bobbin, and the bobbin hangs back a little in its speed, and consequently has the delivered yarn wound on to it. Over-running is prevented by carefully grading the drag, which is obtained by resting the bobbins on some rough surface, such as flannel washers. Yarn made in this way is considered to be of a very superior quality, but the system has now been practically discarded for the ring frame, so that it is unnecessary to enter into any detail as to its working.

Fig. 145 illustrates our general remarks on the ring frame; half the machine is shown in section and the other half in elevation. It is a double-sided machine, *i.e.* each side contains a long line of spindles, suitably spaced, and carried by strong rails, as at H. The spindles are driven by bands from tin rollers T and T_1—(in some cases only one tin roller is used)—the band passing from the wharve G over the top of the nearest tin roller T_1, and on round the farther tin roller T. The entire driving of the machine takes place through one of the tin roller shafts, and in the illustration given, X is the shaft chosen for the purpose; the other tin roller T_1 is frequently driven entirely by the spindle bands, which, passing from T over its upper surface and on to the wharves G, drive it simply by the friction of their contact in going forward. On the driving shaft X is fixed a wheel J, which by a system of gearing ultimately drives the front roller through the wheel P. A compound carrier wheel L M is introduced into the gearing, and at this point any change of speed required in the front roller

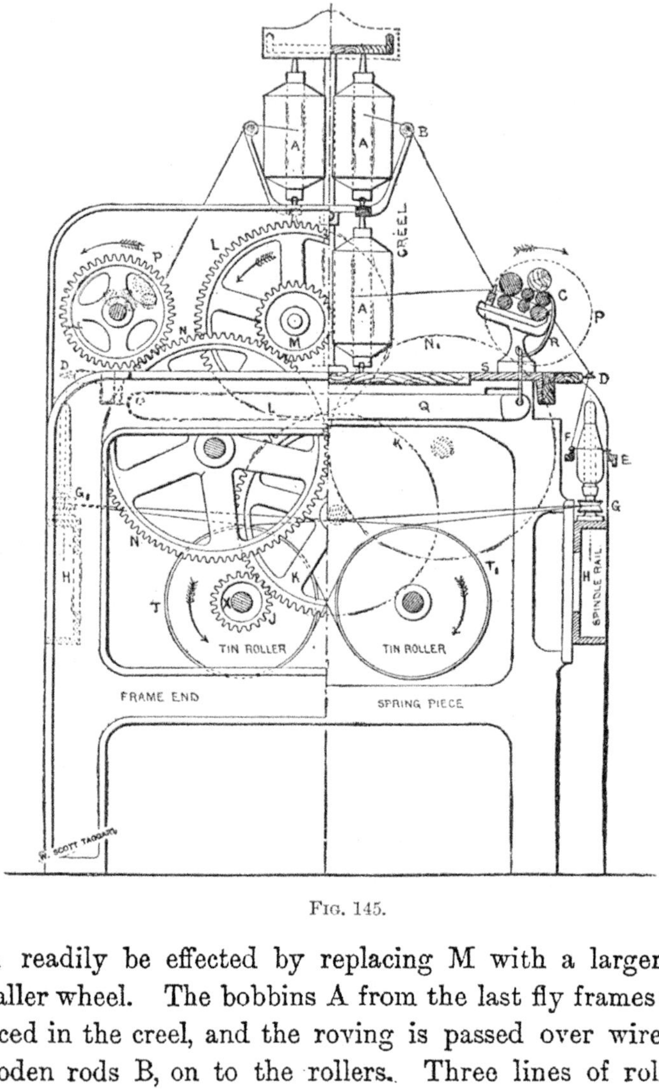

FIG. 145.

can readily be effected by replacing M with a larger or smaller wheel. The bobbins A from the last fly frames are placed in the creel, and the roving is passed over wire or wooden rods B, on to the rollers. Three lines of rollers are generally employed, and, as in all the previous machines, a draft is introduced for drawing out the slivers. From the front rollers the roving is passed through a guide wire, which is placed directly over the centre of the spindle, and on through a small piece of steel, bent in the form of the letter C, called a "traveller," which clips loosely a specially formed steel ring, fixed on a movable plate called the "ring-plate." The spindle and bobbin pass through the centre of this ring, and thus after the yarn is threaded through the above-mentioned steel traveller it is wound round the bobbin. The revolution of the spindle, which is run at a very high speed, in its attempt to wind on the yarn, pulls the traveller round with it, and relieves what would otherwise be a tension in the yarn; at the same time each revolution made by the traveller puts a twist in the yarn, and as the bobbin can only wind on the amount of yarn delivered by the rollers, it follows that the traveller is made to revolve almost as quickly as the spindle, so that we get a most effective twisting operation performed. This is merely a general statement of the action; it will be treated fully in subsequent pages. In order to build up a cop on the bobbin or spindle, the ring-plate is made movable, so that by a special lifting motion it is raised and lowered in a manner suitable for the formation of the cop.

Driving.—Treating in detail the various features of the machine, we shall first briefly mention the driving. Supposing A in Fig. 146 to be the driving pulley of the ring frame, it is possible to drive A in three different ways, namely—direct driving; gallows guide-pulley driving; or driving by half-twisted belt. The last two systems are the ones most generally adopted, and the line shaft is in each case at right angles to the driving shaft of the machine.

With gallows or guide pulley the line shaft may be some distance away, so that the gallows pulleys simply serve to guide the strap on to the machine below.

As already shown, the tin drum on the driving shaft A drives the spindles on the side of the machine marked E. The bands, in passing from the top of the tin drum on A to the spindle, clear the top of the tin drum on B, but the same band from the lower side of A must pass over the top of the tin drum on B, so that it will be seen that the tin drum acts as guide pulleys to the spindle bands. Friction is set up by the large number of bands in a frame to such a degree that whenever two tin drums are used one of them receives no other motion than that obtained through this friction of the bands. Very effective driving is obtained in this way, but a little thought on the matter will lead to several conclusions to the disadvantage of the system. In the first place, the spindle bands, whose object is to drive the spindles, are called upon to also drive the tin drum at a speed of, say, 700 revolutions per minute; the extra strain thrown on them of course soon destroys them, and replacing is a frequent necessity. This is a tangible fault, and requires consideration from an economical point of view. Secondly, we may readily assume that, since the tin drum on B is driven merely by the friction of a number of bands, the spindles on the F side of the machine will exhibit a larger percentage of loss by slippage than the spindles at E. From a practical point of view this objection may be dismissed, for even when, through some local cause, a difference is found, it is so slight that it may without disadvantage be ignored.

The disadvantages mentioned weigh with some people; consequently machine makers are called upon to adopt means to overcome them. In the accompanying sketch,

Fig. 146, a general idea of one way of doing so may be obtained. On each tin roller shaft is placed a band pulley A and B. In some accessible part of the framing is fixed

Fig. 146.

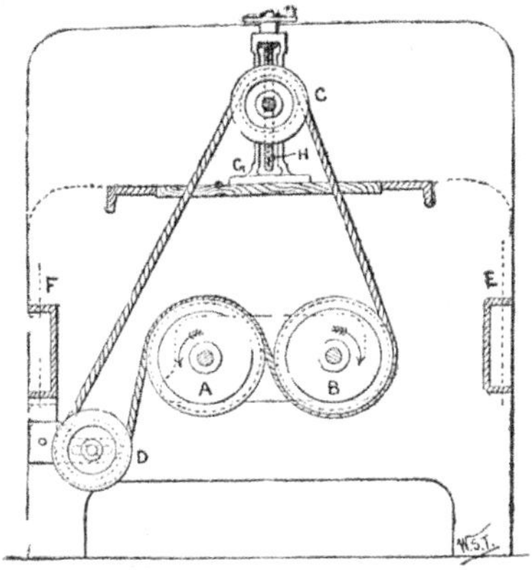

Fig. 148.

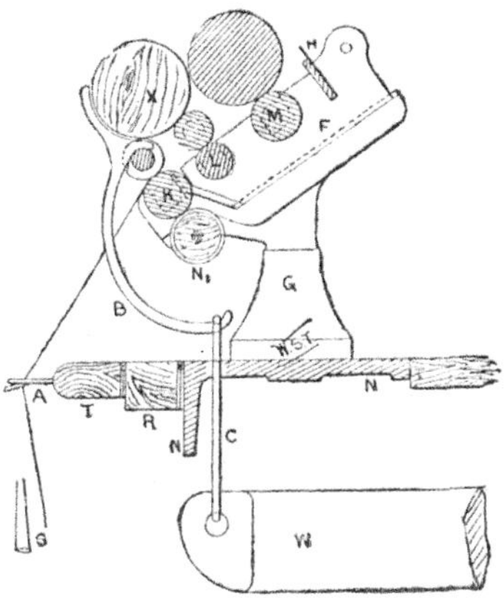

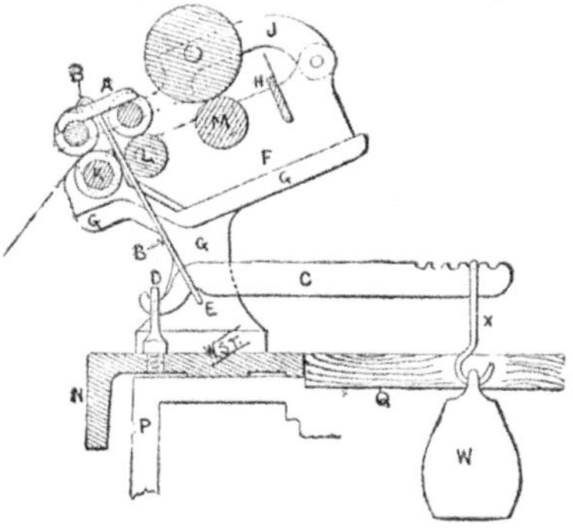

Fig. 147.

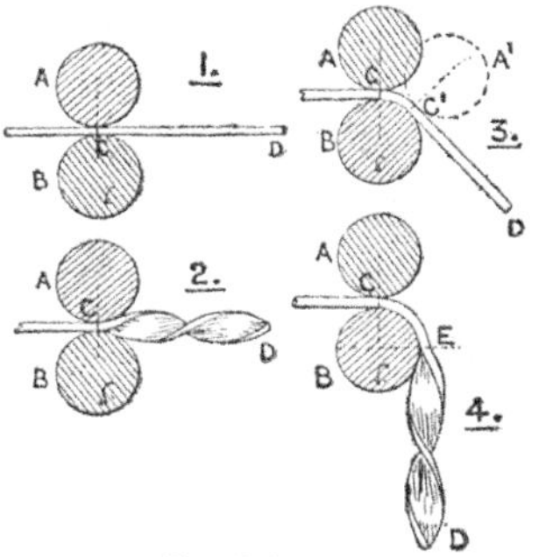

Fig. 149.

a guide pulley C, carried by a bracket G, arranged so that adjustment can easily be made through the screw H whenever the band becomes slack. A guide pulley is also provided at D, with the object of ensuring a good grip of the band on the pulleys A and B. The band is threaded over the pulleys as shown in the sketch, and we can readily understand that B, by such means, can be driven from A with less probability of slippage than by the spindle bands alone. Its practical advantage is apparent in the greater lasting power of the bands, and an economy is at once effected in this direction; but in regard to the speed of the spindles at F a merely fractional improvement is recorded. In one way the application of the band pulleys is a most decided disadvantage, and this in a direction that is very palpable if the trouble be taken to test it. A number of dynamometrical tests in ring frames shows an unmistakable increase in the power required to drive the machine when fitted with the apparatus, from 10 to 20 per cent being no uncommon addition to the usual power. It need scarcely be pointed out that this loss outweighs the economy of spindle bands, and as a consequence many spinners refuse to use such a doubtful improvement.

Roller Stands and Weighting.—Passing from the driving, we shall give some little attention to the roller stands and weighting. Figs. 147 and 148 illustrate the general arrangement of both features. It will be noticed that the arrangement of the rollers follows on the same lines as that of the fly frames and of the mules. One important difference exists, however, as seen in the tilting or inclination of the rollers as a body. The reason for this is a simple one, which can readily be understood; the yarn as it comes from the front roller passes to the thread guide at such an angle that it must pass over a portion of the surface of the front roller before it is clear; also the yarn is in contact with the thread guide (as at A, Fig. 148) all the time it is passing forward to the bobbin. Two points of contact are therefore tending to stop the twists put into the yarn by the traveller from getting up to the nip of the front rollers. Since in the ring frame we have not the agitated movement of yarn during the spinning operation, as in the mule, this interference with the twist would cause a very weak spot to develop at the nip of the rollers, and a great source of breakage would result. The difficulty is overcome by inclining the rollers at such an angle that the yarn is in contact as little as possible with the bottom front roller, so that the twists get right up to the nip of the rollers. The four sketches in Fig. 149 will explain the action very clearly. In each case A B represent the front rollers, C is the grip, and C D a length of yarn delivered from the rollers. If C D be passed out as a flat ribbon, and twisted, say one turn, it would (as in 1 and 2, Fig. 149) become twisted right up to the nip of the rollers, as at C. In the ring frame the yarn passes forward in an inclined direction (Fig. 149), so that it is in contact with the roller B until the point C_1 is passed. Let us notice the effect of this by taking an extreme case (as in 4, Fig. 149), where the yarn goes forward at right angles. In such a case it is in contact with the front roller from C to E, and the twists would tend to stop short at the point E in the manner shown. To obviate this objection the three lines of rollers are bodily inclined (as in Figs. 147 and 148), and the effect of this is to move the top front roller from A to A_1 (as in 3, Fig. 149), so that the nip of the rollers moves from C to C_1, and thus eliminates the objectionable contact surface of the roller B. The angle of the rollers varies from 15° to 35°, according to the cotton being used, but about 25° will be found most general and serviceable.

Referring again to Figs. 147 and 148, two systems of weighting will be found employed—one on the lever system and the other by means of dead weights. It will be noticed, however, that in the first case only the front and middle rollers are weighted, the back being self-weighted by the large iron top roller. A saddle is put across the first rows, and a bridle or link is hooked on to it at a point much nearer to the front roller than to the middle one. This gives a preponderance of weight on the front roller; the lever and weight arrangement is very similar to that shown on the mule roller stand. The dead-weighting of the rollers is illustrated in Fig. 148, and in this case only the front line is weighted, the middle and back rows being self-weighted. The dead weight hangs from a hook placed over the front roller, and it will be noticed that, instead of hanging a weight from each hook, the weight is made twice the necessary size, and long enough to go across the frame, so that its other end hangs from the front roller on the opposite side of the machine. A saddle and bridle lever-weighting, exactly similar to the one illustrated in the mule roller stand, is often adopted on ring frames, so that it is unnecessary to repeat the drawing here. There is one feature that may be of interest to mention, and that is to warn the reader against overlooking the inclination of the weight-hook in the systems mentioned.

It will be sufficient to refer briefly to the matter, without entering upon actual calculations. If a saddle A B (1, Fig. 150) rests horizontally upon two rollers A, and a weight W is hung at C, it is an easy problem to find how much pressure is put upon the roller at A and B. In the same way, if the saddle is inclined as in 2, Fig. 150, the hanging weight W can readily be found to give similar pressures upon A and B as in No. 1, because in such a case the relative horizontal distances of A and B from the direction of the pull of the weight W remain unaltered.

Again, if the saddle is still inclined as at A B (No. 3) and the weight is made to pull in the direction of W, which is at right angles to A B, we should work out the pressures on A and B exactly as in the No. 1 example; but in the

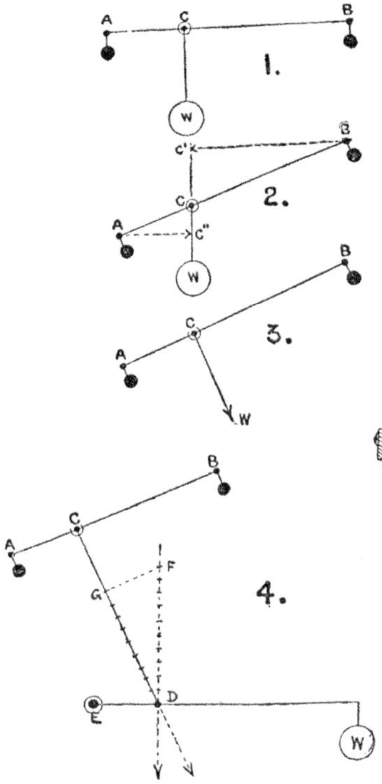

Fig. 150.

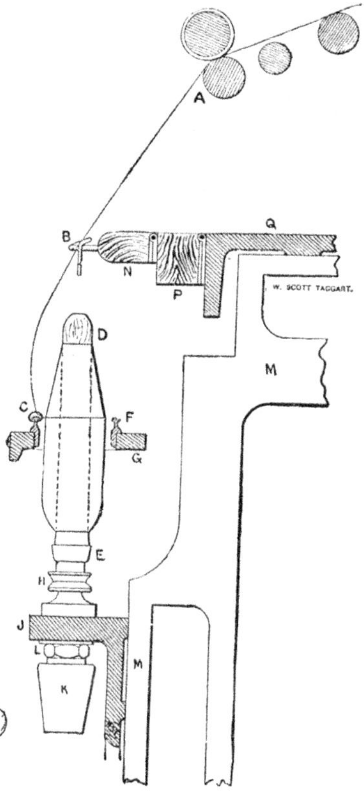

Fig 151.

ring frame we generally find a combination of the arrangements 2 and 3, in which a pull is exercised in the direction of W, this pull being produced by a weight hanging vertically. No. 4, Fig. 150, illustrates the meaning. A B is the saddle, C D a link hooked to the saddle and to the weight lever E W. W hangs vertically and produces a pressure at D in a vertical direction; this pull, however, is exercised along the link C D, so that in consequence of the inclination of C D a portion of the pressure produced by W is inoperative on the saddle A B. It is a comparatively easy matter to find the effective pressure produced by the weight W upon the saddle. Suppose a 2 lb. weight at W gives 10 lb. pressure at D: measure off on a vertical line at D ten given distances, such as quarter-inches; now draw a line, from the upper end F of the divisioned line, at right angles to the line C D, cutting the line C D at G; if we measure off D G and note how many quarter-inches there are in it we obtain the number of lbs. pressure along C D. Among several stands tested it was found that the pressure upon the saddle at C was about ten per cent less than that produced by the weight W at D. If this be duly noted when calculating the pressure on the rollers, the rest of the calculation becomes an easy matter.

Twisting.—After passing through the front rollers A (Fig. 151) the yarn goes forward through the thread guide B and is threaded through a bent piece of steel C, in this form, ⌒, called a "traveller"; from here it passes on to a wooden tube or bobbin D, fitted upon a spindle E, which is driven from the tin roller through the wharve H. The revolution of the spindle begins to wind on the yarn; but since the rollers A only deliver a certain length, and the spindle revolves at a high rate of speed, the tension produced in the yarn acts on the traveller and pulls it round at almost the same speed as the spindle itself. Every revolution of C puts a twist in the yarn, and at the same time the bobbin winds on the amount of yarn given out from the rollers. As winding commences, the rail G, which carries the ring F and traveller C, is caused to rise and fall by lifting mechanism, so that the yarn is wound on in layers, and of a form similar to the cop of the mule.

Thread Guide.—The features mentioned in the foregoing paragraph can now be dealt with in detail. Commencing with the thread guide: this is seen to be a curled piece of wire A, Fig. 152, screwed into V-shaped pieces of wood B, hinged to the thread board proper C. The board C is hinged to the roller beam D. As A is directly over the centre of the spindle, it is necessary to be able to move it out of the

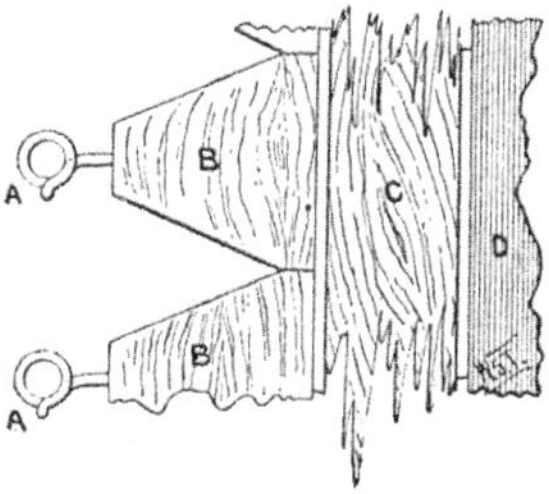

FIG. 152.

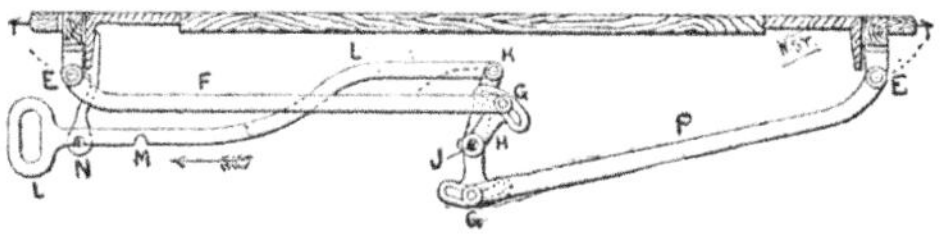

FIG. 153.

way whenever a bobbin is taken off. For this purpose B is hinged so that it can be turned over. For doffing purpose it is found convenient to be able to turn over the whole of the wires and thread board, and arrangements are frequently adopted for doing this. An illustration of one method is given in Fig. 153. To the under side of C is fixed a bracket carrying a pin E, to which are connected links F. The other ends of these links are centred on studs G, carried by a short lever H, pivoted on the shaft J. This shaft J also carries an arm K, to which is attached a handle L, which can be locked in position by the slots M fitting over a projection N. If now the handle L be drawn forward in the direction of the arrow, the thread boards on each side of the frame will be raised bodily out of the path of the bobbins as they are being doffed. To serve the same purpose, an arrangement is sometimes adopted whereby the thread board is moved sideways to the extent of half the space of the spindles.

The Ring.—The ring A in Fig. 154 is made of forged steel, carefully turned and afterwards case-hardened; its general form is similar to that indicated in the diagram, and shown enlarged in Fig. 155. They are carried by and secured to a cast- or wrought-iron plate P, which in the modern machine is now flanged singly or doubly to prevent deflection. The diameter of the ring is its inside measurement, the usual dimensions for the different spaces of spindles and counts spun being as follow:—

For 4's to 20's counts	—$2\frac{3}{4}$ in. space,	$1\frac{3}{4}$ in. dia. of rings
For 20's ,, 40's ,,	—$2\frac{5}{8}$ in. ,,	$1\frac{5}{8}$ in. ,, ,,
For 40's and upwards	—$2\frac{1}{2}$ in. ,,	$1\frac{1}{2}$ in. ,, ,,

If balloon plates are used the space can be reduced a little. The ring is secured to the plate by a set-screw C. Other forms of rings are used, such as the double ring shown in Fig. 156; it is made in this form so that when one flange becomes worn the ring can be reversed. The method of fastening it to the ring-plate is to spring it into the grip E of a special piece of sheet metal C, which is in its turn set-screwed to the plate B. The perfection to which rings are now brought renders it extremely doubtful whether there is any economy in the adoption of this system; but some people still prefer it. An important American firm have introduced slight variations in both the ring and the plate, the plate being made out of sheet steel, while the ring is modified at the point marked A, with the idea of giving better hold to the traveller. Fig. 157 shows the comparison between the old and new form.

From Fig. 154 we obtain an idea of how the ring-plate is carried. At intervals along the frame the plates rest upon the upper part D of a shaft E; these shafts are termed pokers; they slide vertically in bushes F fixed in the spindle rails G, and their lower ends are arranged to be actuated from the building motions in order to give an up-and-down motion to the ring-plate.

FIG. 154.

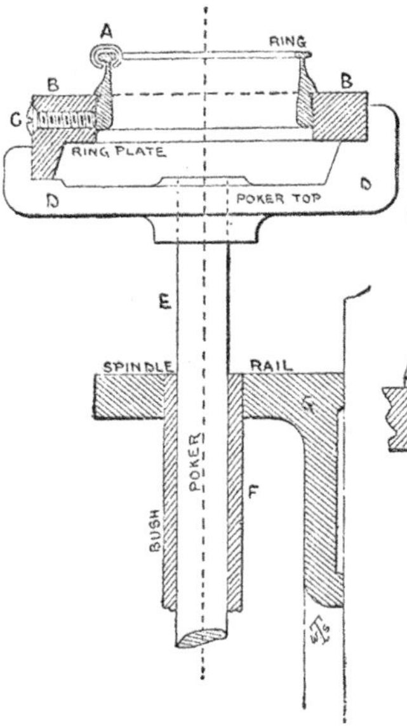

FIG. 155

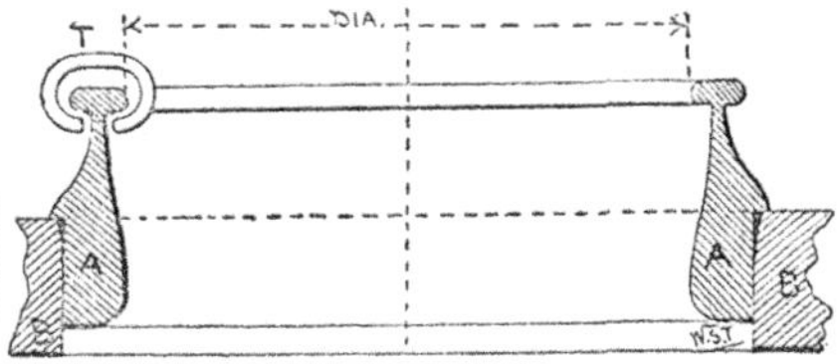

FIG. 156.

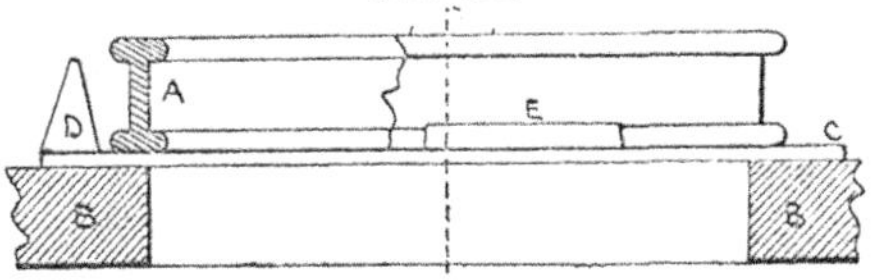

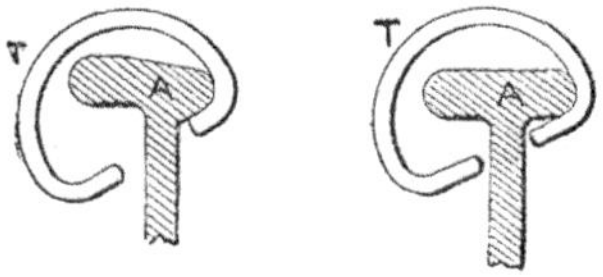

FIG. 157.

Building Motion.—The method of operating the pokers of the ring-plate, so as to give a reciprocating motion for building the cop, is very simple, and in most modern makes of frame there is such a similarity of construction and principle that a single example will be sufficient to explain it.

The principal mechanism employed consists of a cam, actuating a lever, on the end of which is attached a chain, leading to levers that act on the lower ends of the pokers. The speed of the cam is carefully regulated to give a motion to the ring-plate suitable to the counts being spun, and it will be noted that any alteration in the speed of the front roller (consequent on a change in the twist or counts) also results in a similar change in the speed of the cam. This will be pointed out more fully when dealing with the calculations of the machine.

A view of the motion is presented in Fig. 158, which illustrates the essential features. A long lever A is centred on a stud at B, fixed in the frame end; a bowl C is carried by a small dish bracket D, bolted to the long lever, so that by the revolution of the cam E the lever A will be given a reciprocating motion. At the opposite end to B, the lever A carries a bowl G, and round it is wound a chain S, the other end of which is attached to a bowl T. As the lever A is acted on by the cam E, the pull of the chain S will turn T through a portion of a circle; and if to a bowl U by the side of T is connected a chain which leads on to the levers actuating the pokers, we have the motion of E transferred to the ring-plates. From the arrangements so far described we obtain the lift of the first layer of yarn, as at A in Fig. 159. It now remains to consider the further layers B and C. In the first place, the starting-point of each new layer is raised up the bobbin by a taking-up motion, as follows: the bowl G is carried on a short shaft, on which is keyed a worm wheel H, into which is geared a worm V. On the end of the shaft carrying V is keyed a ratchet wheel M, into the teeth of which is engaged a catch carried by a tumbler N. As the lever A is depressed, one end of the tumbler Q is brought against a stop W, and as its further movement downwards is thus arrested it naturally commences to force round the ratchet wheel, and so turn the bowl G, which consequently winds on a small length of the chains S. This action lifts the ring-plates a little higher, so that the next layer must begin higher up the bobbin than the previous one. The same action of the tumbler continues throughout the building of the bobbin, giving

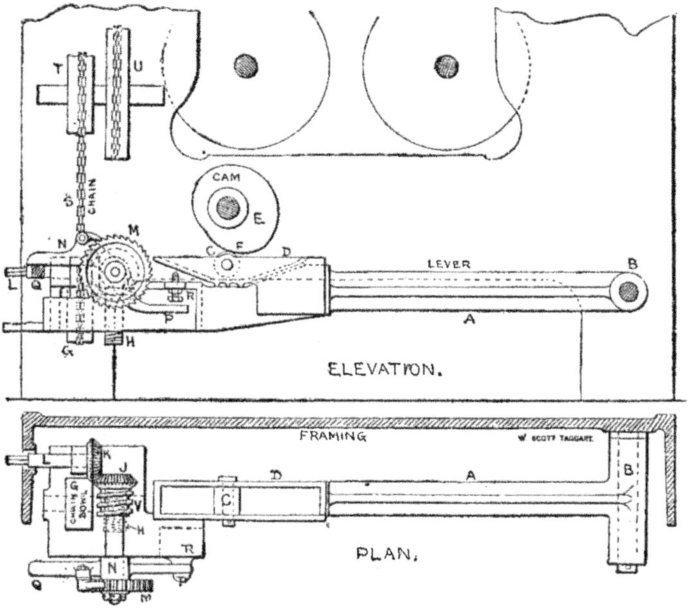

FIG. 158.

whatever lift may be considered suitable. The amount of movement given to the tumbler, and consequently to the ratchet wheel, is regulated by adjusting the tumbler so that its downward movement is not arrested until the stop itself is touched, or (as in the illustration) until the set-screw at P comes into contact with the projection R on the lever A. By careful adjustment of the set-screw we can regulate the number of teeth taken by the catch on the tumbler; or a similar effect may be obtained by changing the ratchet wheel itself for one of a greater or less number of teeth.

Fig. 160 enables us to follow the building motion to its connection with the pokers. The chain N from the bowl U passes to a bowl A, carried by a swing lever B centred on the lever E, whose fulcrum is at F. To the lever E is attached a lever G H, and as the chain S turns the bowl T, the larger bowl U (acting through the chain N) moves the lever E, so that the ends G and F are raised and lowered. (We may add at this point that the ring-plates are not "lowered" by the direct effect of the lever A in Fig. 158; only the "lifting" of the plates is brought about by this means: the lowering is brought about purely by the weight of the plates and their connections, a series of balance weights being so arranged as to permit of this occurring.)

It will be observed that the end G, Fig. 160, has a direct lifting effect on the poker; but since the lifting of G means the lowering of the end H, a chain is used to transfer this movement to a lifting one; a chain is attached to H, and, passing over the pulley J, is brought down and connected to the poker at K, thus producing a lifting action on each poker. The movement given to the lever E is transferred to each poker throughout the length of the frame by means of a rod D, which is coupled-up to similar levers as E at suitable intervals.

The Traveller and its Action.—It will be found convenient at this point to enter upon a discussion regarding the traveller and its action. From a superficial point of view the work of the traveller is comparatively easy to understand, and its effects in spinning and winding offer no difficulty to the observant mind. One reason for this is because the chief actions can be closely watched; consequently experience can be obtained readily and quickly under a variety of conditions. From this statement we are led to remark that the best results are almost invariably dependent upon experiment, and very little reason is called into play. Nevertheless there must be a decided advantage in knowing the reason for a certain line of action, and with this object in view a few remarks will be made explanatory of the functions performed by the traveller in the ring frame.

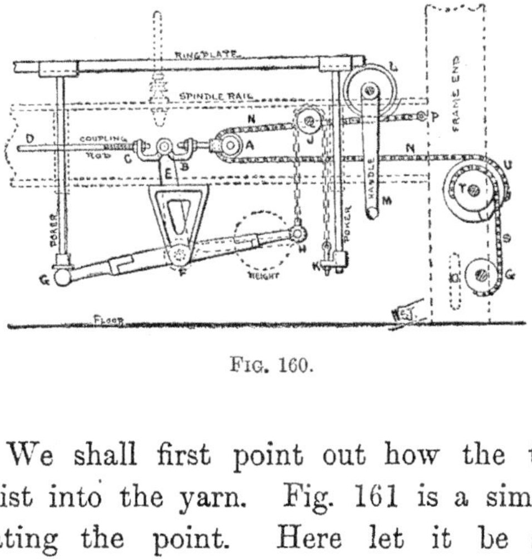

FIG. 160.

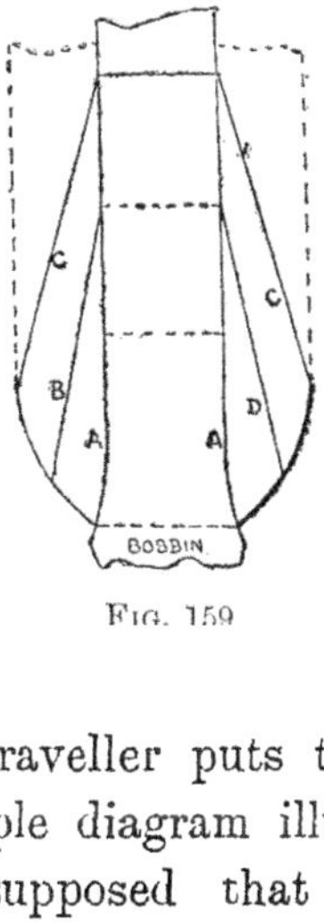

FIG. 159.

We shall first point out how the traveller puts the twist into the yarn. Fig. 161 is a simple diagram illustrating the point. Here let it be supposed that R represents the nip of the front rollers, and that a flattened portion of yarn A R is delivered from them; the end A is carried round in a circle A B C, while the end R is held fast by the rollers. As A is carried round with the small arrows always uppermost, it will be found that by the time B is reached, the tape will have been twisted half a turn; and by carrying the end to C under the same conditions, a full twist will be found to exist in the tape—in other words, one revolution of the end A puts one twist in the length A R. Now, as the traveller performs the duty of carrying the end A of the yarn A R round the circle of the ring, it follows that the traveller is responsible entirely for the twists put into the yarn delivered from the rollers, and that the speed of the traveller regulates this factor in spinning.

Why the traveller revolves? is the next question. In the first place, it must be understood that a traveller is a kind of guide for the yarn; if it were a fixed guide, as at A, Fig. 163, the revolution of the spindle B would simply wind on the yarn as it was delivered by the rollers, and such yarn would be untwisted. On the other hand, if the traveller guide were attached to the spindle, as at A, Fig. 162, the revolution of the spindle B would carry A round with it, and as a result no winding would take place, but every revolution of B would put a twist in the yarn. In the two cases given we have examples of all winding and all twisting; in spinning, these two operations must be performed; so by making the guide A movable and yet unattached to the spindle directly, we get conditions that supply us with the requisite characteristics of the ring frame. The analysis of this action may prove interesting, but we defer it until another feature has been explained. A previous paragraph told us that the spindles are revolved at a constant speed throughout the building of a bobbin; and we now know that the traveller, in addition to putting the twists into the yarn, also winds it on the bobbin. The question now arises—How is the conical part of the cop built up? It is unnecessary to remind the reader that an enlarged diameter of bobbin or cop necessitates, so far as examples in other machines have shown us, a differential speed of spindle, in order to wind the yarn on a varying diameter; but in the case of a ring frame we fail to find any mechanism that performs this apparently required condition of building the bobbin. A few words will make this clear. Let us suppose that the front rollers deliver 528 inches of yarn per minute, and the spindles rotate at the rate of 9500 revolutions per minute, also that the largest diameter of the cone is $1\frac{1}{4}$ inch, and the smallest diameter $\frac{1}{2}$ inch. From these conditions we can readily find the necessary rate of speed of the traveller to wind on the yarn at the two extreme diameters.

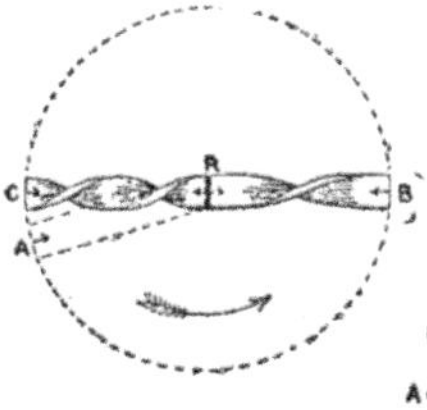

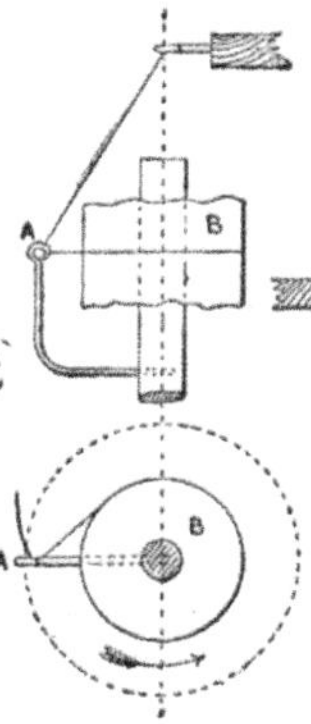

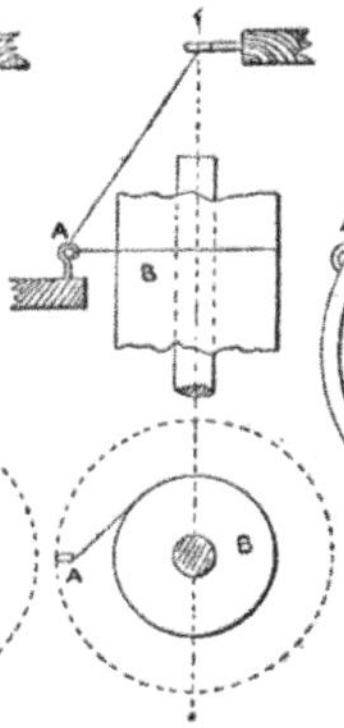

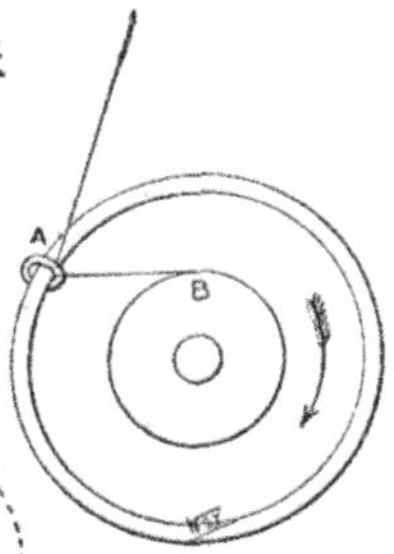

FIG. 161. FIG. 162. FIG. 163. FIG. 164.

To wind 528 inches on to $1\frac{1}{4}$ inch diameter the traveller must make

$$\frac{528}{1\frac{1}{4} \times 3{\cdot}1416} = \frac{528 \times 4 \times 7}{5 \times 22} = 134{\cdot}4 \text{ revolutions}$$

less than the spindle, so that the speed of the traveller will be $9500 - 134{\cdot}4 = 9365{\cdot}6$ revs. per minute when winding on the $1\frac{1}{4}$ inch diameter.

To wind 528 inches on to $\frac{1}{2}$ inch diameter the traveller must make

$$\frac{528}{\frac{1}{2} \times 3{\cdot}1416} = \frac{528 \times 2 \times 7}{1 \times 22} = 336 \text{ revolutions}$$

less than the spindle, so that the speed of the traveller will be 9500 − 336 = 9164 revolutions per minute when winding on the ½ inch diameter. Now, comparing this variation of the speed of the traveller as it winds on the extreme diameters of the bobbin, we find that there is a difference of 9365·6 − 9164 = 201·6 revolutions only, equal to a little over 2 per cent. It is therefore easily understood why we find no apparent arrangement for obtaining differential speed during the formation of the bobbin. This difference, although slight, must be allowed for; and since there is no method adopted for varying the speed of the traveller to the extent noted, resource is had to the lifting cam in causing the lift to vary ever so slightly by a small variation in the shape of the cam that actuates the lifting lever.

In examining the action of the traveller it should first be stated that the yarn is wound on by the bobbin, and that the traveller simply regulates the amount wound on. A bobbin ½ inch diameter revolving 9500 revolutions per minute would wind on 14,928 inches of yarn per minute, while the roller only delivers 528 inches. The consequence is that no sooner does the bobbin commence its revolution than a tension is set up in the yarn, and as this tension is exerted on the traveller, this small piece of bent wire naturally yields, and is pulled round by the bobbin before the tension becomes great enough to break the yarn. Fig. 164 shows this clearly: the bobbin A pulls the yarn in the direction of the arrow; if B was a fixture and the rollers did not deliver yarn fast enough, the yarn would break, but the traveller B, being loose on the ring, gives way, and the yarn drags it round. This is an elementary statement of what occurs, but several important factors enter into the question, and we shall now consider the elements of these.

A thorough investigation of the thread and traveller involves an advanced knowledge of mathematics and some knowledge of theoretical mechanics, and since definite statements of conclusions cannot be thoroughly relied upon without the proof that these sciences enable one to bring forward, a certain amount of what follows must be taken upon trust, as it would be outside the object of this book to use and repeat mathematical formulæ which are not familiar to the average reader. The elements that enter into the question we are about to discuss are as follows:—

(1) The counts of the yarn being spun. This has an important influence, because yarn has weight, and in different numbers the weight varies. For instance, one hank of 840 yards of No. 8's weighs 2 oz., while 840 yards of No. 16's weigh 1 oz. This fact may be expressed by saying that the weight of yarn varies inversely as the counts.

(2) Since the yarn between the traveller and the thread guide has weight, and as it revolves at a very high rate of speed (being almost equal to that of the spindle), it has a tendency to fly outwards from the axis around which it revolves. Centrifugal force is the name given to this tendency of a revolving body to fly away from a centre, and a common example suggests itself in a stone attached to a string, which, on being swung round, will, when the string is set free, fly off for some distance. The result of the centrifugal action of the yarn is such that, instead of the yarn passing in a straight line from the thread guide to the traveller, it flies outward and forms a curve. This always occurs in ring spinning, and the name "ballooning" has been given to the bulging thread.

(3) The amount of ballooning will depend to a certain extent on the counts of the yarn being spun.

(4) The reasoning used in No. 2 is equally applicable to the traveller. Its weight and speed cause it to fly outwards, but being prevented from doing so by the ring, it naturally presses against the ring with a certain degree of force, which is dependent upon the centrifugal force and the tension in the thread. The pressure thus set up produces friction, and this has a retarding influence on the traveller's motion round the ring. From this cause winding is the result.

(5) The pull of the thread between the bobbin and the traveller depends upon the diameter of the bobbin, which varies; upon the weight of the traveller; upon the diameter of the ring; and upon the speed of the spindle.

(6) The pull or tension in the thread will modify the friction between the traveller and the ring, and also modify the ballooning.

These are some of the points that will next be investigated, and from the analysis we hope to deduce conclusions upon which are based the modern practice of ring spinning.

Ballooning.—Ballooning, as we have already noted, is the thread flying away from the centre around which it revolves. The degree to which it takes place, of course, depends upon the counts of the yarn, in other words its weight; the speed and the weight of the traveller; atmospheric resistance to the thread has also some influence. Under normal conditions the curve of the yarn in Fig. 165 represents the shape of the ballooning, and in plan view it will be noticed that another curve is produced showing that the atmospheric resistance has caused the thread to vary from the straight line between the thread guide (over the centre of the spindle) and the traveller. If the tension in the thread diminishes to any great extent, the ballooning will collapse and the yarn become entangled round the spindle, simply through the resistance of the atmosphere forcing it on one side. Under even ordinary conditions slight variations of tension occur, and the result is invariably shown in the effect on the shape of the balloon curve where it changes from the full line curve to the double one. A light traveller is usually the cause of a big balloon curve. It is no difficult matter to prove the conditions of ballooning; but it will be sufficient to indicate that, granted we wish always to have the ballooning the same, whatever counts are being spun, the mass of the ballooned yarn multiplied by its velocity squared and divided by the tension of the thread must equal a constant. This may be represented as

$$\frac{m \times v^2}{t} = \text{constant.}$$

From this we can say that for the curve of the balloon to remain the same, the tension of the yarn must be inversely proportional to the counts being spun, and directly proportional to the speed of the traveller; or, since the traveller's speed varies so slightly from the spindle speed, the speed of the latter might be taken as the speed to work from. Explanatory of this statement, we might say that if the counts are changed and the speed remains the same, the tension must be altered by changing the traveller; or if the traveller is not altered, the speed must be changed.

To obtain the direction of the yarn as it enters the traveller is not an easy matter, but for all practical purposes it will be safe to assume that it is at right angles to the portion of thread which leads on to the bobbin. In dealing with the tension of the thread we come to the crux of the whole question, and the traveller plays the most important part in the matter. We have seen that the traveller is a loose piece of metal capable of gliding over the surface of the ring. Directly it begins to move it is affected by two forces, Fig. 166—a tangential force, x, which tends to make it move in a direction tangent to the ring; a centripetal force, y, which tends to draw it towards the centre of the ring. With these two forces acting upon it, the traveller is compelled to move in a circle. Suppose it moves at a rate of 70 feet per minute, and the mass of traveller and yarn equals ·0000125 lb.; the tangential force $x = m \times v$, and the centripetal force

$$y = \frac{m \times v^2}{r}$$

m = mass, v = velocity, of the traveller, and r = the radius of the ring = $1\frac{3}{8}$ inch. By working these formulæ out we shall find that the centripetal force is about 612 times the tangential force.

If we will now understand that the centripetal force is really exercised by the ring forcing the traveller back, we can easily see that what the ring is doing is being equally done by the traveller in trying to get away; in other words, the centrifugal force of the traveller is equal to the centripetal force of the ring, and to this extent the traveller in moving round the ring is pressing against it with a pressure 612 times greater than the force which tends to cause the traveller to fly off at a tangent. Now the tangential force is due to the weight of the traveller and its speed, and we have seen that this force is a mere fraction compared to the centrifugal force, so from this demonstration we can conclude that the momentum of the traveller, due to its being carried round the ring by the pull of the yarn from the bobbin, is so little that we can afford to completely ignore its weight proper, except so far as it influences its centrifugal force. This must be clearly grasped, as on it depends a right conception of the traveller's action. The chief lesson to be derived from it is that the tangential and centrifugal forces have nothing in common; one is a unit, the other is 612. In ring spinning there is never any attempt of one equalling the other; they are so widely separated in their effect that the tangential force fails to have more than ·016 per cent of influence in ring spinning, and it would require the tangential force of 612 travellers to equal the centrifugal force of one traveller; consequently, outside the mere curiosity of knowing and comparing the two forces there is absolutely no necessity for knowing or mentioning the tangential force or pull of the yarn in dragging the weight of the traveller round the ring.

If a frame was made in which the rings, travellers, and speeds were proportioned so that the tangential force

$$mv = \frac{m \times v^2}{r} \text{ the centrifugal force,}$$

then the radius of the ring would have a dimension in feet equal to the velocity of the traveller in feet per second. Nothing can be more absolutely absurd than this result. The centrifugal force must be and always is, to the extent of over 600 per cent, in the ascendant. Moreover, travellers are so graded in their weight for different counts of yarn, that if the pull of the yarn reduces the centrifugal force below a certain proportion the yarn will break immediately. There is, however, another very important point to consider, namely, the effect this centrifugal force of the traveller has in interfering with its movement round the ring when it is acted upon by the thread from the bobbin—or in other words, the effect the centrifugal force of the traveller has on the tension of the thread. This is a very important point, so we will consider it carefully.

It must be fully realised that the centrifugal force of the traveller, or its pressure against the ring, due to its momentum, is the chief factor to guide us; it is equal, in the example previously given, to a weight of about $2\frac{1}{2}$ oz. resting on the ring. (In passing, it may be observed again that the weight proper of the traveller is so small, compared with this weight due to the centrifugal force, that it may with safety be ignored.) Now this $2\frac{1}{2}$ oz. is pulled round by the thread, and it is the act of pulling it round that causes the thread to be in tension. To find the tension, we must know the coefficient of friction between the traveller and the ring; the ordinary coefficient of friction of polished steel and steel is not applicable to this case. Experimenters have found that it has a wide variation, and depends on such factors as the speed of spindle, diameter of bobbin, diameter of ring, and the dryness or otherwise of the surface of the ring. Professor Escher, of Zurich, found that if

	Oiled ring.	Dry ring.
the bobbin was $\frac{5}{8}$ inch diameter the coefficient of friction was	0·27	0·465
if the bobbin was $1\frac{5}{8}$ inch diameter the coefficient of friction was	0·13	0·272

From his experiments he concluded that the tension will vary the least, the greater the coefficient is between the ring and the traveller; in other words, we might say that the less difference there is between the extreme diameters of the bobbin, the more uniform will the tension be in the yarn. Professor Lüdicke, of Brunswick, found similar variations, and as an example of his researches we give the results of experiments.

Coefficient of frictions for 5000, 6000, 7000, 8000 revolutions
=0·4093, 0·3506, 0·2999, 0·252.

He deduces a convenient rule from his investigations, as follows:—

Coefficient of friction $= 0{\cdot}65 - 0{\cdot}00005 \times$ revolutions of traveller.

It will be noticed that the variable coefficient due to changes in the diameter of the bobbin is ignored in this empirical rule, and it is due to Professor Escher that we can now with certainty rely upon the fact that such variation does exist.

Mr. Bourcart, in a small pamphlet issued some years ago, used for convenience the fraction $\frac{1}{7}$ as the proportion of a weight required to move it round a ring. This of course is much too little, the average being nearer $\frac{1}{3}$ than $\frac{1}{7}$. By taking $\frac{1}{3}$ for our basis as the coefficient, we find that $\frac{1}{3}$ of 2·5 oz. = ·83 oz. will be the tension in the thread required to pull the traveller round. It must be remembered, however, that this tension must be exerted at a tangent to the ring as at T, Fig. 166. If the direction of the pull T varies as at B its force must be increased, because, in addition to overcoming the friction of the traveller, we are now trying to pull the traveller away from the ring, and therefore some of the centrifugal force exists in the thread as tension. This would increase as the pull changes, until, when the direction becomes as at A, the traveller ceases to press against the ring, the whole of its centrifugal force is exerted on the thread, and so the thread would have a tension equal to this force. At the same time that this oblique pull of the thread is taking place another set of conditions exist also, due to the inclination of the pull. We know that if the traveller A, Fig. 166, is pressing against the ring, no amount of pulling in the direction of AC will cause it to move; we also know that the least effort to move A will be along the tangent AT; between these two lines a direction can be found along which, if a pull is exercised, as at AB, the traveller will begin to move. Any pull exercised within the angle BAT will move the traveller, and the amount of the force will become less as the direction of it approaches the line AT. On the other hand, no movement of the traveller can possibly take place if the pull is exercised in a direction that falls within the angle BAC. From this fact we can fix a limit to the diameter of the bare bobbin used on the ring frame, provided we know the diameter of the ring and the coefficient of friction between the ring and the traveller. Assuming the coefficient of friction to be $\frac{1}{3}$, Fig. 167 will give us the size of bare bobbin, while for a weft frame Fig. 168 will represent the conditions. From these diagrams we learn that the smaller the fraction representing the coefficient of friction, the smaller the bare bobbin can be, while if we wish to reduce the size of the bare bobbin, as in the weft frame, the diameter of ring must be reduced.

It has already been shown that there are two forces affecting the traveller that have their origin in the mere fact of its revolution. It has also been shown that the centrifugal force is modified or reduced by the pull of the thread from the bobbin, such thread taking up the force that the traveller loses. A tension therefore exists in the thread, and it is the difference between the tension in the yarn and the remaining centrifugal force of the traveller that regulates the winding. For winding to take place at all, the yarn to the bobbin must always pull against a stronger force than that represented by the tension of the yarn. For instance, if the tension equalled the centrifugal force, the traveller would be in a balanced condition, and it would cease to press against the ring. In such a case it would be carried round as if it were rigidly connected to the spindle, and while twists would be put in, no winding would be taking place. Ordinary observation confirms this remark; for if too light a traveller be used, the balloon flies out directly, because of the slackness resulting from

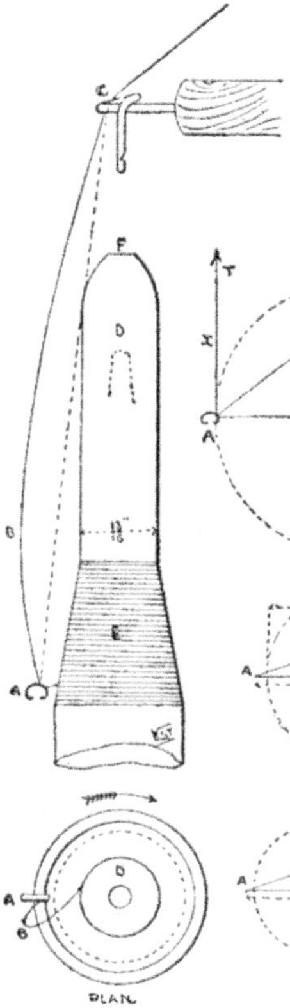

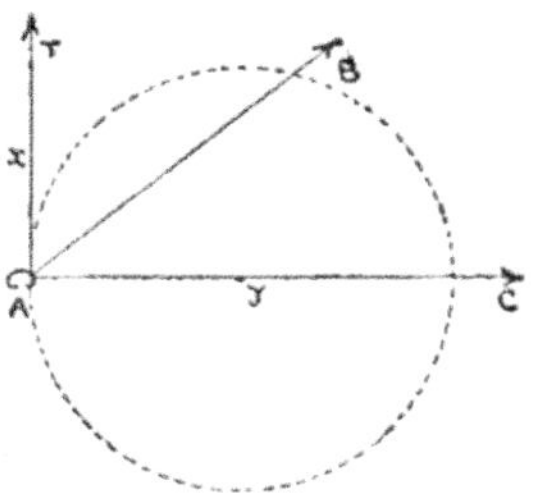

Fig. 166.

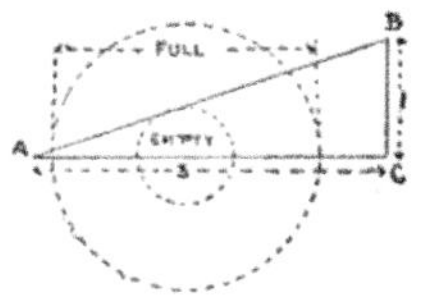

Fig. 167.

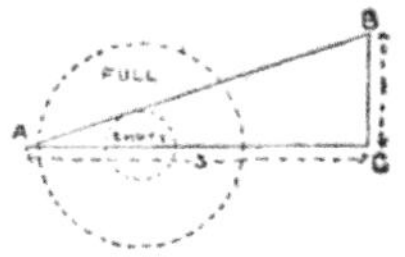

Fig. 168.

Fig. 165.

insufficient winding, due to insufficiency of the centrifugal force of the traveller. In many cases a traveller is used that has only the slightest excess of centrifugal force over the tension; when such conditions exist and circumstances happen that momentarily make the tension equal to the centrifugal force, unsteady ballooning occurs. On the other hand, when too heavy a traveller is used, the centrifugal force is so high that the tension in the thread, in its efforts to move the traveller, becomes so great that all signs of ballooning disappear, and if the tension necessary to do this be equal to or greater than the strength of the yarn, the end breaks.

On examining a traveller that has been working on any ring frame under normal conditions, it will be found that it is worn on that point which touches the inside of the ring; and travellers will last as long as this point resists being worn away by friction. Any traveller that will develop sufficient centrifugal force to cause winding will show signs of wear only on the point which is in contact with the inside of the ring. An interesting experiment will make this clear. Conditions illustrated in Fig. 165 existed on an ordinary frame. Counts 28's were being spun from single roving, and a No. 5's traveller was being used. It was found that all numbers of travellers from 1 to 10 would cause winding, and, moreover, that each new traveller after it had run a short time became worn. An hour's running in all the tests was sufficient to prove the point, but in the heavier travellers a quarter of an hour's spinning showed a comparatively large amount of wear due to the friction on the ring. If rings are somewhat soft, or are irregularly case-hardened, they will also be easily worn out of shape, and for that reason both rings and travellers are made of the best material to resist frictional wear.

From the fact of the centrifugal force regulating the tension in the yarn, several statements might be formulated; for instance, the tension is as the square of the speed of the traveller and in direct proportion to the weight of the traveller and the diameter of the ring. The rule for centrifugal force, namely—

$$\frac{\text{mass of traveller} \times \text{velocity of traveller}^2}{\text{radius of the ring}}$$

will be an obvious proof of the statement. Briefly, the rule means that for a given weight of traveller, if the velocity of the spindle be doubled, the tension in the yarn will be "increased" four-fold; or *vice versâ*, if the tension is to remain the same, after doubling the speed, the weight of the traveller must be "reduced" four-fold. Again, if it be wished to double the tension in the yarn without altering the speed of spindle, the weight of the traveller must also be doubled. Or, if the diameter of the ring be increased and the speed of the spindle and weight of traveller be kept the same, the tension in the yarn will be increased in the same proportion. By similar reasoning we may conclude that the weight of the traveller will vary in direct proportion to the size of the ring. Supposing the weight of the traveller be 1, the velocity of the traveller 2, and the diameter of the ring 2, then

$$\frac{mv^2}{r} = \frac{m \times 2^2}{1} = 4m$$

If, now, the ring be doubled in diameter, the formula would work out

$$\frac{mv^2}{r} = \frac{m \times 4^2}{2} = \frac{m \times 16}{2} = 8m,$$

so that for double the ring we must have double the weight of traveller.

Weight of Travellers.—Some interest attaches to the method adopted in grading the travellers as to their weight and their suitability for spinning certain counts of yarn. Generally speaking, a mill keeps the speed of spindle and diameter of ring the same for a range of numbers spun; the weight of the traveller is therefore altered to suit the changed condition of the counts. The question now is—

How must the weight of the traveller vary as the counts vary? If the "weight" of yarn is taken as a basis, we shall find that 20's yarn is half the weight of 10's, 30's yarn is one-third the weight of 10's, 40's is quarter the weight of 10's, and so on. The following table will present a short series of numbers:—

Counts.		Proportionate increase in weight.		Counts.		Proportionate increase in weight.	
40	is	$\frac{1}{39}$ lighter than	39's	29	is	$\frac{1}{28}$ lighter than	28's
39	,,	$\frac{1}{38}$,,	38's	28	,,	$\frac{1}{27}$,,	27's
38	,,	$\frac{1}{37}$,,	37's	27	,,	$\frac{1}{26}$,,	26's
37	,,	$\frac{1}{36}$,,	36's	26	,,	$\frac{1}{25}$,,	25's
36	,,	$\frac{1}{35}$,,	35's	25	,,	$\frac{1}{24}$,,	24's
35	,,	$\frac{1}{34}$,,	34's	24	,,	$\frac{1}{23}$,,	23's
34	,,	$\frac{1}{33}$,,	33's	23	,,	$\frac{1}{22}$,,	22's
33	,,	$\frac{1}{32}$,,	32's	22	,,	$\frac{1}{21}$,,	21's
32	,,	$\frac{1}{31}$,,	31's	21	,,	$\frac{1}{20}$,,	20's
31	,,	$\frac{1}{30}$,,	30's	20	,,	$\frac{1}{19}$,,	19's
30	,,	$\frac{1}{29}$,,	29's				

From this table we should conclude that the weights of the travellers must vary in the same proportion as the weight of the yarn varies. On the other hand, we might assume that the tension in any yarn is always a fixed proportion of the breaking weight of that yarn. For instance, suppose the tension in the yarn when spinning 40's is one-fifth of the breaking weight of 40's, then there would be reason in assuming that the tension in 20's ought to be one-fifth the breaking weight of 20's. Taking the breaking weight of yarn as published by Messrs. G. Draper and Co., of Hopedale, Mass., U.S.A., as a basis, we get the following:—

Counts.	Breaking weight.	Proportion stronger.	Counts.	Breaking weight.	Proportion stronger.
20	88·3	$\frac{10}{198}$	26	66·3	$\frac{10}{245}$
21	83·8	$\frac{10}{200}$	27	63·6	$\frac{10}{271}$
22	79·7	$\frac{10}{210}$	28	61·3	$\frac{10}{290}$
23	75·9	$\frac{10}{217}$	29	59·2	$\frac{10}{310}$
24	72·4	$\frac{10}{226}$	30	57·3	...
25	69·2	$\frac{10}{238}$	...	...	...

From this table it appears that the breaking weight of the

yarn varies in an "increasing" proportion as the counts get lower. The proportionate increase does not vary so regularly as the weight of the yarn varies; but taking into account the fact that the above breaking weights are from actual tests, the approximate result is near enough to give the assumption strength that the variation in the weight of the travellers might reasonably follow in a gradually increasing proportion as the counts vary. As a matter of fact, the United States and the Scotch standards do vary in the proportion the above reasoning suggests. Although makers of travellers are reluctant to impart information as to the weight of travellers, it is an easy matter to weigh a number of each, say 100, and form a table of the result. Very exact weighings are necessary, but a general idea may be obtained from the few following weights:—

Traveller No.	Weight per 100.	Traveller No.	Weight per 100.	Traveller No.	Weight per 100.
8	200 grs.	3	120 grs.	3/0	60 grs.
7	180 ,,	2	110 ,,	4/0	55 ,
6	160 ,,	1	90 ,,	5/0	50 ,,
5	140 ,,	1/0	80 ,,	6/0	45 ,,
4	130 ,,	2/0	70 ,,	...	...

The system adopted in this table may be seen at a glance, and one can readily understand that as the counts go lower the difference between the weights of each grade can be made greater, just as it can be made less in the higher counts.

In dealing with the amount of twist put into the yarn by the traveller, a little repetition may be necessary. Twist is the result of the traveller lagging behind the spindle. This lagging is due to the friction set up between the traveller and the ring as a consequence of the centrifugal force of the former. While the centrifugal force is practically the same throughout the building of the bobbin, the friction may vary because the coefficient of friction varies, and (as already shown) this has some influence on the lagging, because from this alone it is more difficult to move the traveller when winding is taking place at the nose of the cop than when the yarn is wound on the base. In addition to this, the tension in the yarn required to overcome sufficient of the friction to cause movement is greater at the nose than at the base, because on the nose the yarn is pulling very obliquely to the movement of the traveller. From these two causes, therefore, we conclude that the greatest tension in the yarn exists when winding on the smallest diameter is taking place, and the least tension when winding on the base. Now it must be remembered that whatever the tension may be, and no matter how it varies under normal conditions, it is always simply equal to a part of the pressure of the traveller against the ring; the centrifugal force is always there, though reduced by the pull of the yarn. The direct consequence of this is, that if no delivery of yarn was made, the traveller would be carried round at the same speed as the spindle, whether the yarn was attached to the smallest or the largest diameter; but the tension is greater in the former than in the latter case. On the other hand, if yarn be delivered, the tension will be reduced; consequently the pressure of the traveller against the ring will be increased, and naturally a lagging behind the traveller will be the result until the tension is restored. Continued delivery prevents its restoration, so there is always a lagging behind, as far as the smallest diameter is concerned. As the largest diameter is approached, a uniform continuance of the delivery also relieves the tension at this point; but the addition thus made to the centrifugal force by reducing the tension is a less proportion to the total force than it was at the nose of the cop, and therefore the lagging behind is less at the base than at the nose; in other words, the traveller revolves more quickly when winding on a large diameter than on a small one, and from this we deduce the fact that more twists are put in the yarn as the winding takes place from the nose to the base of the cop. A previous statement, which was used to show that there is no necessity for more than the slightest variation in the lift in order to compensate for the building of a conical cop, might have prepared the ground for this; but another similar example will readily prove it. Let us suppose 530 inches of yarn are delivered per minute, and that the spindles make 9500 revolutions per minute; the traveller must lag behind the bare bobbin of $\frac{3}{4}$-inch diameter 225 revolutions, and behind the full bobbin of $1\frac{3}{8}$-inch diameter 122 revolutions. The speed of the traveller at these two points would therefore be 9275 and 9378 revolutions—a difference of only a fraction above 1 per cent. A difference such as this—indeed if it were much higher—would be impossible to discover; so that from a practical point of view the ring frame is free from any tendency to produce irregularly-twisted yarn as far as its own twisting action is concerned. Irregularly-twisted yarn will naturally exist in ring yarn, just as it does in the mule yarn, owing to the character of the cotton and its previous preparation. Differently coloured rovings spun on the mule and the ring frame will show a remarkable similarity in the irregularity of the twists, which are thrown out very clearly by the contrast of colour.

In regard to the question as to what number of a traveller must be used for any given counts, speed of spindle, size of ring, etc., no definite answer can be given beyond one that depends upon an accumulated mass of practical experience; and even then, local circumstances introduce an element of judgment that compels the use of a traveller which varies from the standard of general experience. The following tables will convey some idea of the general practice, as determined mainly from experience. They are given as a guide only; each user must judge for himself as to how far other conditions necessitate variations from this table.

Counts.	$1\frac{1}{2}$ in. Ring.	$1\frac{5}{8}$ in. Ring.	$1\frac{3}{4}$ in. Ring	Counts.	$1\frac{1}{2}$ in. Ring.	$1\frac{5}{8}$ in. Ring.
4	14's	13's	12's	32	4/0's	5/0's
6	12's	11's	10's	34	5/0's	6/0's
8	10's	9's	8's	36	6/0's	7/0's
10	8's	7's	6's	38	7/0's	8/0's
12	7's	6's	5's	40	8/0's	9/0's
14	6's	5's	4's	42	9/0's	10/0's
16	5's	4's	3's	44	10/0's	11/0's
18	4's	3's	2's	46	11/0's	12/0's
20	3's	2's	1's	48	12/0's	13/0's
22	2's	1's	1/0's	50	13/0's	14/0's
24	1's	1/0's	2/0's	52	14/0's	15/0's
26	1/0's	2/0's	3/0's	54	15/0's	16/0's
28	2/0's	3/0's	4/0's	56	16/0's	17/0's
30	3/0's	3/0's	5/0's	58	17/0's	18/0's
				60	18/0's	19/0's

Note.—The above table is given as a guide to select travellers required, but will, of course, vary according to circumstances.

Travellers of from four to six numbers heavier than stated above are generally required for spinning Egyptian or Sea Islands cotton.

It will be noticed from the above table that as the ring increases in diameter, the weight of the traveller decreases.

The space of spindle and diameter of rings for various counts may be gathered from the following lists:—

			With Ballooning Plates.		
Counts.	Space of Spindles.	Dia. of Rings.	Counts.	Space of Spindles.	Dia. of Rings.
4's to 20's	$2\frac{3}{4}$ in.	$1\frac{3}{4}$ in.	4's to 20's	$2\frac{5}{8}$ in.	$1\frac{3}{4}$ in.
20's to 40's	$2\frac{5}{8}$ in.	$1\frac{5}{8}$ in.	20's to 40's	$2\frac{1}{2}$ in.	$1\frac{5}{8}$ in.
40's upwards	$2\frac{1}{2}$ in.	$1\frac{1}{2}$ in.	40's upwards	$2\frac{1}{4}$ in.	$1\frac{1}{2}$ in.
Weft	$2\frac{1}{4}$ in.	$1\frac{3}{16}$ to $1\frac{1}{4}$ in.			

The traveller tables of well-known makers follow very closely on the tables just given.

The Spindle.—Another important subject to which some space will now be devoted is that of the spindle. A remarkable series of developments have taken place in this feature since the traveller system of spinning was introduced. The root idea of the chief improvements has been to make the spindle work satisfactorily at a high rate of speed with a minimum of power to drive it. It will be interesting to trace out the conditions of work to which the ring frame spindle has had to be adapted before it reached the present type of which Fig. 169 represents an example.

The older form of spindle used on a throstle or flyer spinning-frame was essentially an upright spindle, supported in two bearings, one at the bottom called a "footstep" and the other higher up the spindle, and as near as convenient to the bobbin, such support being called the "bolster bearing." The position of the wharve or driving point on the spindle was generally between the two bearings, but placed much nearer the top support than the bottom one. The mule spindle affords a good example of an arrangement of this kind, and in that machine we see the perfection to which such a system of driving has been brought, and the spindle made suitable for revolving at very high speeds. The conditions of working are, however, different in the mule and the ring frame. In the former a plain spindle is used, upon which the yarn is wound directly; it is not subject to the same tension on the yarn combined with a high speed as in the ring frame, nor is it surmounted by a heavy bobbin whose tendency is to become untrue and out of balance. This latter factor becomes of great importance when the above conditions exist on a spindle running at a very high speed, and it was soon recognised that some improvement of the well-known type was absolutely necessary when greater production was required from the machine. The chief objection to be overcome was, of course, the excessive vibration set up in the spindle when running at a high rate of revolution, which was caused by the spindle or bobbin being out of balance. A low speed does not disclose this vibration to the same degree as a high speed; a bobbin and spindle slightly out of truth might not, at 4500 revolutions per minute, prove very inconvenient, or even show itself clearly; but if the speed were *doubled* to 9000 revolutions per minute, the irregularity of balance would have a *four-fold* tendency to make itself felt, and it shows itself by setting up vibrations in the spindle.

The first attempts at a remedy were made by Rabbeth, in a spindle which dispensed with the lower footstep bearing as such. He extended the bolster bearing in the form of a long tube firmly fixed to the rail, and at the top and bottom of this were bearings for the spindle. Above all was placed the bobbin. This arrangement, it will be seen, was only a slight move in the right direction, but it contained two features that formed the basis of the spindle of to-day, namely, a self-contained spindle, and greatly improved means of lubrication. The next move was made by Sawyer, Fig. 171, who recognised that a greater steadiness of running would be assured if the upper bearing could be raised. He effected this by extending the bush of the bolster bearing, and over this he placed the bobbin. By this means the bolster bearing was placed within the bobbin, and to this extent a decided advantage accrued. He was compelled, however, to still use the lower separate footstep bearing, and to place his wharve between it and the upper support. The early Rabbeth and the Sawyer spindles both contained the elements of a successful spindle to fulfil the requirements of that time, so that

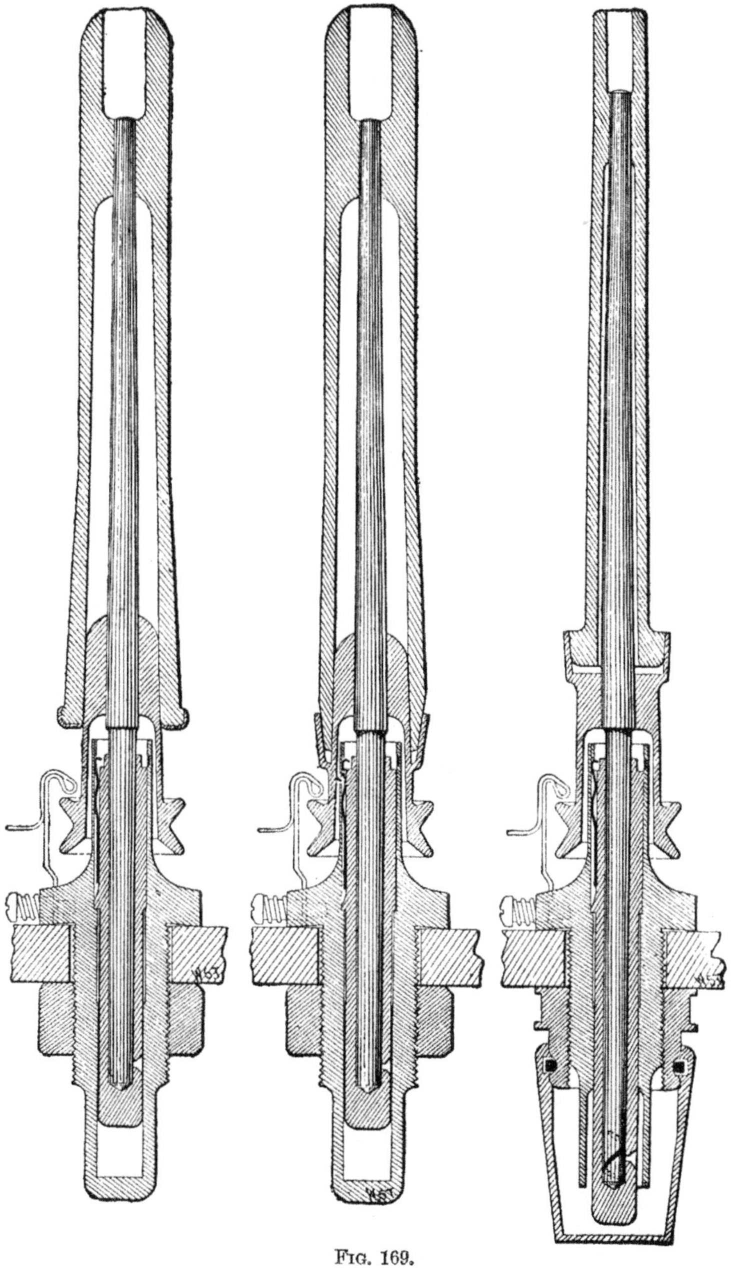

FIG. 169.

eventually both were combined—and in the result the well-known "Rabbeth" spindle was evolved. The improvements all contributed to greatly increased speeds and steadiness in running, and gave the ring frame the opportunity to compete to some extent with other spinning machines.

On reference to Fig. 170 the characteristic features of the Rabbeth spindle will be observed. The steel spindle A is carried by a base or bolster B, which is firmly fastened to the spindle rail C by means of nuts D. The upper portion of the bolster extends to F, where it is bored out to fit the spindle; the lower portion at M is also bored out to fit A. Between these two points the bolster is barrelled or recessed out, so that the open space thus formed serves as a receptacle for oil. The upper bearing at F is usually fitted with a thin bush of some anti-friction metal. A portion of a coarse spiral is left between the ends of the sheet of metal which forms the bush, and by this the oil, if it reaches this part of the spindle, is distributed over the surface of F.

Immediately above the bearing at F, a sleeve G is tightly fitted over the spindle, and is continued in a downward direction to form the wharve H. The position of the wharve is designed so that the top and bottom bearings each bears its share of the strain. It will be noticed subsequently, that in this respect the pull of the spindle band in recent spindles is exercised almost entirely upon the upper bearing. As a rule a brass cup is fitted over the outside of the wharve sleeve at J, which serves for the reception of the lower end of the bobbin K. A loose fit of the bobbin is generally allowed at this point, for a purpose to be explained shortly; but the upper end of the bobbin is made to fit the spindle tightly.

The Booth-Sawyer spindle had a very extensive vogue,

and found great favour with the users of ring frames;

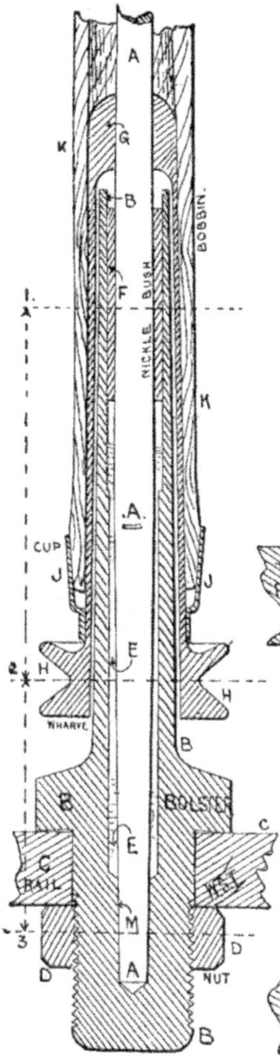

RABBETH

FIG. 170.

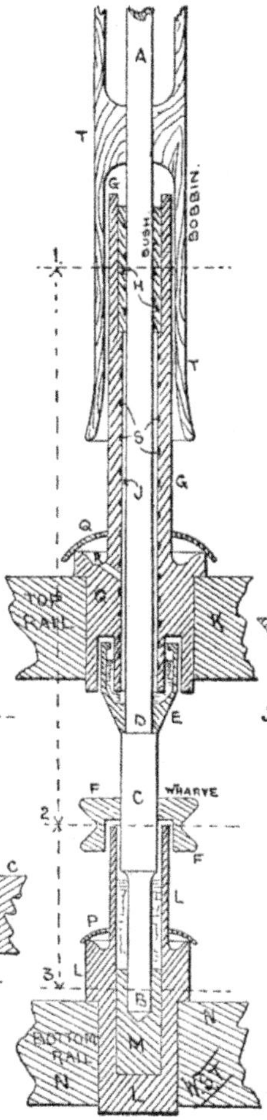

BOOTH-SAWYER.

FIG. 171.

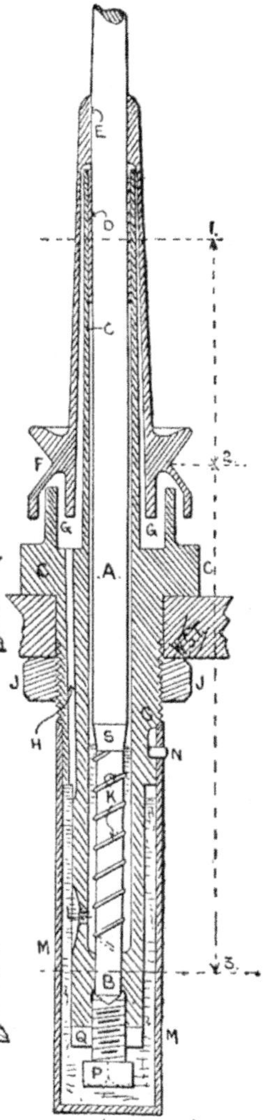

DOBSON-MARSH.

FIG. 172.

but its introduction inaugurated a series of improvements that culminated in the spindle known as the Rabbeth, just described. Its chief characteristics may be summarised as follows: it is self-contained; it has a reservoir or bath of oil in which the spindle works; its upper bearing is within the bobbin; and the pull of the band takes place somewhere between the upper and lower bearings. In practically all modern spindles the two last-named features of the Rabbeth are entirely absent; but the self-contained character and the oiling arrangement is such a basic feature that some authorities classify most recent spindles as being Rabbeth in principle.

The Rabbeth spindle underwent a variety of alterations and improvements, chiefly with the idea of improving the lubrication. The reservoir of oil was very effective in lubricating the footstep bearing, but the upper bearing had to trust to capillary attraction for its oil. The metal bush had no power to take the oil in an upward direction, because the surface of the oil was kept too low for that purpose; and, moreover, if through carelessness too much oil was placed in the spindle, it quickly rose to the top, ran over the bolster, and was dissipated by the wharve, or it ran down and spread over the rail.

A decided improvement was effected when an attempt was made to cause some kind of circulation of the oil within the spindle, whereby the upper bearing might be kept constantly oiled. Another fault showed itself in the fact that when the spindles required re-oiling the old dirty oil had to be pumped out, and indifference in doing this showed itself in accumulations of dirt and gummed oil, which largely increased the power necessary to drive the machine. This defect was also remedied; and in Fig. 172 a Dobson-Marsh spindle is shown in section, which presents an extensively used method of overcoming the objections mentioned. The lower end of the bolster is pierced at a point near to the bottom end of the spindle. Over the bolster is placed, by a detachable bayonet or other means, a cup, carrying a large quantity of oil; on the spindle is placed a spiral of wire, which revolves and forces the oil upwards to the top bearing, and so keeps it constantly lubricated; any oil carried over the top runs down, and, by means of the passage shown, flows back into the cup. When new oil was required after working a month or two, the cup was simply detached without stopping the spindles; and the old oil was poured out, fresh oil supplied, and the cup hooked on or screwed into place again. Another improvement, copied from a still earlier spindle, was incorporated, namely, the set screw P on which the end of the spindle blade rested; as the spindle wore at this point, the screw could be moved upwards to compensate for the wear; the lock-nut Q effectively kept it in position.

We have shown how a spindle revolving at a high speed is subject to strains through being out of balance, and how these strains are augmented through the uncertainty of the bobbin and cop maintaining themselves true. The demand for increased speeds brought about, as indicated, better spindles, in the Sawyer, the Rabbeth, the Dobson-Marsh, and other improved forms. These spindles for a long time served their purpose, but new conditions of speed, etc., began to show weaknesses in their construction, and inventors were thus led on to make further improvements. Eventually a spindle was devised which solved the problem so far as principle was concerned, and the "gravity" or "top" spindle was introduced. Such spindles now assume innumerable forms of construction in details; but, the purpose being the same, a few words of explanation will not be out of place.

If a spindle is perfectly balanced and revolves at a high speed, well supported in bearings, its axis will permanently occupy one position, and any definite strain put upon it will always act in one direction; and there would be little if any vibration set up in such a spindle. If, on the other hand, a spindle is out of balance, *i.e.* heavier on one side of its centre than on the other, there would be two sets of forces at work, and the strain would not be acting equally around the axis of the spindle. Such a condition as this, in which two opposing forces are at work, interferes with the uniform motion of the spindle round its axis, and there is a constant struggle going on to revolve round an axis which would be common to the two unequal sides. Vibration is set up as a consequence of rigid bearings, and, together with considerable wear and tear, the spinning operation is performed under disadvantageous circumstances. The object of the improvement was to arrange the spindle so that it could revolve round its own axis and also round the real centre of its movement. A spinning-top is sometimes used to illustrate this principle, and from the example the new spindle was formerly referred to as a "top" spindle. "Gravity" spindle was a name also applied. Either name, however, is only partially correct, and while a top may enable some idea to be obtained of the principle involved, the word "gravity" is entirely a misnomer. If a perfectly balanced top be set spinning at a high speed, it will revolve with its axis vertical; but if it be moved out of that position, it will continue to revolve round its own axis and at the same time revolve in a circular path forming the outline of a cone whose apex is the point where it touches the ground. Now this is not what occurs with a spindle: an unbalanced top would represent the action much better. In such a case the top could not revolve with its axis vertical; it would certainly revolve round its axis, but at the same time its free position would permit its axis to become inclined and a bodily movement to take place round the axis of a cone whose apex would be some distance in the ground. It will be readily seen that although this example is a better illustration than a balanced top, still it does not approach the actual conditions of a spindle; in the top, the axis is not supported in any way, while in a spindle we are compelled to have such support.

If a bar of iron be taken, and a pound weight placed on one end and an ounce on the other end, the middle point of the bar is clearly not the centre round which the bar could be set revolving; neither would the bar revolve at a high speed if the point were taken, on which the bar and weights would be balanced. We require to know such a point in the bar that the energy developed by each weight would equal one another. This point is known as the "centre of gyration," and in the case of an unbalanced spindle it is this centre or axis round which the spindle must be capable of revolving at the same time as it revolves round its own axis. For most practical purposes, pulleys, etc., are balanced on the principle of making their "centre of gravity" correspond with the centre of their rotation, but in important organs great care is taken to balance them so that the "centre of gyration" corresponds to the axis of the shaft on which they revolve. In a spindle this cannot be done, so the balancing effect is obtained by leaving sufficient room in their bearings to permit them to occupy and revolve round their natural centres. Spindles constructed on the above-mentioned principle are termed "elastic" or "flexible" spindles.

Fig. 173 represents five of the prevailing types of flexible spindles used by machine-makers in this country. They are all self-contained, on the principle of the Rabbeth. Referring to A, it will be observed that the spindle D is fitted with a bolster or pillar E; this bolster is itself fitted within the pillar F, which is firmly bolted to the spindle rail. The inside bolster E is only permitted to fit the fixed pillar F at its upper end, and even then the fit is very easy; the length of the bearing is shown as from B to C. The lower end, it will be noticed, is quite free from contact with the lower part of F, so that if the upper part of the spindle and the bobbin fitted over it become unbalanced, it may be deflected from an exact vertical line, and revolve, as already pointed out, round its centre of gyration. Every possible precaution is taken to prevent or eliminate the tendency in the spindle and bobbin to become unbalanced, and consequently never more than the slightest tendency makes itself evident, even with the highest speeds. Allowance need only be made, therefore, to a limited extent, and this accounts for the small clearances shown in the drawings.

It is quite evident that the pull of the band must be exercised on some part of the upper bearing, and this is marked clearly in the sketches, B and C representing the top and bottom of the bearing, while A is the centre of the wharve. In this connection it may also be remarked that it is advisable to arrange the pull so that it may be somewhere near the centre of the bearing.

A variety of means are adopted for keeping the inside bolster in position and preventing it from revolving. Pins and slots, screw caps, and spring catches are the usual methods; in the example B, a square end is provided on the bottom of the bolster, which rests within a corresponding but slightly larger hole in the outside pillar.

There is one great inconvenience associated with the

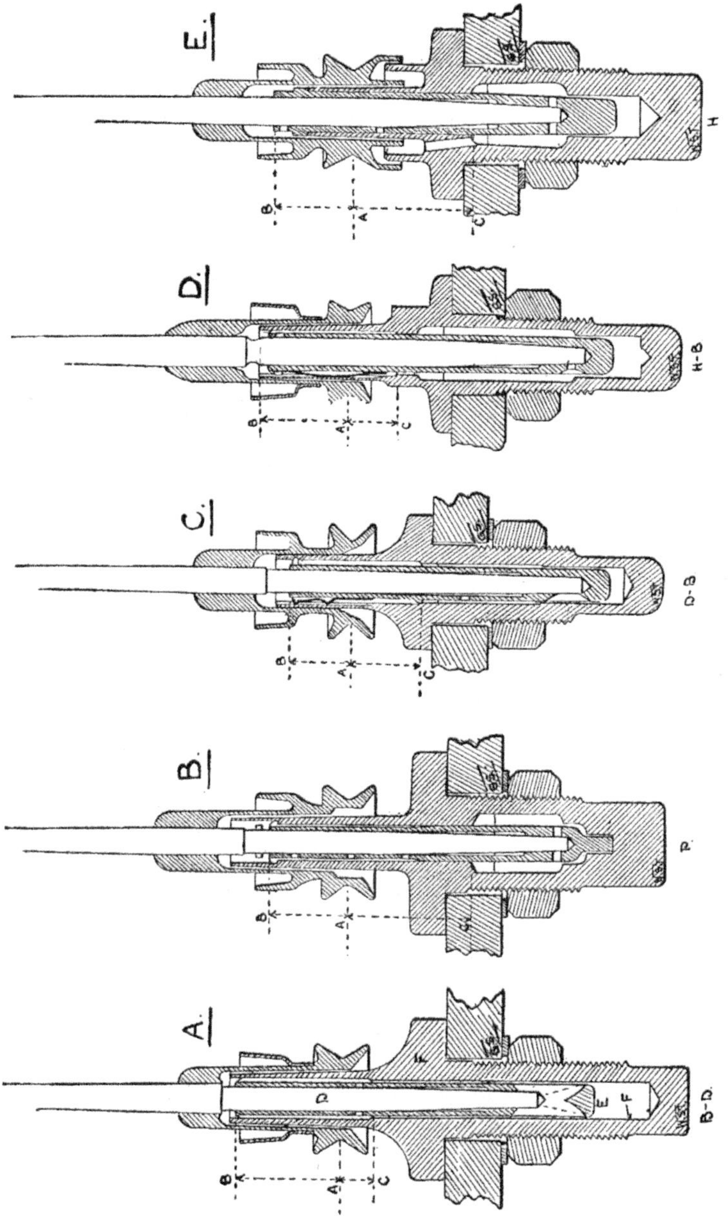

FIG. 173.

spindles just illustrated: that is the method of renewing the oil. The machine must be stopped, all bands be taken off, the spindles taken out, and a pump used to extract the dirty oil; then the whole operation must be performed again in the reverse order. All this means a waste of time, as well as offering an opportunity for carelessness to show itself. This objection was overcome by making the lower end of the outside pillar open, and attaching thereto a cup containing oil.

The Dobson-Marsh spindle illustrated this method, but as modified, in its present construction, the oil cup is attached to the pillar by means of a spring ring fitting within a recess. The oil is circulated the full length of the spindle blade by means of a spiral cut on the lower end of the spindle, and it returns to the cup by grooves cut in the inside of the outside pillar; these can easily be traced in the drawing. Re-oiling can be performed without stopping the machine or touching the bands, all that is necessary being to take the cups off, empty the old oil out, put in the new, and replace the cups. The work is done quickly, and saves a deal of time. An arrangement of this kind has such decided advantages that the method has been applied, with slight variations in detail, to several makes of spindles, one of which is represented in Fig. 174. The cup C in this case is fastened on the inside of the pillar by means of a quick-threaded screw, as marked at B. The lower end C is made square, so that a key can be used to fix it firmly in position and make it perfectly tight. Similar inside cups have been used for some time by attaching them to the pillar in various ways, such as by means of clip rings, hook-and-slot, and bayonet joints.

On reference to Figs. 175, 176, another improvement will be noticed. In order to prevent the spindle from lifting up from its position, a catch is so arranged over the wharve that such an action is impossible; before the spindle can be moved, the catch must be moved on one side, and it is moreover necessary that the catch be so constructed that on the replacement of the spindle it will permit the spindle to fall with certainty to its place. Fig. 169 and also Figs. 177 and 178 illustrate similar catches. The lid shown in the oil cup

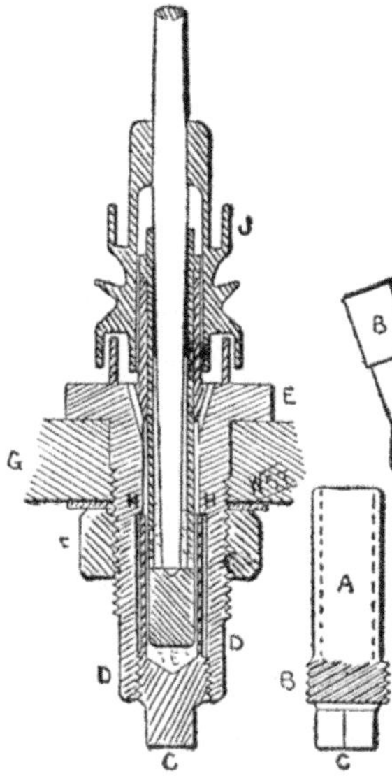

FIG. 174.

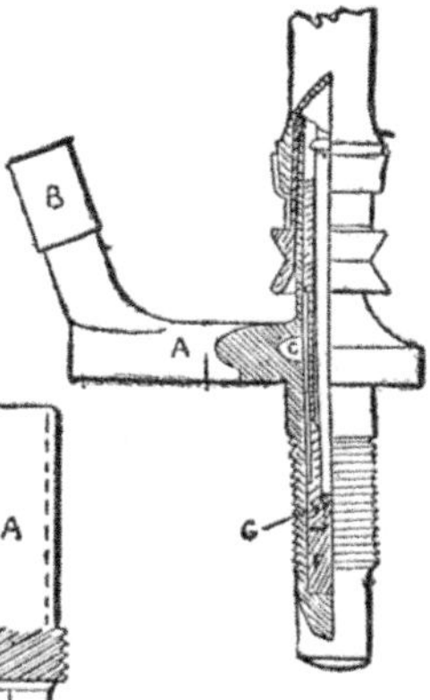

FIG. 175.

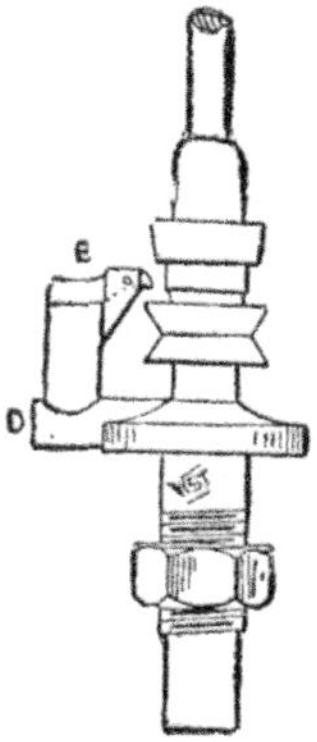

FIG. 176.

in Fig. 176 is to permit oil to be supplied to the spindle to compensate for any evaporation that may take place.

It will readily be understood that the best-made spindles will wear very little indeed during the course of years, and if the oil be of the best quality (as it ought to be) there will be no gumming nor will it become very dirty, and evaporation therefore will reduce it in quantity only. The passage to the oil cup is thus a decided convenience to those who use the best lubricating oil obtainable.

Fig. 175 illustrates a method of obtaining the same effect on a self-contained spindle. An extension is made to the outside pillar at A, and it is bored out for the passage of the oil. A cap B prevents the entrance of fly, dirt, etc. An improvement on this is shown in Fig. 176, wherein the cap is replaced by a hinged lid D, so arranged that it serves also the purpose of a catch to prevent the lifting of the spindle. An indispensable adjunct to the ring rail is to be noticed in what is called the "traveller clearer." While the machine is working, a good deal of dirt and fine fibre is always flying about, which settles upon the frame, and some of it naturally rests upon the ring itself. In course of time accumulations would occur, which would interfere with the action of the traveller by clogging its action. A small projection is therefore placed on the ring plate by screwing or other convenient means, in such a position that the traveller in its revolution just misses it. In consequence of this any fibres adhering to the traveller are caught up by the projection, and the traveller passes on cleared of its encumbrances. A catch of this kind will be noticed in Fig. 156 at D.

THE BALLOONING EFFECT has already been explained; it only remains to point out that under some conditions it has a tendency to cause the space between the spindles to be greater than is desirable in order to avoid the adjacent threads coming into contact with one another. To prevent this, balloon plates, or, as some species of them are called, "separators," are adopted. Such appliances are only really necessary during the formation of the lower part of the bobbin, so that when this stage is passed they are generally arranged to be automatically moved out of the way. Different ways for doing this have been introduced, but in essentials they consist of the introduction of projecting

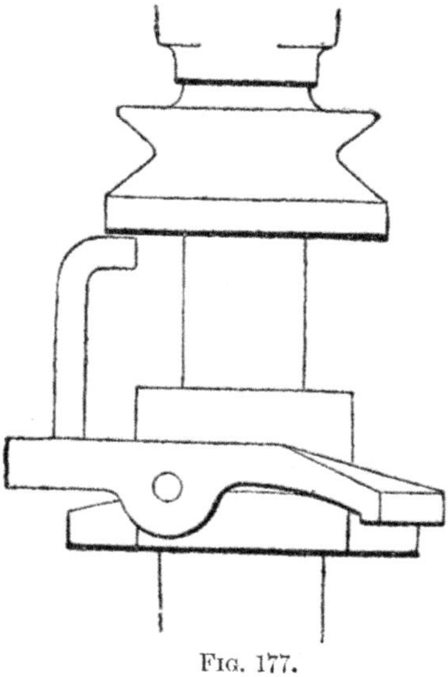

Fig. 177.

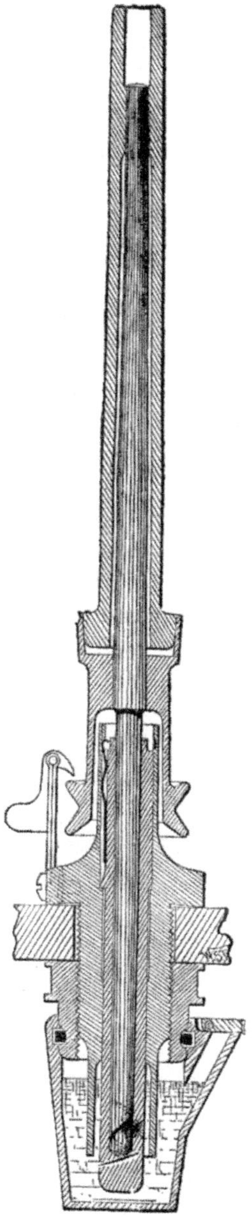

Fig. 178.

pieces of metal between the spindles, so that the bulging thread is kept from coming into contact with neighbouring threads. A vertical plate of sheet metal is a favourite method; its only disadvantage is that the open back and front causes the yarn to bulge out at these points, so that as it passes the sides it strikes against the plates, and of course such an action is a disadvantage. This can be neutralised to some extent by using plates that are closed in at the back, so that as far as practicable the yarn is always kept moving in a circle. Complete rings have even been adopted for anti-ballooning purposes, but they introduce difficulties in doffing and piecing, so have therefore not been very successful.

It has already been intimated that the yarn spun on a ring frame must be wound upon a bobbin whose diameter is relatively large. This has always been a great drawback and has prevented the machine competing with the weft yarn made on the mule. Weft yarn is of course made on the ring frame, and in large quantities; but it is not done under the best conditions, and the size of the bobbins made is a great disadvantage. Many attempts, therefore, have been made to spin on the bare spindle for both twist and weft purposes, by getting rid of the bobbin, so as to obtain the greatest amount of yarn in the smallest space, as in the mule cop. A surprising amount of ingenuity and exertion has been put forth to solve the problem of spinning on the bare spindle, and, so far as making a cop is concerned, it may be added that the problem has been successfully solved in several ways. Commercial success, however, is another matter, and in this direction nothing but failure has rewarded the efforts that have been made. To be successful, a machine for spinning on the bare spindle must have a production equal to the present ring frame; it must make a compact cop equal to the mule, which must possess the quality of "readying" to an equal degree; the stopping and starting of the machine must present no difficulties, and the travellers or guides must be as permanent as possible; the strains in the yarn must be uniform, especially in soft-twisted yarn (as in weft); and elasticity must be a quality possessed by the yarn produced.

The chief difficulty, that of causing the traveller to approach the spindle as the smaller diameters are being wound, has not proved insurmountable; but most of the other points mentioned above have hitherto not been attained, and until these have been overcome, spinning on the bare spindle can only be said to be in its experimental stage. Bearing this in mind, it would be inadvisable to present the reader with the numerous methods that have been tried unless some claim to success could be made out for them. Every machine-maker is, more or less, devoting considerable time and money to bring it to a successful issue, and no doubt something will be done soon to make the ring frame a satisfactory cop spinner.

These notes upon the ring frame would be incomplete without some reference to a comparison between the mule and the ring frame. There is such a divergence of opinion upon the matter that the subject can only be briefly touched upon, and it is done without the slightest idea of treating it controversially. Thus far practical experience points to a limit beyond which ring yarn cannot excel yarn made on the mule. Between 60's and 70's might be taken as this limit—though the writer can point to a firm where as high as 100's is spun equal to anything in strength and quality that the mule produces. Weft yarns are not so easily produced on the ring system as on the mule, but improvements in the machine and conditions of working enable weft up to 40's to be very successfully spun; beyond this, practical difficulties arise, which prevent commercial success being attained. It is frequently stated that the ring frame requires more power to drive than the mule; but a considerable practical experience, extending over both machines, suggests no great disadvantage in this respect in the ring frame (especially with our high-class modern flexible spindles). The ring frame is much the cheaper yarn spinner, in some cases exceeding the mule by as much as 40 per cent. In medium and finer counts no advantage in this respect can be claimed, but below, say, 40's there is a decided gain.

A comparison of the strength of yarn produced on the two systems gives to the ring yarn the claim to superiority, in some cases rising as high as 40 per cent. In regard to regularity and elasticity, there is room for doubt as to which claims the advantage, especially the latter quality; but the mule appears to attain a higher degree than the ring frame in the elasticity of the yarn made, and is more uniform in that respect. The ring frame has the advantage over the mule in the space occupied, an economy of 50 per cent being claimed for it. The cheapness of the labour and the ease with which the ring frame can be learned and attended to are economical advantages to be considered.

The horse-power required to drive a ring frame is a very variable quantity, depending upon a number of conditions that can scarcely be found alike in two machines. The spindles, of course, absorb the greatest proportion of the power, and differences in spindles account for much of the variation in power between one machine and another. Anything affecting the spindle, such as its speed, the pull of the band, the size of the bobbin, the length of the traverse, the lubrication and condition of the oil used--all are factors in the problem of the power. The construction of a machine, its erection, gearing, rollers, and weighting, are conditions which more or less must be considered in relation to the power. Therefore it is no easy matter to set up a standard by which the power of a ring frame can be gauged. From practical dynamometrical tests made by the writer, extending over scores of machines of all the best makers, the number of spindles per indicated horse-power has ranged from 60 to as high as 110. The following table, taken from one of Draper's publications, presents in a convenient form the results of tests which were conducted to indicate the power absorbed by different parts of the machine. Speed of spindles, 9300. Diameter of ring, $1\frac{1}{2}$ inch. Nos. spun, 43's.

Power taken by rollers, traverse motion, and gearing	11	per cent.
Power taken by weight of bobbin and yarn	11	,,
Power taken by the pull of the traveller	17	,,
Power taken by cylinder and bare spindles	61	,,
	100	,,

The pull of the band has a very deciding effect in the power, and 20 per cent may easily be added by banding too tightly. To those who use a band tension scale a pull of 2 lb. is strongly recommended by the best authorities, and in no case ought it to exceed 3 lb.

Another feature which is not sufficiently observed by many users of ring frames is the lubrication of the spindles. It has become an axiom among those who have devoted attention to the subject that only the very best oil it is possible to get ought to be used on a ring frame. The price of such an oil is an apparent objection, but when it is considered that large percentages of power are saved—which means a great saving in the coal bill, a longer life to the machine, and far better work—it is not difficult to see

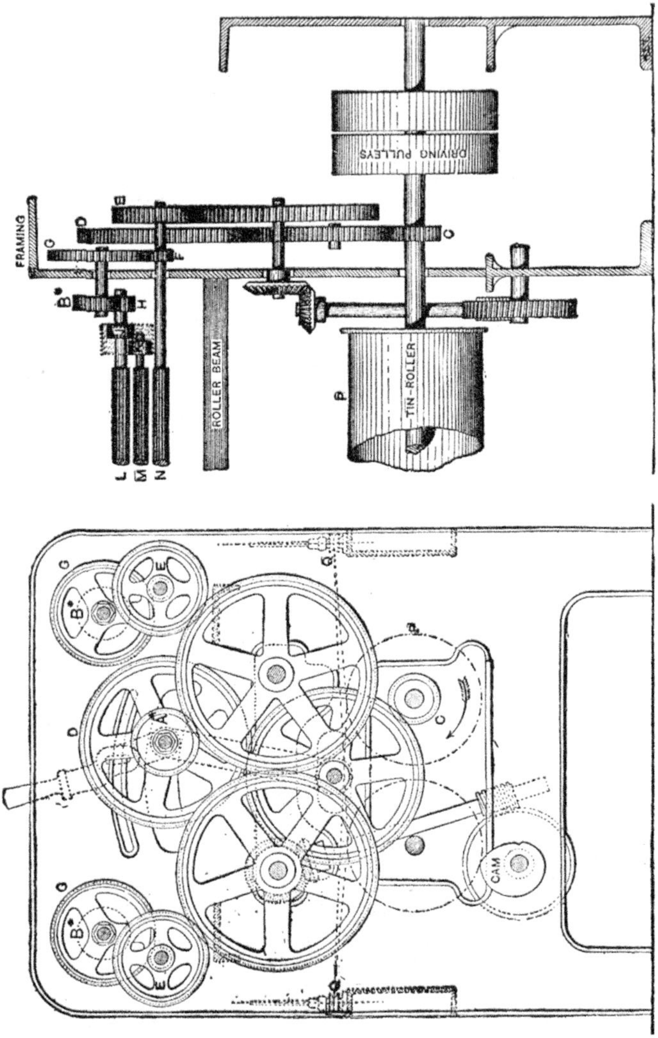

FIG. 179.

that lubrication is too important a matter to be ignored or lightly dealt with.

Calculations.—Speed of spindles $=\dfrac{\text{Revs. of P} \times \text{dia. of P}}{\text{Dia. of Q.}}$.

$$\text{Revolutions of front roller} = \frac{\text{Revs. of C} \times \text{C} \times \text{A}}{\text{D} \times \text{E}}.$$

$$\text{Turns of spindle for one of front roller} = \frac{\text{E} \times \text{D} \times \text{P}}{\text{A} \times \text{C} \times \text{Q}}.$$

$$\text{Twist per inch} = \frac{\text{E} \times \text{D} \times \text{P}}{\text{A} \times \text{C} \times \text{Q} \times \text{N} \times 3{\cdot}1416}.$$

$$\text{Twist wheel} = \frac{\text{E} \times \text{D} \times \text{P}}{\text{Twist per inch} \times \text{C} \times \text{Q} \times \text{N} \times 3{\cdot}1416}.$$

$$\text{Constant number for twist} = \frac{\text{E} \times \text{D} \times \text{P}}{\text{C} \times \text{Q} \times \text{N} \times 3{\cdot}1416}.$$

$$\text{Twist wheel} = \frac{\text{Constant number}}{\text{Twist per inch}}.$$

$$\text{Twist per inch} = \frac{\text{Constant number}}{\text{Twist wheel}}.$$

$$\text{Twist wheel} = \frac{\text{Present twist wheel} \times \sqrt{\text{Present counts}}}{\sqrt{\text{Required counts}}}.$$

$$\text{Twist wheel} = \sqrt{\frac{\text{Present twist wheel}^2 \times \text{Present counts}}{\text{Required counts}}}.$$

$$\text{Draft} = \frac{\text{H} \times \text{G} \times \text{N}}{\text{B} \times \text{F} \times \text{L}}.$$

$$\text{Draft wheel} = \frac{\text{H} \times \text{G} \times \text{N}}{\text{Draft} \times \text{F} \times \text{L}}.$$

$$\text{Constant number for draft} = \frac{\text{H} \times \text{G} \times \text{N}}{\text{F} \times \text{L}}.$$

$$\text{Draft} = \frac{\text{Constant number}}{\text{Draft wheel}}.$$

$$\text{Draft wheel} = \frac{\text{Constant number}}{\text{Draft}}.$$

$$\text{Ratchet wheel} = \frac{\text{Present shaper wheel} \times \sqrt{\text{Required counts}}}{\sqrt{\text{Present counts}}}.$$

$$\text{Ratchet wheel} = \sqrt{\frac{\text{Present shaper wheel}^2 \times \text{Required counts}}{\text{Present counts}}}.$$

Fig. 179 will enable all the above calculations to be easily followed.

It may be observed that the above rule for twist is only approximate; but it differs from exactness by such a small fractional amount that it may be used in all circumstances.

CHAPTER IV

BOBBIN WINDING FRAME

It will only be necessary to briefly describe the uses to which cotton yarns are put after coming from the spinning machines, and these may be summed up in two chief purposes, namely, weaving and doubling. The former term includes all forms of cloth manufacture into which cotton enters, whether as the only material used or simply as the warp of the cloth, some other substance, such as silk, wool, linen, etc., being used as the weft. The latter term, doubling, signifies the twisting together of two or more strands of single yarn in a simple or compound form, for the purpose of making sewing thread, lace, embroidery, knitting, crochet, hosiery, netting and other fancy yarns. For the purpose of weaving, the weft yarn is generally used in the form in which it comes from the spinning machine, while, on the other hand, the warp threads require to be put through several important operations to fit them for their purpose. The only one of these operations which concerns our subject is also employed in the doubling series of operations, so that by confining our attention to this branch a repetition will be unnecessary.

The cops from the mule or bobbins from the ring frame

are brought to this machine to have their yarn wound on to large double-flanged bobbins, similar to those shown in the sketch Fig. 180 at A; they are generally termed warpers' bobbins, because yarn is practically always wound into this form before being transferred to the warping beam. The reason for this form of bobbin is that when the yarn has to be again unwound and placed on a beam or otherwise, it will come from the bobbin at a fairly uniform tension, because of the parallel layers; from a cop or bobbin this would be impossible on account of the constantly changing diameter of the conical ends. In addition to being used for weaving purposes, the bobbins are largely used for the doubling frame for the same reason which prompts their use in warping. This point will be treated more fully in a subsequent paragraph.

A general idea of the machine can be obtained on reference to Fig 180, which represents the upper portion of one side of the frame. Two rows of spindles (B) are carried in bearings D and E on each side, and driven from a tin drum in the centre of the machine through the wharves C. The upper ends of the spindles carry the bobbins A. The cops or ring frame bobbins are supported by small brackets at N, and from here the yarn is led forward over a flannel-covered board L, where it is likely to be cleared of any loose fibres or dirt adhering to it. On its way to the bobbin it passes through the bristles of a brush K, where a further cleaning will naturally occur, and then on through a guide H and on to the bobbin. The guide H is generally of a special form for the purpose of clearing the yarn of any persistent motes, slubs, etc., which stick to it, or to prevent the passage of badly-formed piecings or knots due to carelessness in the previous processes. This clearer is, therefore, an absolutely essential feature, and it has afforded innumerable opportunities for the display of ingenuity in so arranging its parts as to obtain the greatest usefulness

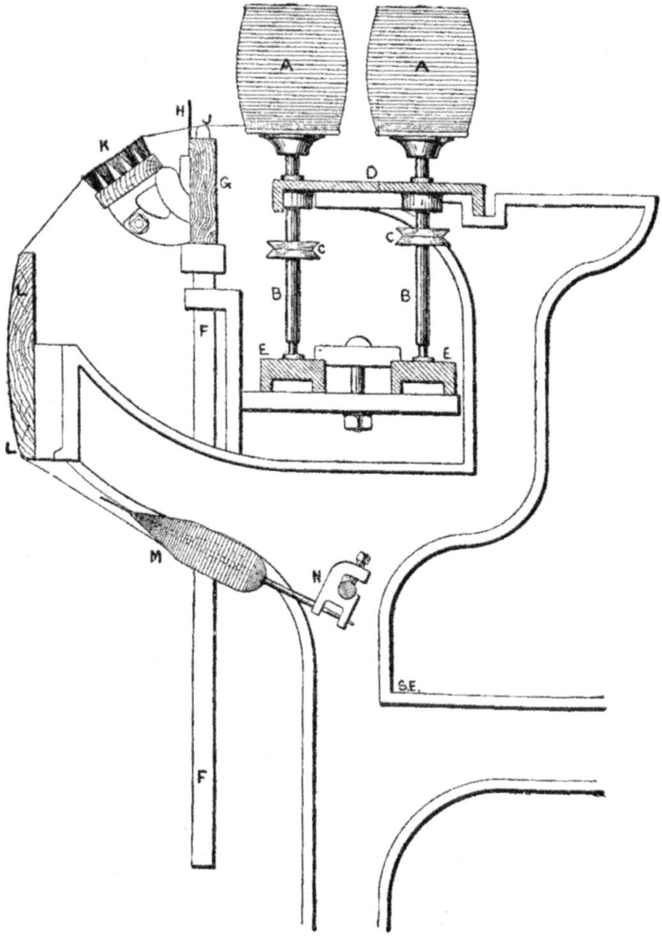

FIG. 180.

out of it; it also adapts itself well to different counts and classes of yarn. This remark applies to other machines through which yarn passes and where such guides are employed; in some cases a kind of winding machine is used simply for the purpose of clearing the yarn, so that the clearer is the chief feature. The guide H is carried on the top of a lifting rod F, which may be operated in its upward and downward movement by a cam or the well-known mangle-wheel arrangement. The resulting bobbin can be made barrelled, as in the sketch, or perfectly parallel. Bobbin boxes are placed under the row of cops M, and in the middle of the machine is a receptacle for the full bobbins after doffing. In most machines an arrangement can be applied, in the form of an endless band or apron running down the middle of the frame, which carries the full bobbins to the end of the machine and deposits them in a large box or skip. Instead of winding from cops, arrangements can be substituted in order to wind from hanks. In Fig. 187 will be found a fuller and better idea of the machine just described.

Quick-traverse Winding Frame.—When yarn is to be used for doubling purposes, that is, a combination of two or more ends into one, the yarns from two of these bobbins are passed together through the rollers of a doubling frame and then twisted together as one strand. This system, however, is not now so general as formerly, though it is still practised, and in the case of high numbers winding is dispensed with and the cops placed directly in the creel of the doubler. The usual method of doubling the ends together for the doubling machine is to take the cops or bobbins to what is called a quick-traverse winding frame, where a bobbin is made upon an ordinary paper tube without flanges. A section of such a machine is given in Fig. 181. It is a double-sided frame, and, like the one just described, it will wind bobbins of any diameter at the same time. Two shafts run the full length of the machine, and on them are threaded and keyed drums M, whose lengths are suitable for the lift of a bobbin required. Resting on the drums are wooden rollers N supported in

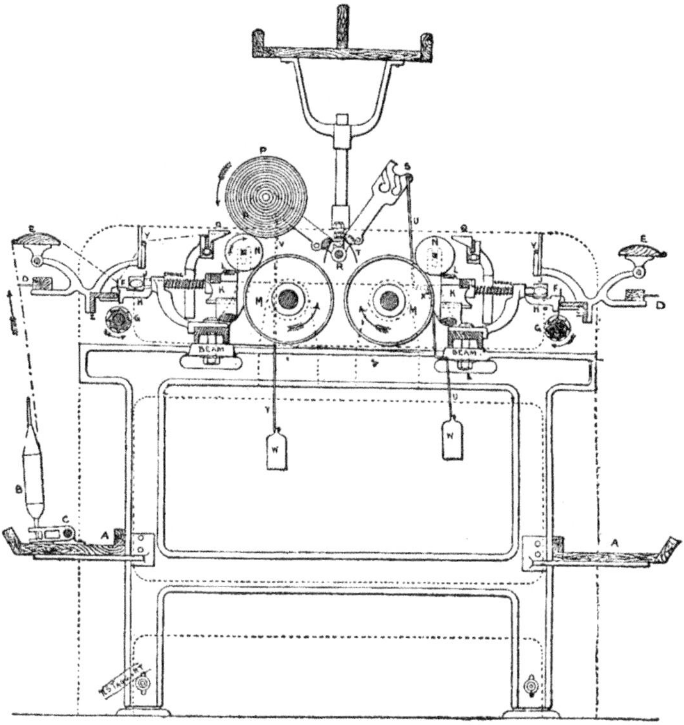

FIG. 181.

the slotted bracket carried by the beams, and on the rollers N rest the steel spindles upon which the bobbin is to be wound. The rollers are driven by friction, so that the bobbin is built up through frictional driving, and as each bobbin is driven from a distinct wooden roller the diameter of the bobbin as it is formed does not interfere with its correct shape. The bobbin boxes or creels contain a series of supports C for the cops or bobbins, and the yarn is led through a guide D and over a covered clearer E; from here it passes through the drop needles F and over a stationary guide Y and on to the bobbin through a spoon Q. This spoon can be placed very near to the nip of the bobbin and wooden rollers N, and it receives a very quick to-and-fro movement from a cam or other suitable mechanical motion. The quick traverse of Q causes the yarn to be wound on the bobbin P in a series of very coarse spirals, and these are such that the quick return motion enables a bobbin to be formed without the usual wooden ends, thus saving both weight and space.

In this machine any number of ends from one to six can be doubled together and wound on one bobbin, and each end will have exactly the same tension, the arrangement of the parts being such that the tension can be readily adjusted. The importance of maintaining the exact number of ends continuously is obvious, especially if for doubling purposes, when only two ends are to be twisted; it is therefore natural that an automatic stop motion is necessary, and for this purpose the needles at F are employed. Each end passes through a separate needle; when the end breaks, the needle falls and comes into contact with a revolving grooved roller G. Since the needles are carried by a swivelled catch-lever H, the consequence of F falling into the path of G is to move the lever on one side, and in doing so it releases the catch end of H from a projection J which it has held in position. When J is held by H a slide K is drawn back so that its upper surface at L is out of contact with the wooden rollers N, and at the same time the spring is put into tension. When the catch H is released, the spring forces K forward and the part L impinges against the wooden roller N and lifts it clear of the drum M, thus stopping the bobbin. This stoppage enables the broken end to be at once pieced. The improbability of more than one end breaking at once or more than one cop becoming empty at the same time, enables the piecing-up to be done without bunch knots occurring; it prevents waste and overrunning, and in keeping the yarn always at the same tension obviates the great fault of corkscrewing when the bobbins are taken to the doubler. The frames carrying the bobbins M are centred on the rod R, and wire hooks supporting weights W are added to give grip and steadiness. A further improvement in this respect is obtained by the use of the spring T, especially when the bobbin is small. The hooks U and V are often formed with bent portions, so that the bobbin itself can be lifted out of contact with N and kept so by resting the bent portion of the wire upon a convenient projection, as at X.

Another well-known and successful quick-traverse winding frame is illustrated in Fig. 182, where half the machine is shown in section. A shaft L drives a series of drums A, whose outer surface is in the form of a thin shell having a fine double helical slit piercing it all round; this slit corresponds to the usual cam which gives the quick traverse to the other makes of winding frames. The spool or bobbin D carried by a lever B which is centred at F presses against the under side of A, and the yarn is led from the cop or bobbin through the usual detector wire G, over K, the roller H, and through the slit in the drum A on to the spool. The revolution of A will naturally cause the yarn which passes through it to travel backwards and forwards the full length of the cam slit in its surface, and as the spool is driven at the same time through friction by being in contact with A, the yarn is wound on D in a series of coarse spirals in such a manner that the ends of the spool are built up solidly and squarely, and are capable of being handled and transported safely and economically. A steel blade at E serves to keep the yarn always at the bite of the spool and drum, and the lever B is so fulcrumed that, as the spool enlarges, the point of contact with the drum remains always the same. The stop motion is sufficiently interesting to merit description. On the breakage of an end, the needle G falls into the path of the revolving wiper L, and is with its carrier at once moved backwards; this pulls M with it and lifts up the catch at "m"; "m," it will be observed, has resting against it a finger "a," which is centred on the supporting lever A^1, which carries the drum A. Directly "m" is moved out of the way of "a" the drum A falls back against a fixed brake N and is at once stopped in its revolution; the broken end can at once be pieced, and if it is necessary to draw the spool away from the drum a catch fulcrumed on B^1 enables this to be done by allowing the other end of B^1 to come against a stop on the beam; on pressing down at B^1 the spool will fall immediately against the drum ready for work. The pressure of the spool against A is carefully regulated by the weight W, and is practically the same from the empty to the full bobbin. It may be observed that this machine is essentially a quick-traverse winding frame, and it could not be used for slow winding without great loss of production; alteration in the speed of the traverse cannot be made irrespective of the speed of winding, and as a consequence the character of the winding practically remains the same. Where the traverse is independent of the revolution of the spool, half the pitch of the spiral can be obtained without altering the production, and a slow-traverse bobbin can even be made with the same effect.

Fig. 183 gives another full sectional view of a frame made by a well-known maker of this kind of machine, and the following remarks will enable the working to be under-

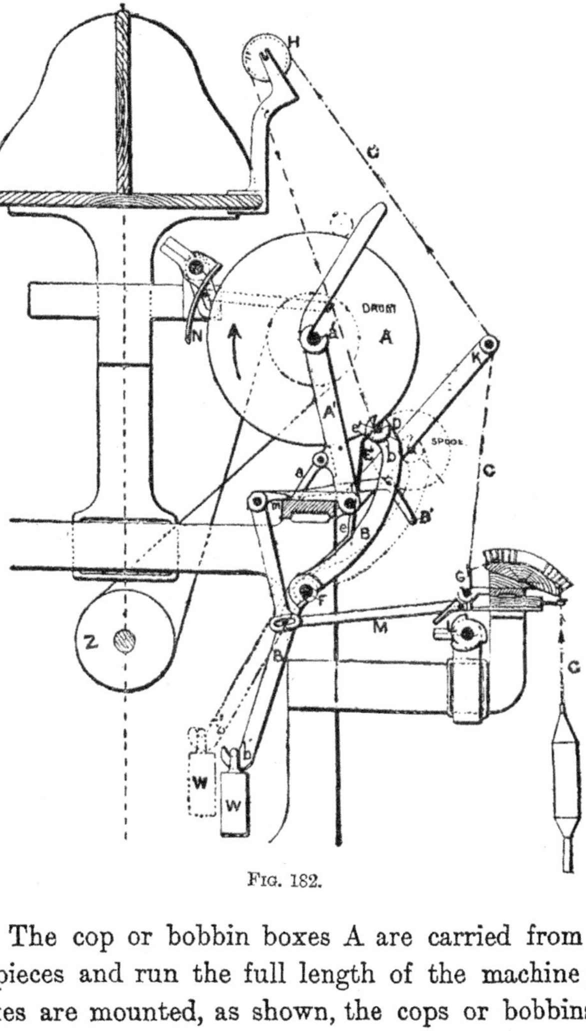

FIG. 182.

stood. The cop or bobbin boxes A are carried from the spring pieces and run the full length of the machine; in the boxes are mounted, as shown, the cops or bobbins B. The yarn from these is passed through wire guides C and on over an adjusting drag board D, the regulation of which is effected through the screws E. After leaving the drag board, the yarn is passed through detector needles F carried by a short swivel cradle, which rests upon one end of the swivel frame G. The yarn is now taken in an upward direction over the wooden guide rollers R, and from here it passes direct to the flanged wooden winding bobbin M.

The bobbin M is supported upon the upper end of a lever J fulcrumed on the swivel frame G; the lower end of J is connected by a cord or chain to a weight K, which keeps M pressed against a central revolving drum N, so that the bobbin has a constant pressure and an unvarying surface speed. The threads Q are fixed to the traverse rod O, which is actuated by a slow-motioned traverse, either of the cam or mangle-wheel type. The automatic stoppage of the machine when an end breaks is brought about through the medium of the needles F; these are kept out of contact with the revolving ratchet shaft I when the yarn is passing forward, but on an end breaking, the needle falls and is at once moved aside by one of the wings of I. This at once frees the end of G which carries the needle box, and it rises, thus lowering the other end, which carries the lever J; in this way the bobbin is lowered from its normal position, and in doing so it is kept out of contact with the drum N by a brake L, whose knife edge projects almost to the nip of the drum and bobbin. This naturally stops the revolution of M; but to make this stoppage absolutely certain, a projection on J comes against an extension of the brake L, and the weight of J forces the upper end of L against the bobbin M, and so stops its further motion immediately. For the purpose of piecing-up, the lever J and bobbin can be pulled forward and automatically hooked in a convenient position; when all is in order, the setting-on

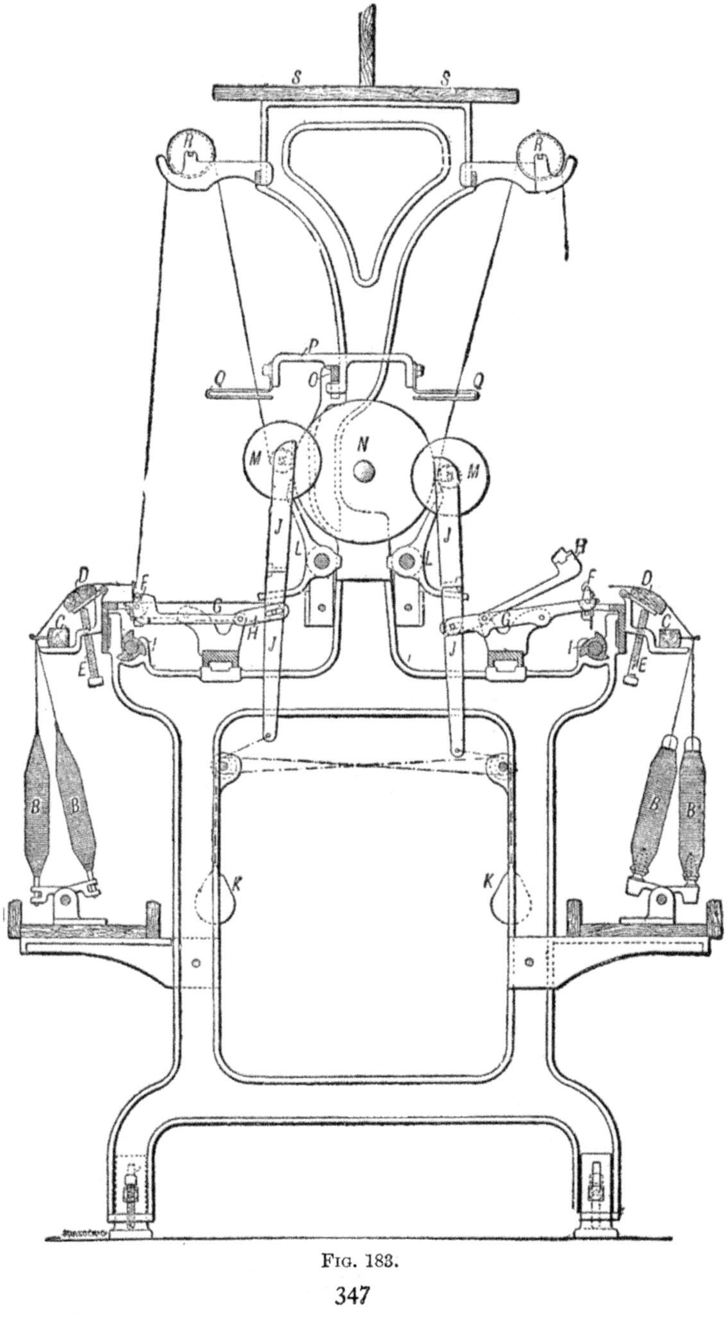

FIG. 183.

handle H is depressed, and this action at once puts the whole arrangement in correct position for continuing the winding. One side of the machine is shown as when winding is being performed, while the right-hand side represents the altered positions taken up by the various levers when an end breaks. No difficulty whatever is experienced in finding and piecing a broken end, and any number of ends from one to eight can be wound together.

The machine just illustrated is used, with very little alteration of construction, for quick-traverse winding; a change in the method of giving the traverse being all that is necessary.

The motion in this machine for making the "cheeses," as the quick-traverse bobbins are termed, is an interesting example of mechanism; instead of the usual cam, an attempt has been made to use a crank. Readers will know that although a crank gives a to-and-fro movement, such a motion is not a uniform one, the middle of the throw giving a quick movement while the ends produce a slow one. In the example, this irregular action of the crank is overcome, or rather modified, by the introduction of a special cam groove, so formed that the crank pin travelling in this groove is made to give to a traverse rod an absolutely uniform motion.

Fig. 184 gives a section through the traverse motion. Upon the drum shaft A are keyed a series of drums B; the traverse rod J is connected to crossheads I, which slide to and fro in guides. A crank pin H on the top side of the crank-plate slide F fits in a groove of the crosshead, the slide F sliding within a groove on the crank plate E. It will be seen that if F and E are fastened together and revolved, the pin H would impart to the traverse rod J a simple crank movement which is incapable of building a straight bobbin. However, F is made so that it can slide along guides carried by E, and by means of a bowl G which it carries, and which fits in a cam groove cut in a revolving plate D, the pin H is made to move in a special manner towards and away from the centre of the crank plate E, so that the irregularities of the crank motion are neutralised and a uniform traverse is the result. From the sketch an idea of how the arrangement is driven may be obtained. The shaft K is driven from the gearing C; on K is keyed

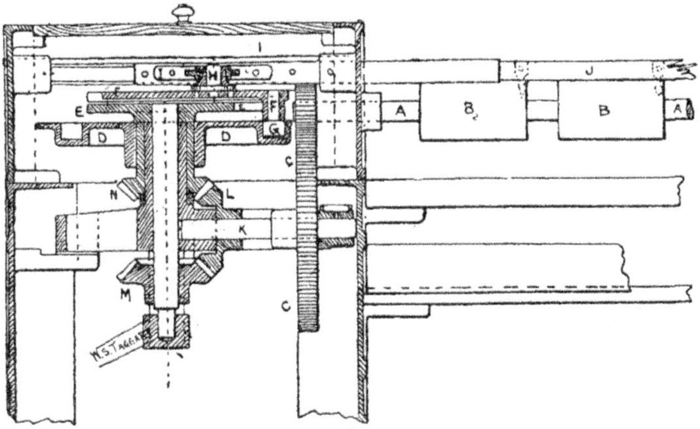

FIG. 184.

a bevel L, which gears into two bevels M and N. The lower bevel wheel M is keyed to the crank shaft P, and so drives the crank plate E; the upper bevel wheel N is keyed to the boss of the cam D, and from it the cam receives its motion.

Two partial views of a quick-traverse frame are given in Figs. 185 and 186. This frame is fitted up, as usual, with automatic stop motion, but the drawings will serve their chief purpose by showing the most common form of cam. The revolution of the cam A moves the pin B to and fro, and with it the traverse rod C, which runs the full length of the frame and carries the guides D.

A drawing has been prepared, Fig. 187, to illustrate a more modern example of the ordinary winding frame for making warpers' bobbins than that given in Fig. 180. This machine is sometimes called a **clearing frame.** The ring

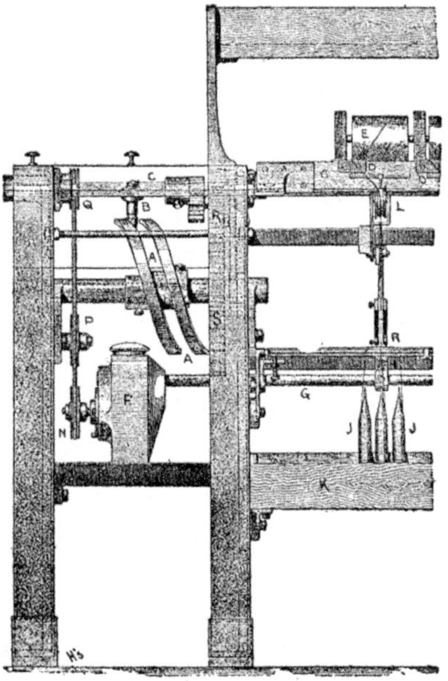

Fig. 185.

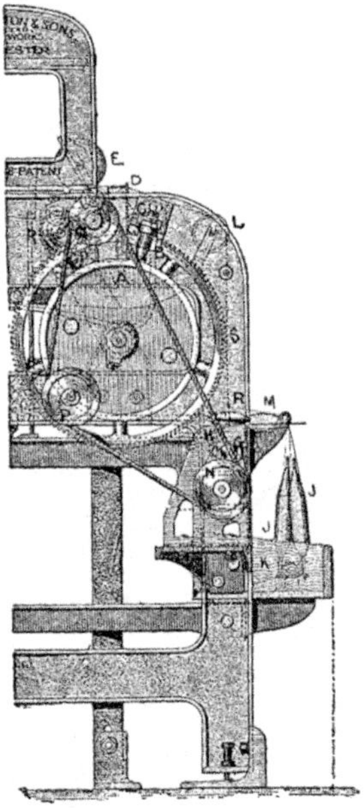

Fig. 186.

frame bobbins G are mounted, as shown, upon a box arrangement F, which serves also the purpose of a receptacle for bobbins. The yarn is led upwards through guides over a drag board E and through another guide H; from H the yarn is taken over a rod J, and when used as a clearing frame it passes through an adjustable yarn clearer C. This clearer is really a series of narrow slits, so arranged that the slits can readily be made wider or narrower according

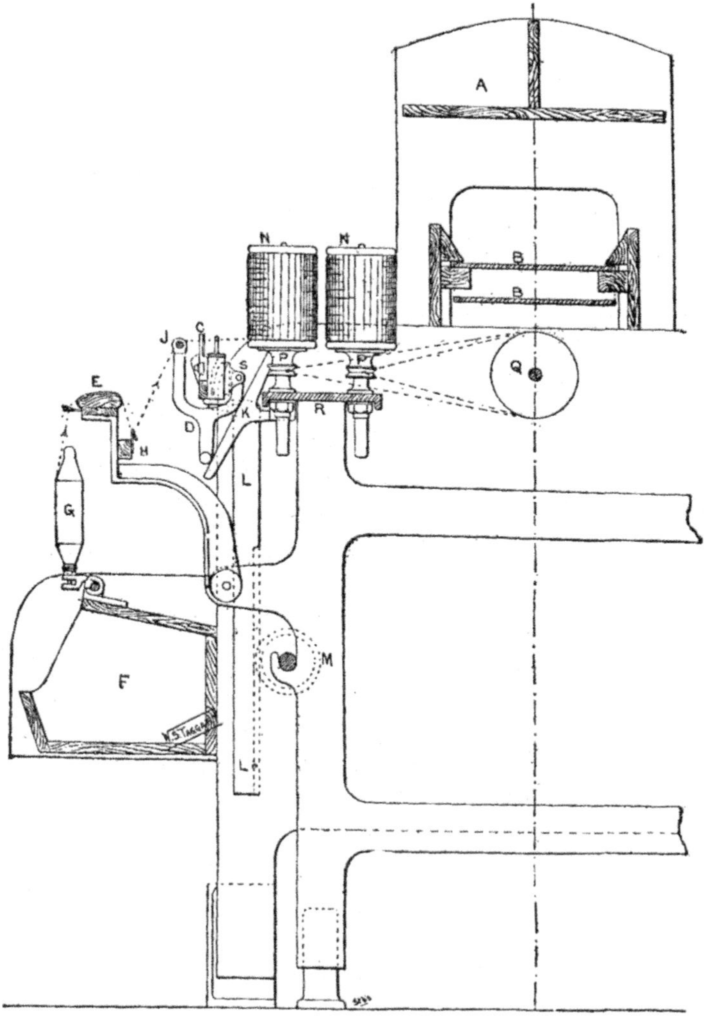

FIG. 187.

to the counts or condition of the yarn, the object being to prevent the passage of knots or other imperfections in the way of "slubs," etc. The yarn now passes on to the bobbins N, which are mounted upon Rabbeth spindles P carried from the beam R; the spindles are driven from the tin roller Q, the diameter of which is usually about 5 inches.

To economise time when doffing, the full bobbins are taken from the spindles and put upon a travelling apron B, which carries them to the end of the machine and deposits them into a basket or box. The building motion takes effect through the wheel M gearing into a vertical rack L, on the upper end of which is mounted the clearer arrangement. An ingenious contrivance is introduced at D: it is very desirable not to have the yarn always passing through the clearer C at the same spot, so the rod J is carried by a lever D, whose centre is at S; a projection on D rests upon an incline K, so that as L is raised and lowered the lever D will receive an oscillating movement, and the rod J will guide the yarn through the clearer C in a constantly varying position.

CHAPTER V

DOUBLING

THE bobbins from the winding frame are now taken and placed in the creel of a **doubling frame**, or, as it is sometimes and more correctly called, a twisting frame or **twister**, where the ends are twisted together into one thread.

The doubler bears a general resemblance to the ring spinning frame, and its twisting operation is exactly the same, but an observer would notice in most machinery three points of difference, namely—the bobbins in the creel are different, as already explained; the feed rollers are not as in the ring frame—we find no drawing rollers at all in the doubler, for the reason that no drawing effect can be obtained from threads already so well twisted; instead of three lines of rollers we only find a single line. The other difference noted would be the character of the bobbins built on the machine; as a rule these are built up as parallel layers on double-ended bobbins and not in conical layers, as in the ring frame, though it may be observed that ring spinning frames are sometimes made with parallel lifts and doubling frames are made with conical lifts.

Fig. 188 will convey some general idea of the doubler. As in the ring frame, there are two rows of spindles HH, one on each side of the machine; these are driven from the

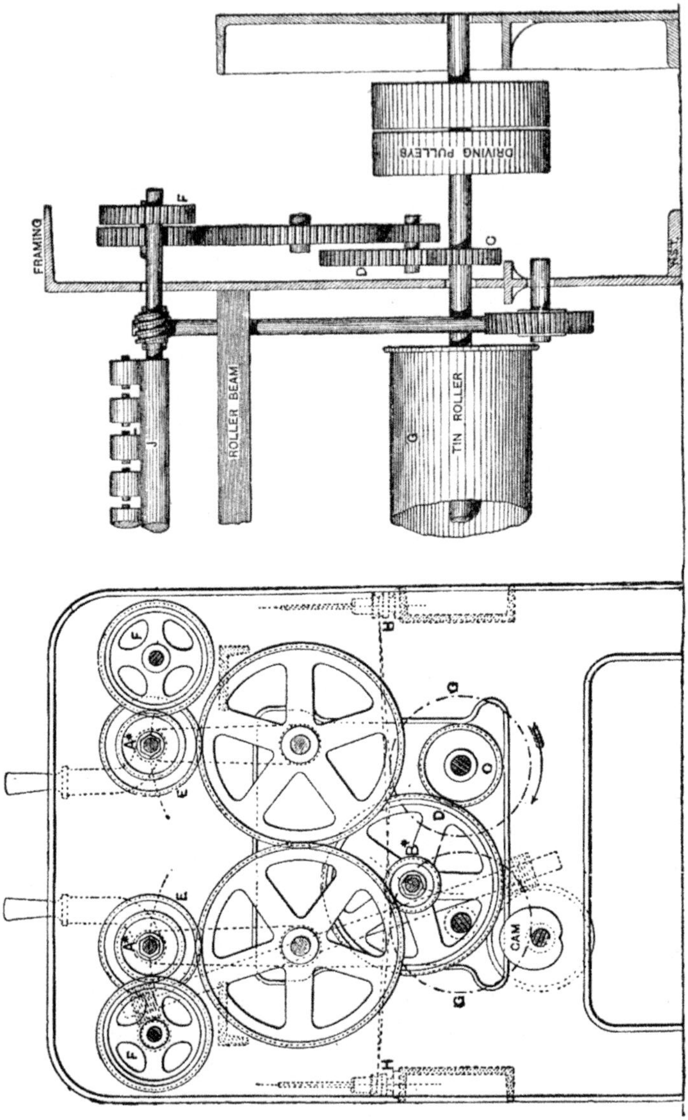

FIG. 188.

tin drums G. The driving of the single line of rollers J starts at C on the tin roller shaft and through the gearing

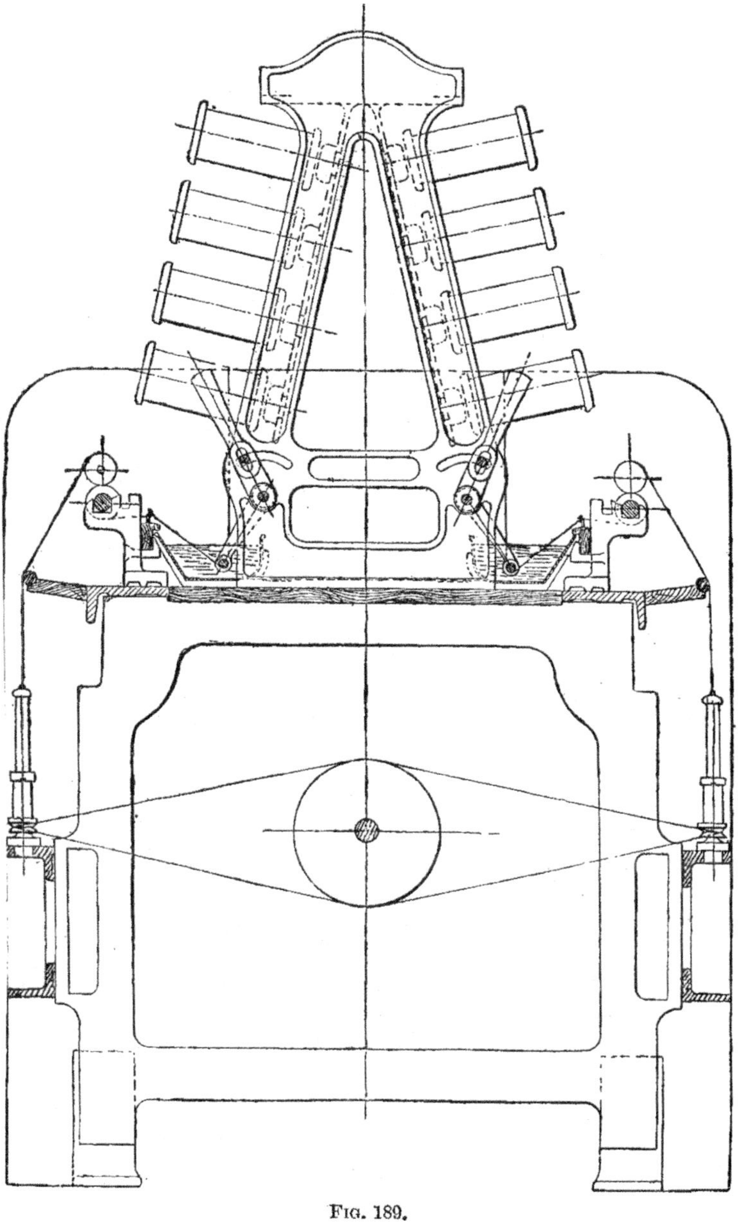

FIG. 189.

shown, to F. On account of the wide range of twists put into the various doublings, two change places are introduced at A and B to enable this to be readily obtained. The lifting cam, it will be noticed, is an equal heart-shaped one, giving the up and down motion of the rail a uniform

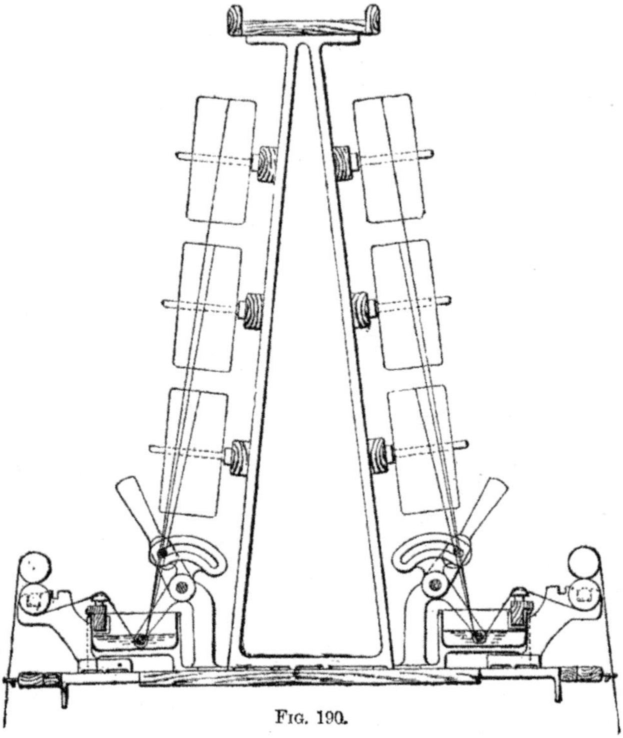

FIG. 190.

traverse; it is driven generally from the feed roller J in the manner illustrated.

Creels.—Illustrations are given in Figs. 189, 190, and 191 to show the method of doubling from flanged bobbins, from quick-traverse winding drum bobbins, and from cops; in all cases the system of doubling is called wet doubling, from the fact that the yarn before reaching the rollers passes through a trough of water. Dry doubling is practically the same system with the exception of the water trough, and iron rollers are used instead of brass covered ones; dry doubled yarn is used chiefly for warp threads in

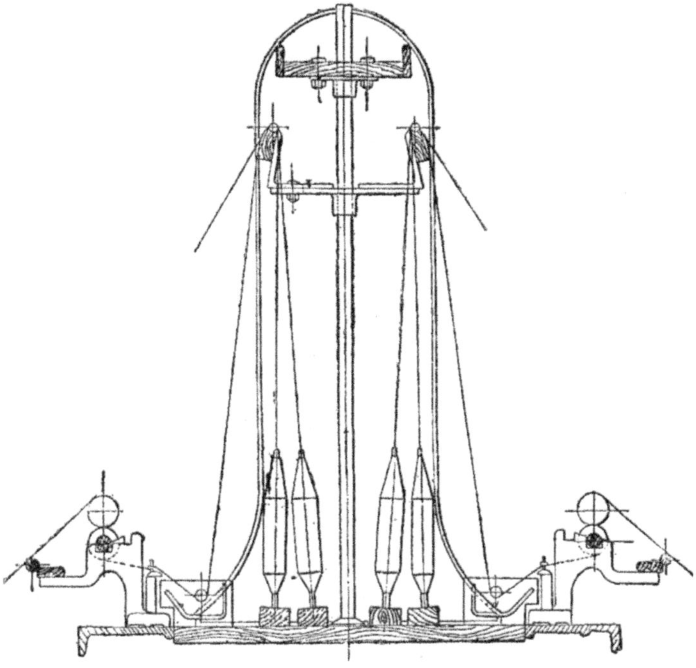

FIG. 191.

weaving, and also in many cases simply for the selvedges in cloth where the general warp is single yarn.

English and Scotch Systems.—The two illustrations in Figs. 192 and 193 will serve to illustrate the details of the trough in wet doubling. Two systems are employed, namely, the Scotch and English. Fig. 192 shows the English system; here the yarn coming from the bobbins passes down into the trough of water and under a glass rod

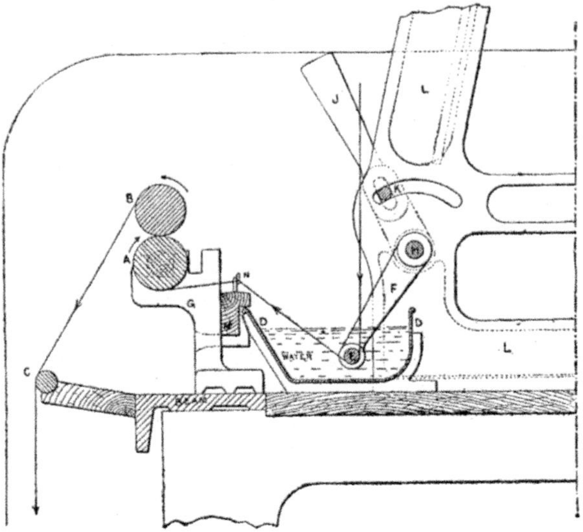

Fig. 192.

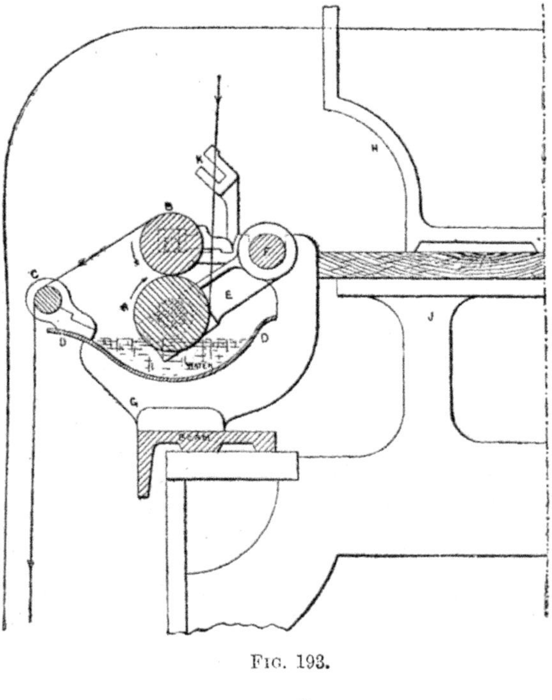

Fig. 193.

carried by a series of short arms centred as shown. On emerging from the water the yarn passes through a guide wire and on to the rollers; these rollers are covered with brass so that the weft yarn has no corroding effect on them. Apart from the fact that the rollers deliver the yarn at a fixed rate, they serve the purpose of pressing the surplus water from the yarn, the top roller being heavy and driven entirely by friction from the bottom roller. For cleaning purposes, etc., the glass rod can be lifted out of the trough by the handle shown in the illustration, and the trough itself can be emptied by a tap placed at one end of the frame. The effect of water on the yarn is to give it a solidity, and strength is added from the fact that all loose fibres are smoothed down and readily incorporated in the double thread when twisted. The Scotch system is given in Fig. 193; here the rollers themselves are located within the trough, so that the yarn passes direct from the creel to the underside of the bottom roller. The top roller is out of contact with the water in the trough, but it is constantly wet through its contact with the bottom roller, so that in

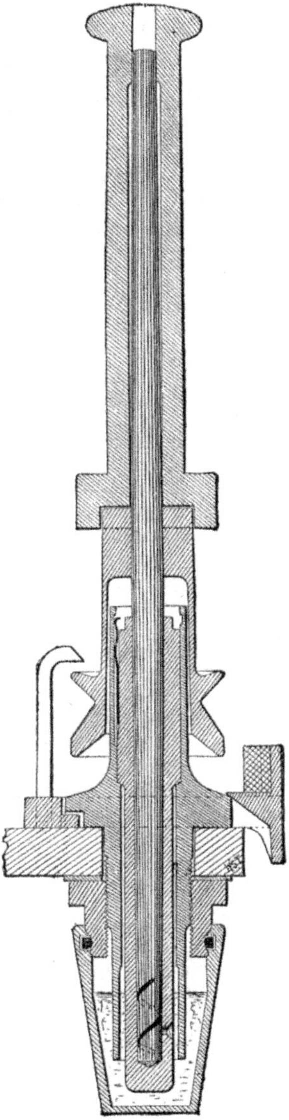

FIG. 194.

this case the yarn passes to the spindles in a far wetter condition than in the English system.

Spindle.—The spindles of a doubling frame are practically the same as those used on a ring spinning frame. Fig. 194 will enable a comparison to be made, and it will be noted that stronger and heavier spindles are required for doubling than for spinning. The bobbin is double-

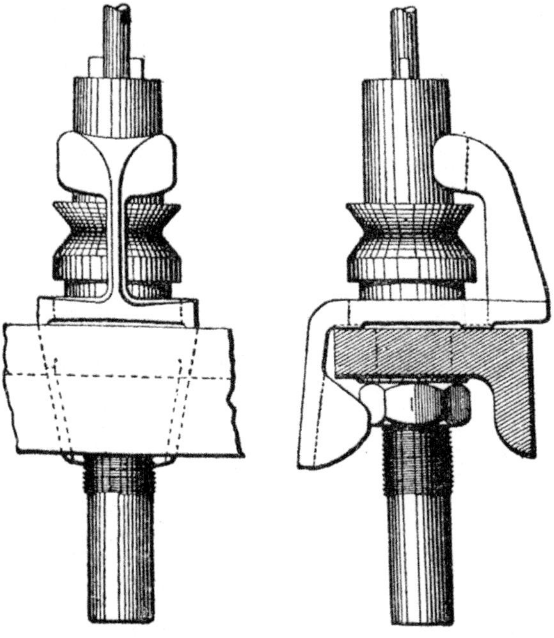

FIG. 195.

ended, and the yarn is wound on in layers the full length of the lift, no crossing effect being given.

Knee-brakes.—An appliance called a knee-brake is frequently applied to a doubling spindle. This is done so that the attendant can, by pressing the brake with the knee, stop the spindle while the end is being pieced. Fig. 194 shows the brake, and attached to it is a projecting wire used as a catch to prevent the spindle lifting out of its bearing. The brake illustrated encircles the base part of the spindle carrier, and on the underside of the front is cut an inclined surface resting upon the edge of the pillar base. A piece of leather is shown in the hatched part of the drawing, and it is so arranged that if the knee presses against the front of the brake, the inclined surface permits the brake to slide up until the leather comes into contact with the lower part of the spindle wharve; the friction resulting from this pressure stops the spindle, and piecing can be performed quickly and conveniently. In withdrawing the knee the brake by its own gravity falls out of contact with the wharve and takes up its normal position.

There are innumerable types of these knee-brakes, but one more example will be sufficient; this is illustrated in Fig. 195, where a single casting, as in the first case, rests on the spindle rail and is prevented from any side play by small projections fitting round the pillar base. Pressure applied by the knee to the front part at A causes the upper leather-padded end C to press against the spindle and so stop it.

Stop Motions.—Two examples are given of stop motions; the first is a very cheap and simple, but at the same time a very useful, type. Its object is to prevent the delivery of yarn when an end breaks; it will readily be understood that twisted yarn, if it continues to be delivered, will either be wound on the roller in the form of a lap and will require some trouble to cut away, with the possibility of damaging the roller, or it will be delivered and fly around in the path of other ends and cause several breakdowns in addition to its own. Apart from the trouble involved, much waste is caused by it, and the necessity of a stop motion will be obvious. In the motion illustrated (Fig. 196) a metal holder D is loosely centred on the pivot of the top roller A; attached to D is a wire F curved around so that its lower end G, which is curled, allows the yarn to pass through on its way to the spindle. At E is fixed a strip of leather, so that when the end breaks the wire F instantly falls, and the leather E passes into the nip of the two rollers A and B, and is carried a little forward; while the leather strip is in this position, it is impossible

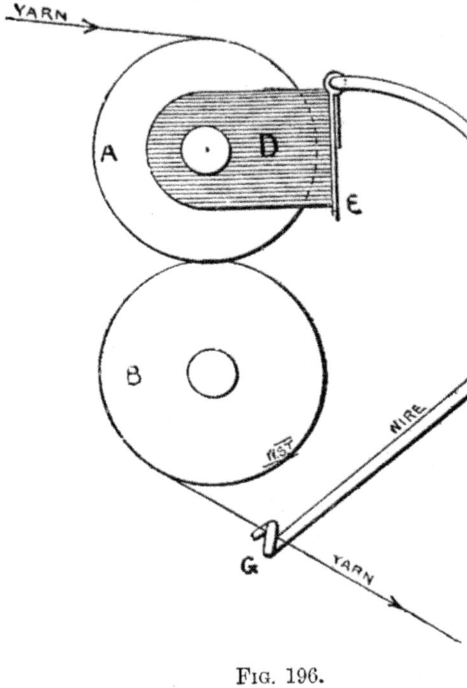

FIG. 196.

for yarn to be further delivered, so both waste, laps, and additional breakages are prevented. The second example is a well-known and well-tried arrangement, and, as will be seen in Fig. 197, it stops both the rollers and the spindles. The stoppage of these two organs was formerly considered an altogether unnecessary act for twofold yarn and even for threefold; this opinion is still held by many, but experience is proving that substantial advantages accrue even when twofold yarn is being doubled, and the success of the method shown in the diagram is a strong proof of the efficacy of such an arrangement. The drawing is almost self-explanatory, and can be easily followed in its action,

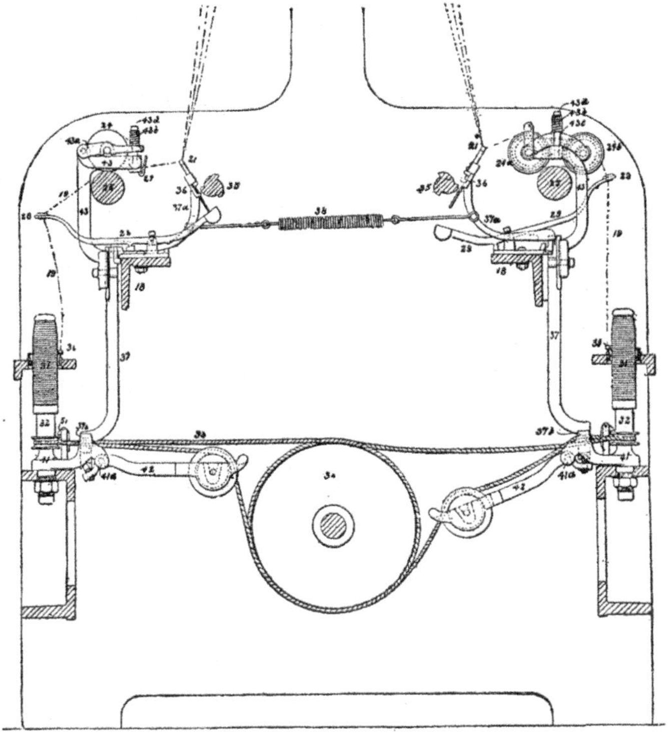

FIG. 197.

the connection between the stopping of the rollers and the spindles being clearly depicted.

Twisting.—The twisting action on the doubler requires no special description, because it depends upon the same principles as in the ring frame. In the dry doubler, ring and traveller are precisely as in the ring frame. In wet doubling, however, a slight variation is introduced in the form of the traveller, with the object of obtaining a larger frictional surface between it and the ring. This will be observed in Fig. 198, where A is the ring and B the traveller. To prevent wear, doubler rings are oiled or greased; several very ingenious methods have been tried to do this without hand labour, but so far the greasing in most places is purely a manual task. A point to notice in connection with a doubler ring and traveller is that any wear that takes place will be at the part A on the under side of the ring. It is a large surface, and wear must be considerable to become inconvenient. From this fact we find little effort made to use double rings in doublers, though they are by no means unknown. A list of suitable travellers for 2, 3, and 4-fold yarns for wet and dry doubling would be too long to insert here, but any machine firm of repute would, no doubt, willingly supply the reader with the information.

An interesting subject is presented to us when we come to consider the twisting together of two or more yarns to form a cord. As this is the chief purpose of the machine, a brief mention of the operation will be made. If a twisted thread of single yarn be taken from a cop that has been spun "twist way," the spirals will have the same direction as the threads of a left-handed screw (see Fig. 199. If this thread is now allowed to sag until it becomes doubled, it will be observed that the parts of the doubled end immediately begin to twist themselves together, as shown in the sketch, Fig. 200. The peculiarity in this action lies in the fact that the twist of this doubled thread is opposite to the twist of the single thread which composes it. Moreover, the single thread, if left alone, would begin to untwist itself, while the double part has no such tendency;

on the contrary, its tendency is to become more tightly twisted and to remain so. The action just described is a perfectly natural one, and has been performed entirely by the forces within the yarn itself. Its explanation is simple enough when we remember that a thread twisted "twist way" will untwist itself by turning "weft way." If two threads twisted "twist way" are put together, each

Fig. 198.

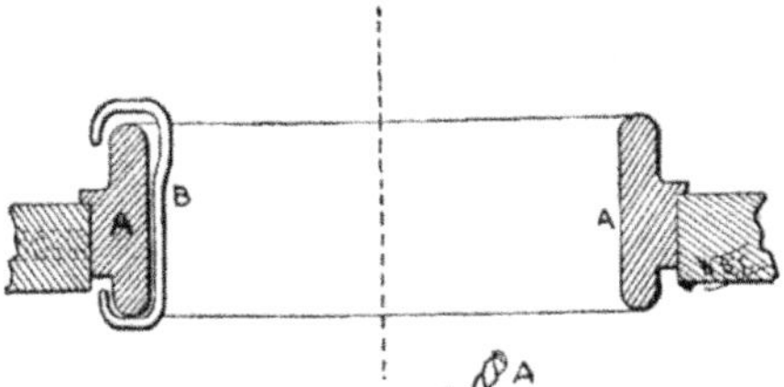

Fig. 199.

Fig. 200.

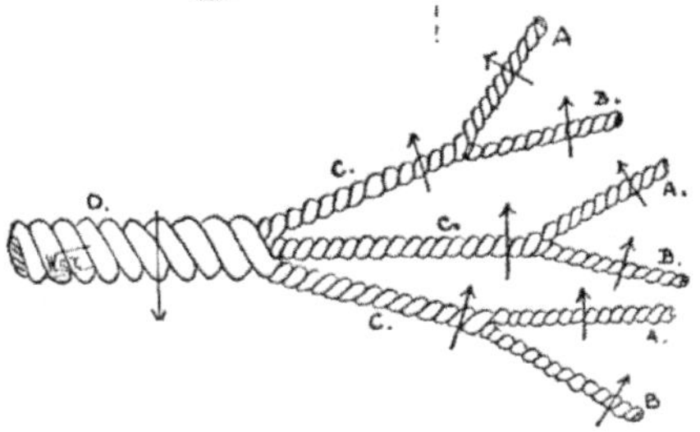

Fig. 201.

tries to untwist "weft way," and consequently they wind round each other and form a combined thread which is twisted "weft way," as the sketch illustrates.

This example supplies us with the foundation upon which to base our doubling operations. In doubling two or three-fold, the twist must be opposite to that of the single yarn. In four and six-fold two operations are necessary. First, two ends are made into one; these are then re-wound on a winding machine and two or three of them are twisted together again on the second doubler into a single cord: such a cord would be denoted as a four or six-fold thread. Fig. 201 will illustrate how these ends must be twisted in order to obtain a thread that will be well twisted and will remain so without a tendency to untwist. Note, that since a six-fold is to be the object of our doubling, the three two-folds of which it is composed are not to be twisted as if they were to stand as simple two-fold yarn, but rather they must be so twisted as to make a permanent six-fold yarn. In the first place, single yarns A and B, both with the same twist, are taken and twisted together into one thread, as at C. The two are twisted in the same direction as the twist in the single yarn; this twist is not a permanent one, for, as already mentioned, this two-fold thread would at once untwist itself if allowed to be free. We have, therefore, in the doubled yarn at C two forces at work—the twist in the single yarns A and B tending to untwist, and the same twist in the double yarn tending to untwist in the same direction. If three of these threads C are put together, they would among themselves twist into a cord in the opposite direction to the twist of the component threads; they must, therefore, be twisted together in this direction if we desire a permanently twisted six-fold yarn. This is shown at D in the sketch, which, it will be noticed, has its twist opposite to that of the threads C and A and B. A thread made by this method is said to be "cable laid," to distinguish it from some threads, or rather cords, which are made by simply twisting six or more ends at one operation into a single cord. Commercially, doubled yarns are denoted by first stating the number of folds and then following with the counts of the single yarn of which it is composed—for instance, six-fold 120's means that the

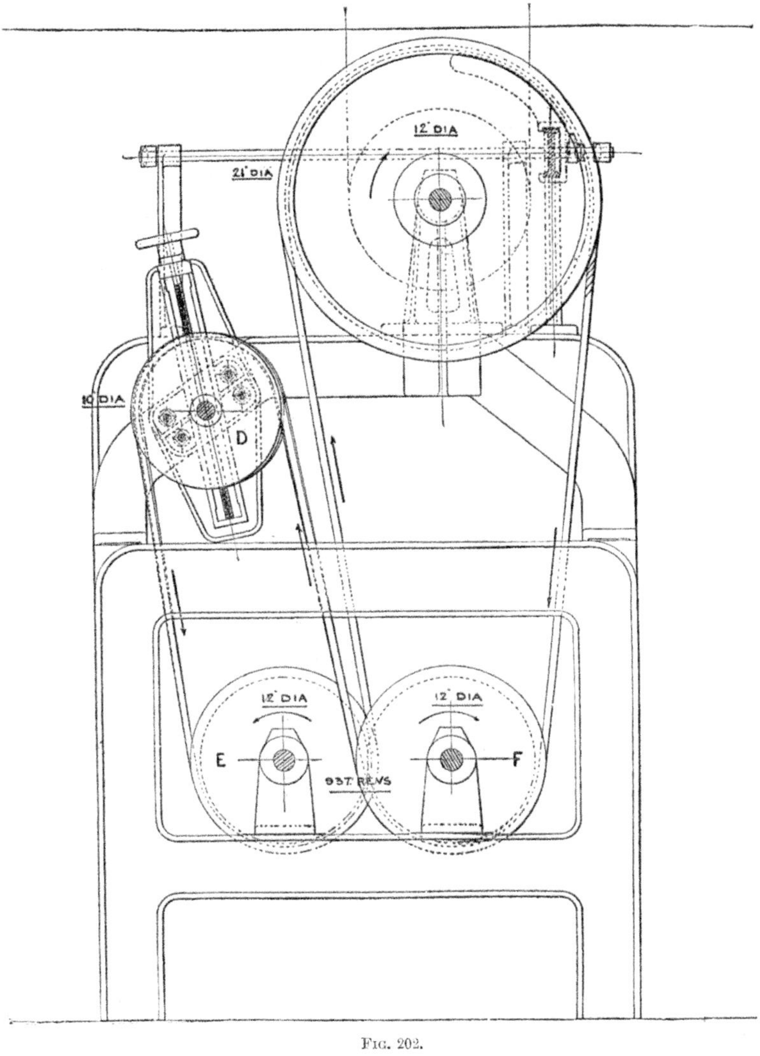

Fig. 202.

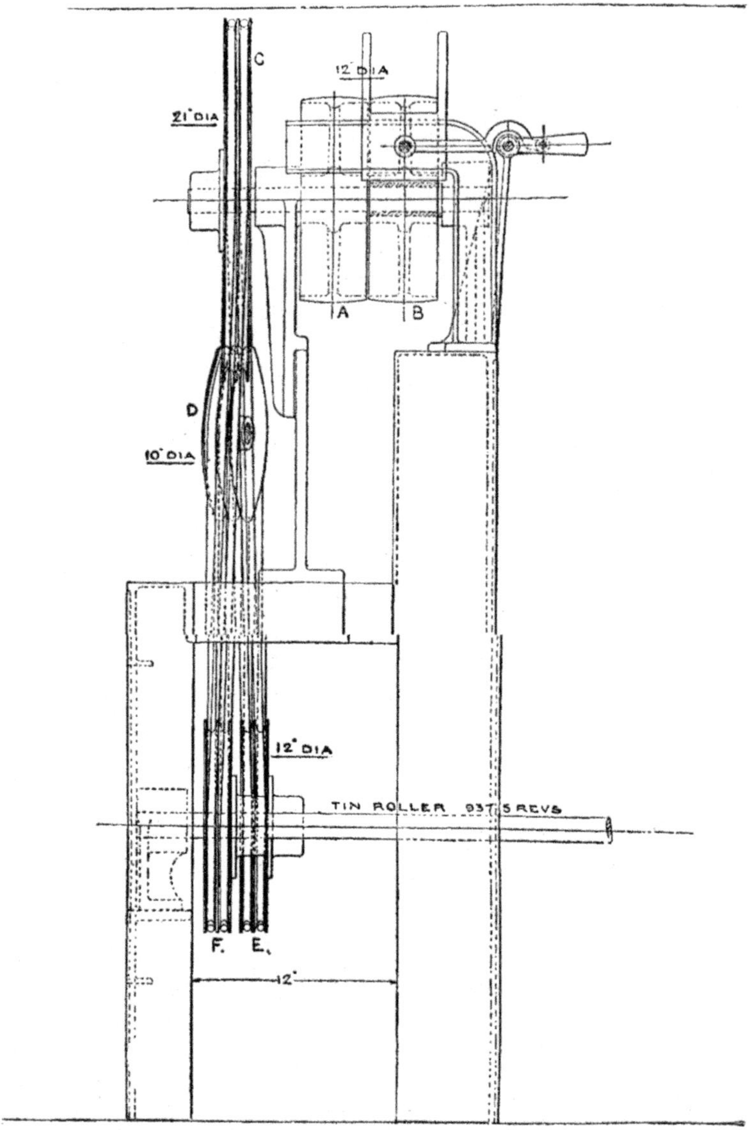

Fig. 203.

combined thread has six strands of 120's in it, and is represented frequently as 6/120's.

Rope Driving.—A feature now frequently seen on doubling frames and sometimes on ring spinning frames is an arrangement for enabling the speed of the tin roller to be altered. In each case, however, the necessity for its use only arises in case of a wide range of twists being worked in the same mill on the same machines. Several systems are adopted, an example of which is given in Fig. 202. The driving pulleys, instead of being placed on one of the tin drum shafts, are carried by suitable supports above the frame end, as at A and B. These two pulleys, it will be observed, are placed between two strong supports, but the short driving shaft is extended, and on it is fixed a rim band pulley C. An endless band passes around this pulley, and is threaded round pulleys E and F keyed to each of the tin drum shafts, and on over a guide pulley D. The passage of the band over the pulleys can be easily understood from the drawings, Figs. 202 and 203; a point to observe is that no crossing of bands is necessary. The pulley D is carried by a slide bracket containing a screw, through which the band can be tightened and kept at a suitable tension. The change pulley C is exactly similar to those used on the mule, and the method of fastening it on the shaft is also the same, so that it is quite a simple matter to change it. From some points of view there is also an advantage in this arrangement of driving, inasmuch as both tin drums are driven alike.

Calculations.—Fig. 188 will enable the following calculations to be readily understood:—

$$\text{Speed of spindles} = \frac{\text{Revs. of G} \times \text{dia. of G}}{\text{Dia. of H}}.$$

$$\text{Revs. of front roller} = \frac{\text{Revs. of } C \times C \times B \times A}{D \times E \times F}.$$

$$\text{Turns of spindle for one of front roller} = \frac{F \times E \times L \times G}{A \times B \times C \times H}.$$

$$\text{Twist per inch} = \frac{F \times E \times D \times G}{A \times B \times C \times H \times J \times 3{\cdot}1416}.$$

$$\text{Twist wheel } A = \frac{F \times E \times D \times G}{\text{Twist} \times B \times C \times H \times J \times 3{\cdot}1416}.$$

$$\text{Twist wheel } B = \frac{F \times E \times D \times G}{A \times \text{Twist} \times C \times H \times J \times 3{\cdot}1416}.$$

In regard to the twist required in doubled yarns, the common practice is to find the counts of the combined thread and put in the twists that would be required for a single yarn of the same counts; a slight allowance is to be made for the difference produced as a consequence of contraction due to twisting.

Two threads of		40's	doubled	= 20's.
Three	,,	60's	,,	= 20's.
Four	,,	80's	,,	= 20's.

The following rules, relating to doubled yarns when the numbers of the single yarns are not the same, are only occasionally required:—

For two yarns of different counts, say A and

$$\frac{A \times B}{A + B} = \text{doubled thread.}$$

When we know the counts of the doubled thread, and it is desired to know the counts of another thread to use with a known yarn to produce it, all that is necessary is to proceed as follows:—

$$\frac{A \times B}{A + B} = \text{doubled thread.}$$

We know the counts of A, and we want to know the counts of B, therefore—

$$\frac{A \times \text{doubled thread}}{A - \text{doubled thread}} = B.$$

If A = 40's and the doubled thread = 20's, what must B equal?

$$\frac{40\text{'s} \times 20\text{'s}}{40\text{'s} - 20\text{'s}} = \frac{800}{20} = 40\text{'s}.$$

For three yarns of unequal counts, doubled together, the usual rule is as follows:—

Weigh a lea of each of the counts;
Divide the weight of each lea by 1000;
Add the quotients together, and
Divide the sum of them into 1000.

Or, Divide the highest count by itself; then
Divide each of the other counts into the highest;
Add the three quotients together, and
Divide their sum into the highest count.

Or, Let three mixed numbers be *a*, *b*, and *c*, then the mixed counts equal

$$\frac{a \times b \times c}{ab + ac + bc} = \text{counts.}$$

Ex.—When Nos. 20, 40, and 60 are doubled, what is the resultant numbers?

$$\frac{20 \times 40 \times 60}{20 \times 40 + 20 \times 60 + 40 \times 60} = 10{\cdot}9 \text{ counts.}$$

Note.—1 per cent is generally allowed when doubling different numbers.

CHAPTER VI

YARN PREPARING MACHINES

Reeling.—When yarn has to be dyed, bleached, and probably shipped abroad, it is usually made into as loose a condition as possible. Within recent years means have been adopted — by using perforated skewers and forcing dye or some bleaching liquid through the cop from its interior as it is immersed in vats or kiers—to bleach and dye the yarn while in the cop and bobbin form. In spite of the progress that has been made in this direction, however, the process of reeling is still largely used for unwinding the cops and bobbins, and, while the yarn is in the unwound state, bleaching and dyeing it by the usual methods.

In some processes the cop and bobbin are not most convenient for the yarn. This is the case especially in the hosiery trade : as the yarn used for this purpose is generally bought, it is first reeled and then made up into bundles for safety, convenience, and economy in transit. On its arrival at the mill it is re-wound into the special form of bobbin or cop that is most convenient for the purpose.

Reeling is performed on a machine called a reel, the chief element of which consists of a "swift." This swift is built up of six longitudinal staves of wood, arranged in the form of a hexagon, each stave being supported throughout its length by arms at an equal distance from the centre of a shaft. Fig. 204 illustrates this feature. Upon the shaft

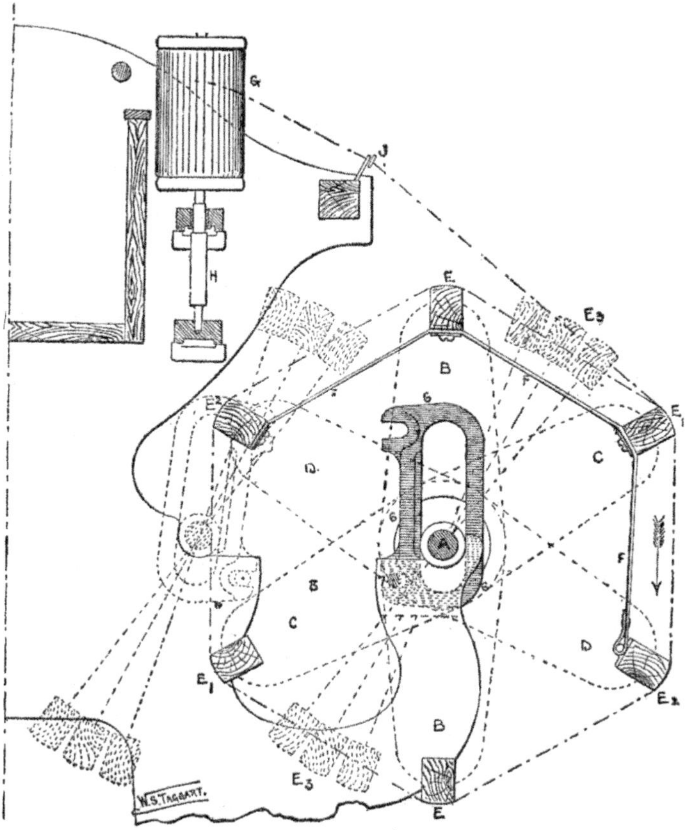

FIG. 204.

A are mounted arms B, C, and D; one of the arms B is fixed to the shaft (or, instead of a shaft, a tin cylinder is used, as being lighter and larger in diameter). The staves E, E_1, and E_2 are in pairs, and are carried by the arms B,

C, and D. Two of the arms (C and D) are loose on the shaft, and can be moved round so that the staves can be brought together, as shown by the dotted lines at E_3; a strap F is generally used to keep the staves in their correct positions when working. All that is necessary, when doffing, is to unhook one end of the strap at G, and the whole swift can be closed up as at E_3.

The circumference of the swift is $1\frac{1}{2}$ yards = 54 inches; the diameter, therefore, across two opposite staves will be approximately—

$$\frac{52 \times 2}{6} = 18 \text{ inches.}$$

Reels assume a variety of forms, according to the work they have to do, but, speaking generally, they may be divided into cop and bobbin reels—or, in other words, the differences between most reels consist of variations in the creel and method of driving. There are single reels for cops or bobbins arranged to be driven by hand or power, these having only one length of swift; a double form for the same purpose with a swift on each side of the machine; a special form for reeling from the cheeses made on a quick-traverse gassing frame, and other types or variations. It will be sufficient to illustrate the essential features of one type only, this being the bobbin reel shown in Fig. 204. The bobbins G are placed upon spindles H, and the yarn led from them through guides or clearing plates at J on to the swift; the swift is driven from the end of the machine (see Fig. 205) by the pulley K, L being the loose pulley. There are several methods of winding the yarn on to the swift so that it will be in a loose condition when taken off again. It must be remembered, however, that while the loose condition is essential, the winding must be so performed that no entanglement will occur in the subsequent processes; therefore some definite method must be adopted that will facilitate re-winding. This, in conjunction with the fact that it is frequently necessary to have very exact lengths wound on the swift, introduces traverse and measuring mechanism, which form a chief feature of most reels.

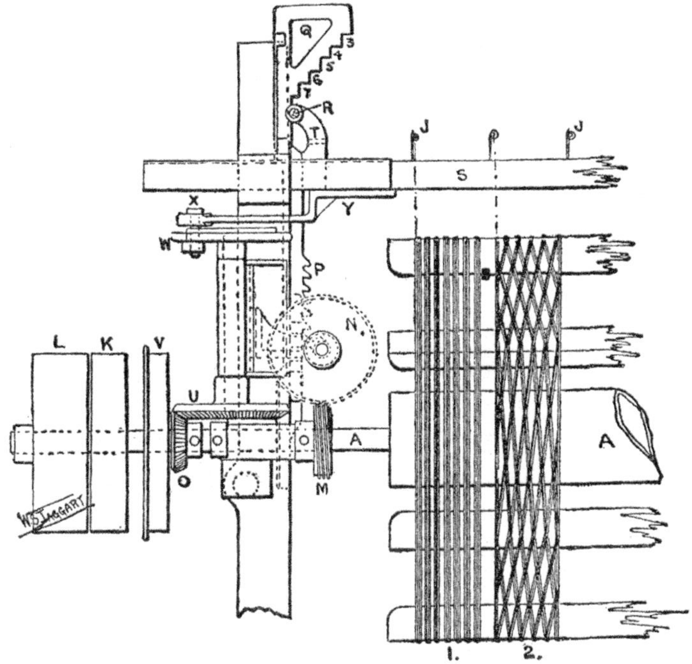

Fig. 205.

When the yarn is wound on the swift in a definite length, the basis of the English system of numbering yarns—namely, 840 yards—is taken as the standard, and as 840 yards of any yarn is termed a "hank" we have a very simple method of obtaining the necessary revolutions for the swift to wind any given weight of yarn. For instance, the swift in one revolution winds on $1\frac{1}{2}$ yards; it will therefore require

$$\frac{840}{1\frac{1}{2}}=560 \text{ revs.}$$

to wind on one hank. Knowing the counts of yarn being wound—say 20's—there would be twenty of these hanks to 1 lb. The usual practice is to make the hanks into bundles of 5 or 10 lb., so that it is an easy matter to say how many of the hanks from a reel are required for a bundle. The hank (or 840 yards) wound on a swift may be put on by laying one-seventh of a hank—that is, 120 yards—on one part of the swift, and then moving the guide wire a little to one side and laying another 120 yards—or "lea," as it is termed—by the side of the first ones. If this is done seven times we get the hank divided into seven parts (or leas), as shown at 1 in Fig. 205. Another method is to arrange the traverse motion to guide the yarn over a space on the swift equal to that occupied by the seven-lea motion. By doing this quickly the yarn is wound on in a "crossed" condition, as shown at 2 in Fig. 205; when definite lengths are wound on in this way the yarn is said to be "skeined," though it is as well to bear in mind that the system can be and is adopted for winding on any length other than a hank. The "Grant" system of winding is a modification of the crossed form, the yarn being crossed in a special manner, which enables a thread to be passed through the openings between the crossings, so as to tie the whole together and prevent entanglements. The seven-lea arrangement also permits the seven divisions to be easily tied together. Fig. 205 will enable the traverse mechanism to be understood. On the shaft A is keyed a worm M, which gears into a worm wheel N. This wheel carries a finger which in the course of its revolution comes against a tooth in a vertical rack P, and lifts the rack bodily. The upper end of the rack P carries a stepped bracket Q, against which a pin R, carried by the finger T, is kept pressed by a spring. The finger T being fixed to the guide rail S, the guide wires J will cause the yarn to be wound on the swift at the points they happen to be opposite. Now suppose the rack is lowered and the pin R is on the step at 3, then when the finger on N lifts the rack P the pin will shoot into the next step at 4, and the guide plate S will move the depth of the notch away on one side of its previous position. This continues until the seven leas have been laid on the swift; at the termination of the last length an automatic arrangement moves the strap on to the loose pulley, and a brake comes into contact with the pulley V and stops the frame instantly. A catch which drops into teeth on the side of the rack P prevents the rack falling back after it is lifted by the finger on N.

The same illustration shows the method of obtaining the crossing motion. In this case the traverse is continuously moving to and fro, so that the parts M, N, P, and R are not used; instead, we have a wheel O driving another wheel U, on the upper end of which is a crank W carrying a pin X, which also engages with a finger Y fixed to the traverse rod S. The revolution of W will move X backwards and forwards, and according to the "throw" of the crank will give a traverse of the yarn on the swift. The pin at X is the same as that used at R, so that if a machine is fitted up with both motions, the change from the "lea" motion to the "crossing" motion can be effected by simply changing the pin R to X.

The doffing of the yarn from the swift is done as follows:—First, the staves are drawn together, as shown at E_3, Fig. 204. As there are generally 40 hanks wound on one swift, all this requires to be moved to one end of the machine and then taken off. This is not a simple matter, for the swift and the yarn it contains are heavy, and the support of the swift if fixed prevents taking off the yarn; the end support is therefore modified, and in its place the old style of bearing was that shown at Fig. 206. Better systems are now adopted, one of which is shown in Fig. 204; this is called the "bridge" doffing motion. The shaft A is carried by the framing, but surrounding it is a loose guide 6, which is loosely hinged on a pin 7; when the staves are brought together the yarn is drawn over the end and dropped into the opening 8. This done, the guide 6 is turned over from the pin 7 as a centre, and then occupies the position shown in the dotted lines, leaving a clear opening for the doffed yarn to be lifted out.

FIG. 206.

Coleby's Reel.—A machine sometimes used instead of the long 40-hank machines illustrated in Fig. 204 is given in Fig. 207. It consists of four independent "swifts" (B, C, D, and E), all of which are driven from the middle of the machine by one belt. The swifts are of the ordinary construction, and each one is generally made to reel ten hanks, so that a complete machine will work forty hanks at a time. Every thread has an automatic stop motion, but only that section or swift is stopped which is at the time reeling the thread; this is a decided advantage, as it saves considerable time and material; only ten hanks are stopped for piecing instead of forty hanks, as in the previously described machines.

The off end of each swift, as at J and H, is supported by a bracket K, which is provided with a stud J and a sliding bush H. When doffing, the bush H is moved along the shaft of the swift and a clear opening is left for the reeled hanks to be slipped off the closed staves of the swift. The drawing shows the machine ready to be doffed, and one swift is represented as at the point ready for the hanks to be taken off. Any system of reeling can be applied to the machine and any required length wound on the swifts with automatic stop motion to each section when the correct quantity is put on. Cops, bobbins, cheeses, etc.,

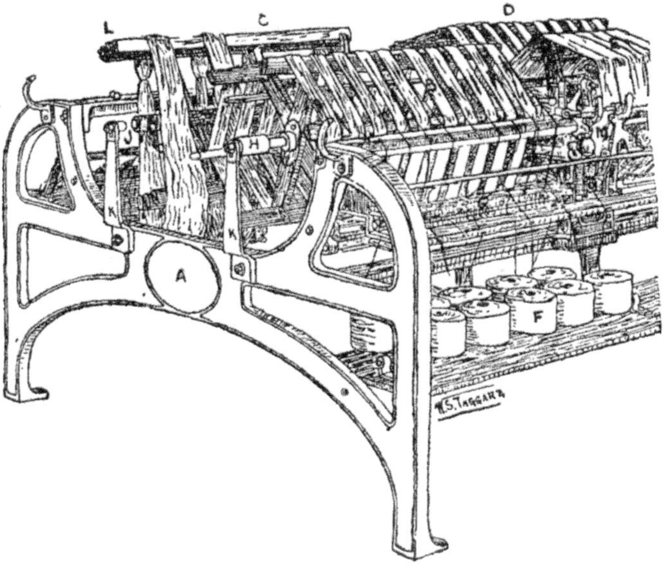

FIG. 207.

can be reeled, and one, two, or more ends reeled together. As in the previous machine, a change from "lea" to "cross" or "Grant" reeling can be effected by simply changing a stud. The gain in production, through saving in time alone, over the ordinary reel is as high as 25 per cent, and one child with any smartness can take care of one side of a machine—that is, twenty cops—and produce as much as an average reeler from an ordinary 40-hank reel.

Fig. 208 shows a sectional view of the machine, and it

will be observed that the strap A drives both lines of swifts. This is not done directly, but through the arrange-

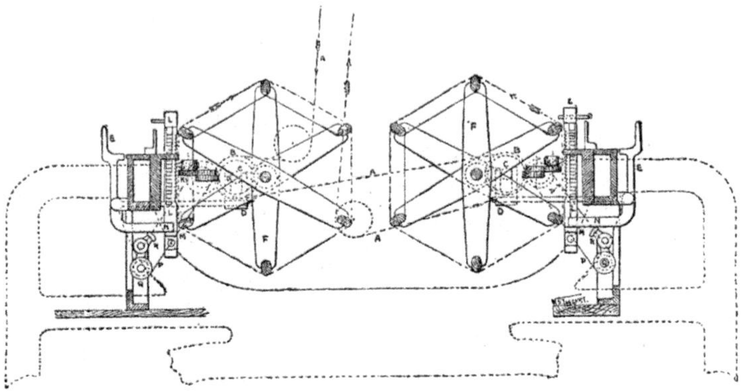

Fig. 208.

ment shown in Fig. 209, which represents a plan view of the gearing. The strap drives the pulley B which runs loose on the shaft C; each side of the pulley has one part of a clutch wheel which gears with the other half connected to the shaft C by a float key. The boss of this half of the

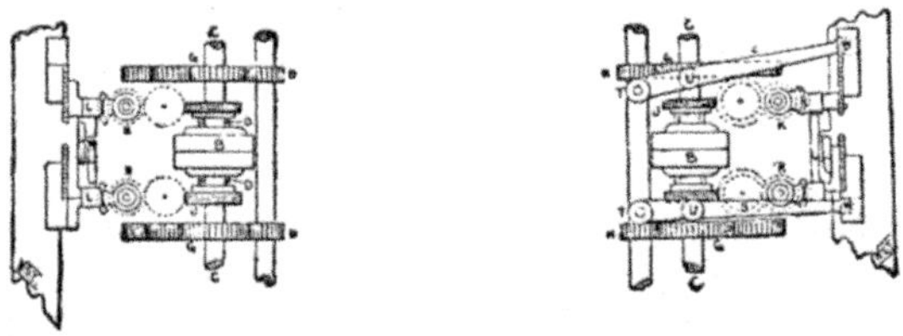

Fig. 209.

clutch wheel has a groove for the fork D, so that by moving aside the handle E any one or all the swifts F can be stopped through disconnecting the clutch wheels. The shaft C drives the swift through the wheels G and H. The regulating and measuring mechanism is driven from the worm J; the finger on the wheel K lifts the vertical rack L, and so permits the traverse rod to escape a tooth after one-seventh of a lea has been wound on at one spot. The stoppage of the swift after it has revolved a sufficient number of times is brought about by a stop M on the vertical rack coming against a projection N on the setting-on handle E and releasing it from a catch which holds it in position. The automatic stop motion when an end breaks is worked by the band P, which drives the shaft Q, on which are the spiders; a needle falls on the spiders when an end breaks, but their continued movement forces the needle box on one side and causes the lever R to lift up and release the handle E.

The right-hand sketch in Fig. 209 will illustrate how the crossing motion is obtained; a lever S centred at T is actuated by a cam at U; this gives a to-and-fro motion to the end of the lever at W; its connection to the traverse bar produces a quick reciprocation motion, and so crosses the yarn on the swift.

Gassing.—Gassing is a process in which yarn is passed rapidly over a light for the purpose of burning off the numerous ends of fibres which stand out from the body; it is a very necessary operation for all purposes where yarn is required to be as round, smooth, and solid as possible. In general, it consists in taking the yarn from the cops or bobbins A (Fig. 210), and after passing it through tension guides B and C, threading it backwards and forwards over small grooved pulleys D and E, and from here on to a quick-traverse drum-winding arrangement at F. Between the two pulleys D and E, just under the point where the yarn is crossed, is placed a gas-burner G, having a number of very small jets of flame; in many cases the burner is of the atmospheric kind known as the Bunsen burner. The pale-blue flame, devoid of carbon, is intensely hot, and performs the function of singeing most effectively. The yarn passes through the light at from 200 to 250 feet per minute, and it does this from 7 to 11 times before being

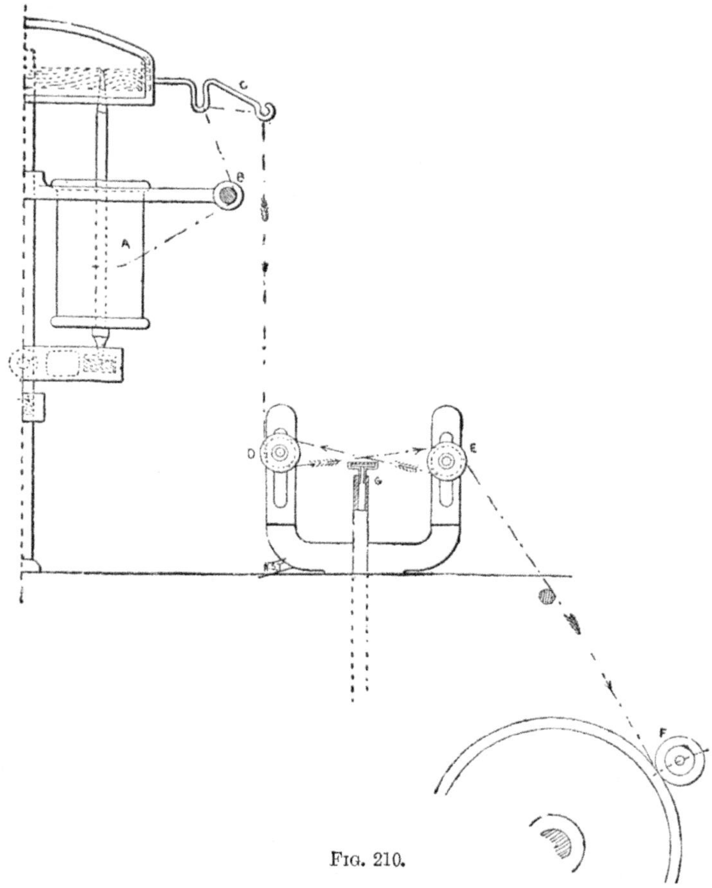

Fig. 210.

wound upon the "cheese." The coarser numbers—say, 30's twofold—go through the light at the slower rate and the higher number of times, while the higher counts—such as 200's twofold—pass over the light at the quickest rate and the least number of times. Extra folds of yarn go slower still, and the higher the counts the quicker the passage of the yarn must be.

There is, of course, a loss in gassing yarn, the amount depending upon the extent of the gassing and the quality of the yarn—7 to 8 per cent representing the average, this meaning that if 100's yarn is gassed the resulting yarn will be about 108's. The gas is supplied to the burners by a pipe which runs the full length of the frame on each side; the burners are connected to it by a swivel joint, and arrangements are made so that when an end breaks and piecing is being effected, or the machine is stopped, the burners move aside from underneath the yarn. On setting on the machine, the winding commences before the burners are moved, so that scorching or burning the threads is entirely avoided. The traverse motion is now almost universally an adaptation of the quick traverse motion, practically the same mechanism being used.

Bundling Press.—As its name implies, this is a machine for pressing a number of hanks of cotton into a smaller compass, and while under pressure the compressed yarn is tied up into bundles by hand. The size of the bundles varies, but generally 5 or 10 lb. bundles are formed in the machine. The illustration (Fig. 211) will convey a very good idea of the press; its upper part is formed by a series of strong bars projecting above the table in two sets; a narrow space separates each bar, through which string is passed; packing paper is placed over the string, and on this are placed the hanks of the number to make up the weight to 10 lb.; another paper is placed over the top, and the machine set in motion. The action is as follows:—The driving pulleys, through strong gearing, turn a pair of eccentrics or cams; these lift up the sliding base upon which the hanks are placed, and as this base rises, the top of the yarn box is formed by a series of strong bars hinged to one of the sets of projecting bars above mentioned, being automatically caused to swivel down and become locked in the opposite set of bars; a strong top is thus formed, and against it the bundle is pressed to the required

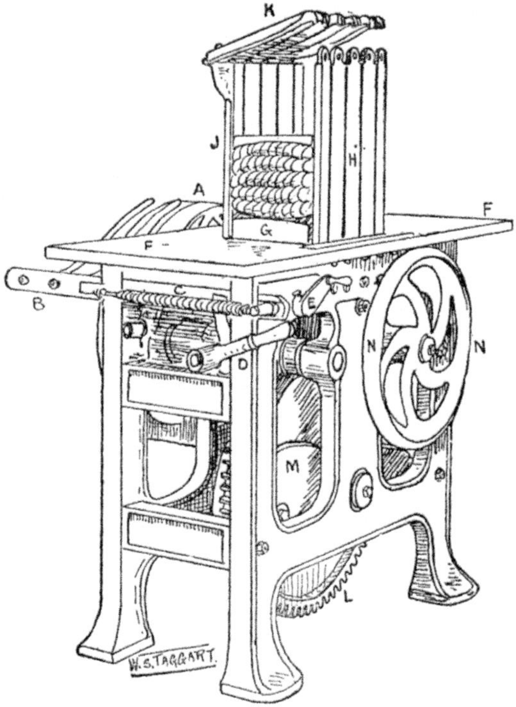

Fig. 211.

dimensions. An interval is allowed for completing the packing and tying by hand, after which the base lowers and the top bars move upwards, leaving a clear space again for the withdrawal of the now complete bundle. The machine is generally made to be worked either by hand or power, and 180 ten lb. bundles can be made per day of 10 hours.

CHAPTER VII

MILL PLANNING

MILL planning is essentially a branch of a student's work in the subject of cotton spinning, and for that reason it will be noticed in these pages. It is, however, so emphatically a distinct branch, from a practical point of view, that outside of giving an intelligent idea of how mill planning is done, it is not our purpose to do more than give sufficient information to enable one to use it as a basis for further practice, and as an aid in understanding the work of others. To plan a mill with all due regard to room, lengths of machines, arrangement of machinery, passages, driving, economy in carriage from process to process, speeds, hanks, productions, drafts, etc., etc., requires years of practical experience, and only those whose sole occupation it is can do it thoroughly.

The total weight of yarn required to be produced and the counts spun are the two chief items it is necessary to know before commencing to plan the machinery; with these as a basis, we will proceed to give examples of various mills spinning different classes of cotton.

Example.—A 10,000 spindle mill, spinning average 20's on mules.

A suitable space of spindle will be $1\frac{1}{8}$ inch.

A suitable length of mule will contain 1000 spindles, so that there will be ten mules.

A mule spinning 20's will produce about thirty-two hanks per spindle per week.

The ten mules will produce—

$$\frac{1000 \times 32 \text{ hanks}}{20 \text{ counts}} = 16{,}000 \text{ lb. of yarn.}$$

Proceed now to make a table as follows, filling in the necessary data according to experience; the data given in the tables are good average results.

Machine.	Hank Roving.	Draft.	Hanks and lbs.	Total Weight.
Card . . .	·138	100	700	16,800
Draw Frame . .	·138	6	1000	16,680
Slubber . . .	$\frac{1}{2}$	3·6	50	15,560
Intermediate . .	$1\frac{3}{8}$	5·5	44	16,440
Rover . . .	$3\frac{1}{4}$	4·72	40	16,320
Mule . . .	20	6·15	32	16,000

Note.—Allow five per cent waste between the card and the mule. Of this, allow two per cent in the mule, the rest being divided among the other machines, this being sufficient for practical purposes.

On referring to the table we note that the card produces 700 lb. per week, so that

$$\text{No. of Cards} = \frac{16800}{700} = 24$$

$$\text{No. of Draw Frame deliveries} = \frac{16680}{1000} = 16{\cdot}68$$

$$\text{No. of Slubber Spindles} = \frac{16560 \times \frac{1}{2}}{\cdot 50} = 165$$

$$\text{No. of Intermediate Spindles} = \frac{16440 \times 1\frac{3}{8}}{44} = 514$$

$$\text{No. of Roving Spindles} = \frac{16320 \times 3\frac{1}{4}}{40} = 1326$$

$$\text{No. of Mule Spindles} = \frac{16000 \times 20}{32} = 10000$$

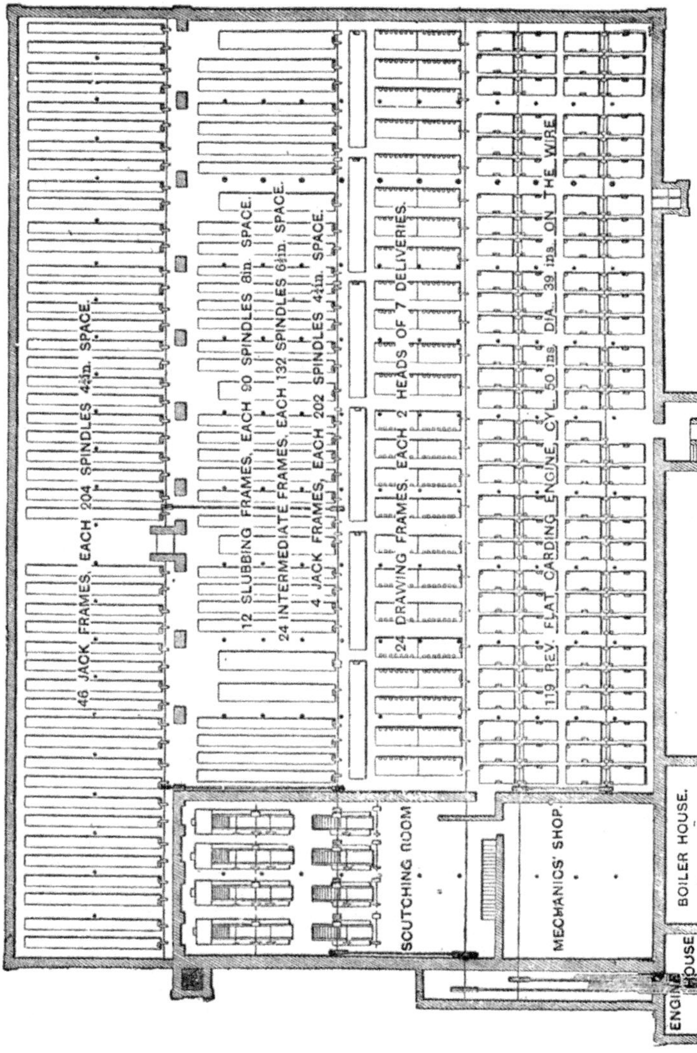

PLAN OF THE CARD ROOM FOR A MILL OF 80,648 SPINDLES.

FIG. 212.

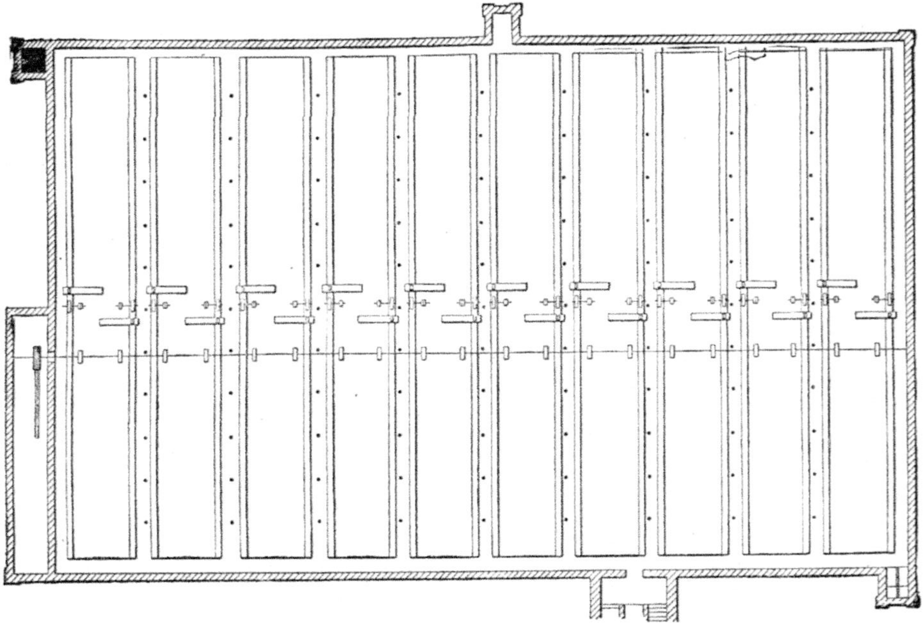

4th SPINNING ROOM OF MILL Spinning 40's Twist.

1st Room	contains	20 mules,	each	1050 spindles,	$1\frac{3}{4}$	in.	space	=21,000	spindles.
2nd	,,	,,	16	,,	,, 1058	,,	$1\frac{11}{16}$	,,	,,	=16,928 ,,
3rd	,,	,,	20	,,	,, 1066	,,	$1\frac{21}{32}$	,,	,,	=21,320 ,,
4th	,,	,,	20	,,	,, 1070	,,	$1\frac{5}{8}$	,,	,,	=21,400 ,,
										80,648 ,,

FIG. 213.

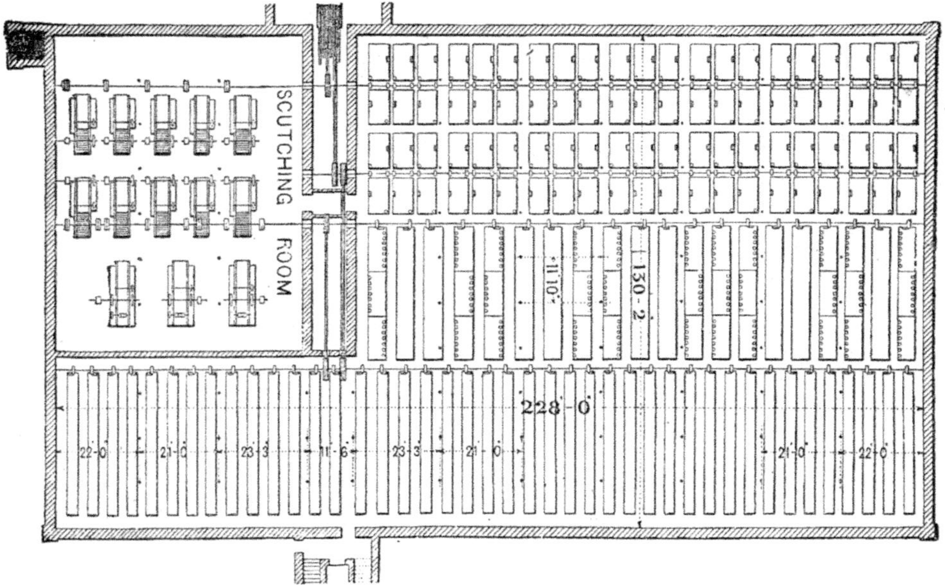

CARD ROOM OF 5-STOREY MILL

Containing 78,384 Mule Spindles, Spinning American Cotton, Nos. 32's Twist and 45's Weft.

1 Bale Breaker and Lattice Mixing Arrangement.
3 Horizontal Exhaust Openers.
10 Single Scutchers.
84 Carding Engines.
11 Draw Frames, each 3 heads of 7 deliveries.

9 Slubbers, each 98 spindles, 8 in. space.
16 Intermediates, ,, 132 ,, $6\frac{1}{2}$ in. ,,
26 Rovers, ,, 172 ,, 5 in. ,,
14 Rovers, ,, 172 5 in. ,, in 1st Spinning Room.

FIG. 214.

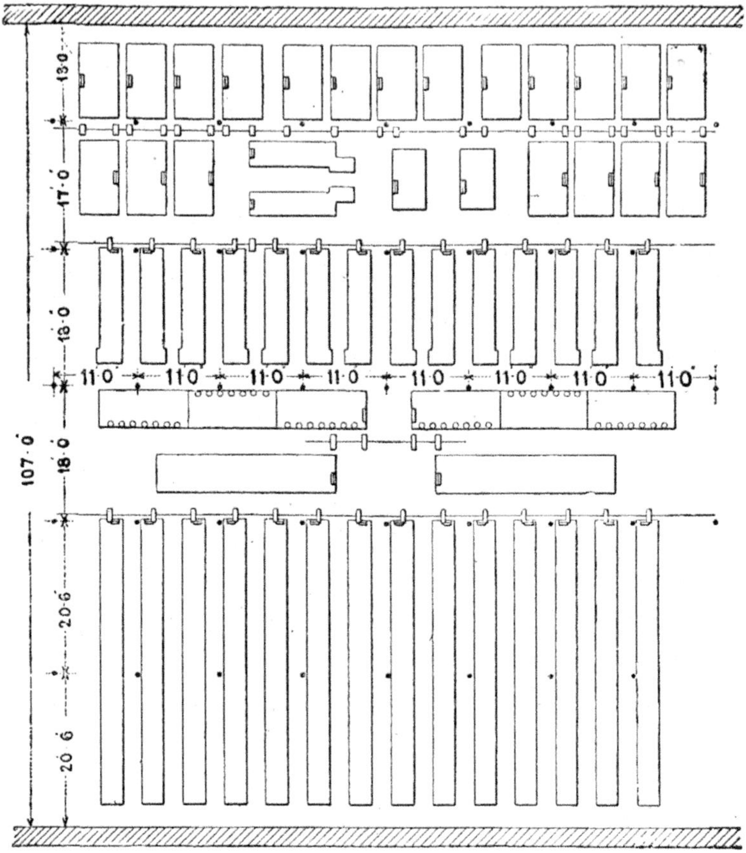

PREPARING MACHINERY

For 20,000 Mule Spindles, Spinning No. 80's Combed Yarn.

20 Carding Engines.
2 Silver Lap Machines.
2 Draw and Lap Machines.
14 Combing Machines of 8 heads each.
2 Draw Frames, each 3 heads of 7 deliveries.
2 Slubbers, 64 spindles each, 7 in. space.
4 Intermediates, 130 ,, ,, $6\frac{1}{2}$,, ,,
10 Jacks, 200 ,, ,, $4\frac{1}{4}$,, ,,

FIG. 215.

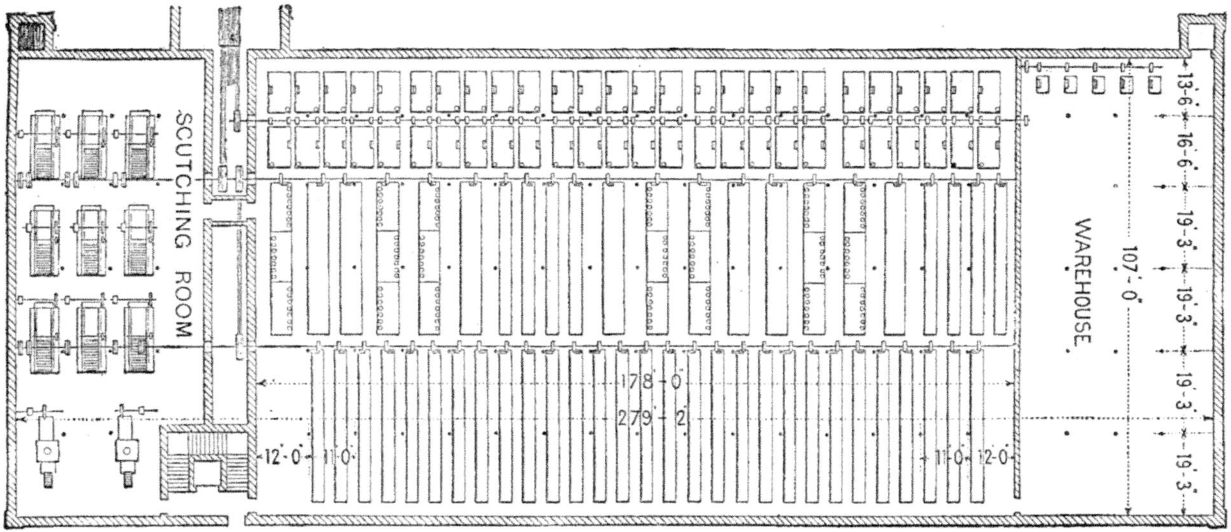

CARD ROOM OF INDIAN MILL,

Containing 20,000 Ring Spinning Spindles, Spinning No. 20's.

2 Single Vertical Openers, with Porcupine Feeder.	8 Intermediates,	120	spindles each,	$6\frac{1}{2}$	in. space.
9 ,, Scutchers.	4 ,,	122	,, ,,	$6\frac{1}{2}$	,, ,,
52 Carding Engines.	25 Rovers,	160	,, ,,	5	,, ,,
7 Draw Frames, each 3 heads of 7 deliveries.	5 Bundling Presses.				
7 Slubbers, 78 spindles each, 10 in. space.					

FIG. 216.

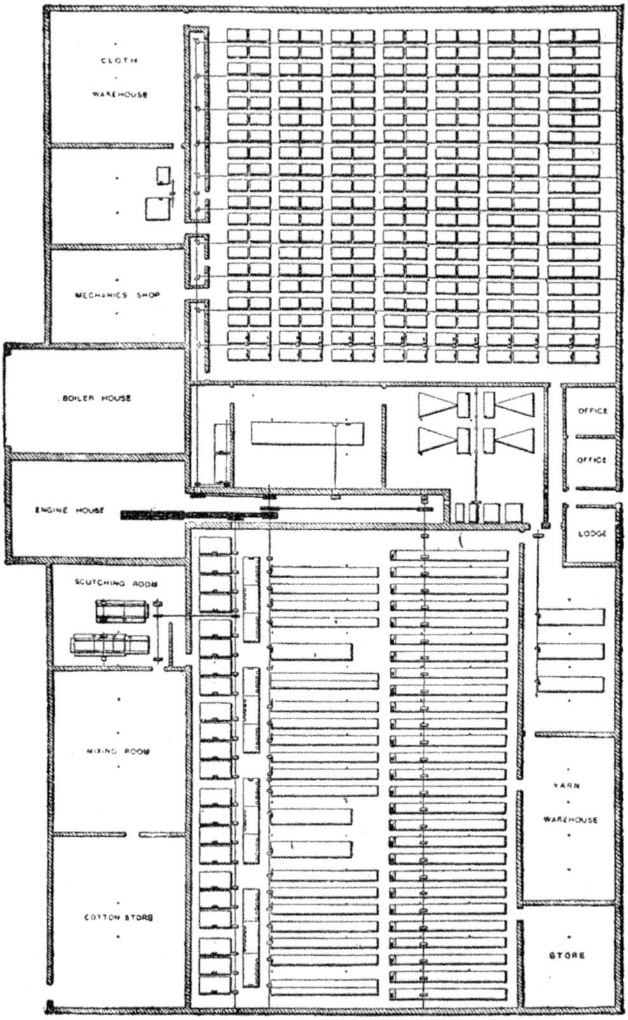

SPINNING AND WEAVING MILL.

Containing 9316 Ring Spinning Spindles, Spinning 16's to 36's Twist and 20's to 36's Weft, and 280 Looms.

Fig. 217.

From these results we decide upon suitable lengths of frames, etc., and draw up a table of the machinery necessary for the mill, first noting that a vertical opener will produce 30,000 to 40,000 lb. per week, and a single scutcher 15,000 to 20,000 lb. per week.

1 Double Vertical Opener.				
3 Single Scutchers.				
24 Cards.				
4 Draw Frames, 3 heads of 5 deliveries.				
2 Slubbing Frames,	86 spindles,	8 in. space,	10 in. lift.
4 Intermediate ,,	132 ,,	$6\frac{1}{2}$,, ,,	9 ,, ,,
8 Roving Frames,	170 ,,	5 ,, ,,	7 ,, ,,
10 Mules,	1000 ,,	$1\frac{1}{8}$,, ,,	

From this example the general method adopted to obtain the number and dimensions of the machines will be easily understood. They are then planned out to scale to the best advantage, and to illustrate this planning several examples are given in Figs. 212 to 217.

The following particulars may prove useful as a guide to the planning of a mill, together with other information given in various articles that have already appeared :—

PRODUCTION OF THE CARD.

Nos.	Kind of Cotton.	Weight of Lap per yd.	Hank Carding.	Lbs. per Card in 10 hours.	Usual Draft.
16	Indian or American .	$13\frac{1}{2}$ oz.	·138	178·2	93
18	,, ,, .	$13\frac{1}{2}$,,	·138	168·15	93
20	,, ,, .	$13\frac{1}{2}$,,	·138	159·3	93
30	American . . .	13 ,,	·154	141·6	100
40	,, . . .	12 ,,	·173	119·05	104
40	Egyptian . . .	12 ,,	·189	80·53	113
50	,, . . .	$11\frac{1}{2}$,,	·208	71·5	119
60	,, . . .	11 ,,	·208	61·95	114
70	,, . . .	11 ,,	·231	58·4	127
80	,, . . .	11 ,,	·231	52·1	127
80	,, Combed .	11 ,,	·231	55·3	127
90	,, ,, .	10 ,,	·277	53·5	138
100	,, ,, .	10 ,,	·277	50	138

For a week of $56\frac{1}{2}$ hours multiply the above productions by 5·65 for a week's production.

$$\text{Production in 10 hours} = \frac{\text{min. in 10 hours} \times \text{revs. of calender roller} \times \text{dia. of calender roller} \times 3{\cdot}1416 \times \text{weight of sliver in grains per yard}}{36 \times 7000}$$

PRODUCTION OF DRAW FRAMES.

Dia. of F. Roller.	Revs. of F. Roller.	Weight of Sliver per yard.	Hank.	Lbs. per Delivery in 10 hrs.	Nos.	Cotton.
$1\frac{1}{8}$	400	66	·126	180·54	10's to 20's	Indian or China
$1\frac{1}{8}$	400	60	·138	164·25		
$1\frac{1}{4}$	350	60	·138	158·76	20's/24's	Indian or China
$1\frac{1}{4}$	350	54	·154	142·83	24's/32's	American
$1\frac{1}{4}$	350	48	·173	126·9	32's/40's	American
$1\frac{3}{8}$	300	48	·173	120·18	32's to 40's	American or Low Egyptian
$1\frac{3}{8}$	300	44	·189	110·27		
$1\frac{3}{8}$	300	40	·208	100		
$1\frac{1}{2}$	280	48	·173	122·48	30's/40's	Egyptian
$1\frac{1}{2}$	280	44	·189	112·21	40's/45's	,,
$1\frac{1}{2}$	280	40	·208	102·12	45's/50's	,,
$1\frac{1}{2}$	250	40	·208	91·15	60's	,,
$1\frac{1}{2}$	250	36	·231	81·95	70's	,,
$1\frac{1}{2}$	200	40	·208	72·74	80's	,,
$1\frac{1}{2}$	200	36	·231	65·49	90's	,,
$1\frac{1}{2}$	200	30	·277	54·51	100's	,,

$$\text{Production in 10 hours} = \frac{\text{min. in 10 hours} \times \text{revs. of F.R.} \times \text{dia. of F.R.} \times 3{\cdot}1416 \times \text{grains per yard of sliver}}{36 \text{ in.} \times 7000 \text{ grains}}$$

PROPORTIONS OF MACHINERY

Nos.	Cotton.	1 Card.	1 Comber.	1 Draw Frame.	1 Slubber.	1 Intermediate.	1 Rover.	1 Jack.	Sliver Lap.	Draw and Lap.
		Number of Mule Spindles to								
16 T	Indian or American	465	...	476	47·6	26·4	8·7	...	...	...
18 T	,, ,,	494	...	476	52·9	30·7	8·7	...	...	...
20 T	,, ,,	519	...	519·4	59·5	30·3	8·5	...	...	...
20 T	Indian	500	...	500	50·7	18·77	6·56	...	...	...
30 T	American	714	...	714	77	29·8	9·9	...	...	...
40 T	,,	952	...	952	101	38·4	11·9	...	...	...
40 T	Egyptian	645	...	833·3	92·59	26·8	...	7·09	...	...
50 T	,,	763	...	1111	106·38	32·25	...	8·18	...	...
60 T	,,	869·5	...	1333	119	38·4	...	9·18	...	...
70 T	,,	1025·6	...	1428·5	121·4	38·4	...	9·09	...	...
80 T	,,	1111	...	1428·5	156·25	38·46	...	10	...	...
80 T	,, Combed	1000	1428	1428·5	156·25	38·46	...	10	10,000	10,000
90 T	,, ,,	1176	1818	1666	294	135	41·6	10·4	10,000	10,000
100 T	,, ,,	1333	2000	2000	311·9	138·8	40·16	11·45	10,000	10,000
		Number of Ring Spindles to								
16 T	Indian or American	285·6	...	292·4	29·12	16·24	4·68	...	...	...
20 T	,, ,,	313	...	313	35·8	18·25	4·62	...	...	...
20 T	Indian	384	...	400	38·5	14·25	5·00	...	...	...
30 T	American	485·8	...	485·8	52·5	20·3	5·65	...	...	...
40 T	,,	631	...	631	66·9	20·7	6·57	...	...	...
14 T	China	319	...	303	31·5	14·4	5·6	...	...	...
14 T	Japan	277	...	277	27·7	12·6	4·6	...	...	...

The foregoing Table is based upon the following data:—

Nos.	Cotton.	Machine.	Hank Roving.	Hks. per Spindle in 10 hrs.	Lb. per Spindle in 10 hrs.	Draft.	Ring or Mule Yarn.
16	Indian or American	Card	·138	...	178·2	93	Ring and Mule
18	,, ,,	,,	·138	...	168·15	93	,, ,,
20	,, ,,	,,	·138	...	159·3	93	,, ,,
30	American	,,	·154	...	141·6	100	,, ,,
40	,,	,,	·173	...	119·05	104	,, ,,
40	Egyptian	,,	·189	...	80·53	113	Mule
50	,,	,,	·208	...	71·5	119	,,
60	,,	,,	·208	...	61·95	114	,,
70	,,	,,	·231	...	58·4	127	,,
80	,,	,,	·231	...	52·1	127	,,
80	,, Combed	,,	·231	...	55·3	127	,,
90	,, ,,	,,	·277	...	53·5	138	,,
100	,, ,,	,,	·277	...	50·0	138	,,
16	Indian or American	Draw Frame	·126	...	181·07	5·45	Ring and Mule
18	,, ,,	,, ,,	·138	...	161·07	6	,, ,,
20	,, ,,	,, ,,	·138	...	158·0	6	,, ,,
30	American	,, ,,	·154	...	140·5	6	,, ,,
40	,,	,, ,,	·173	...	118·2	6	,, ,,
40	Egyptian	,, ,,	·189	...	103·36	6	Mule
50	,,	,, ,,	·208	...	102·4	6	,,
60	,,	,, ,,	·208	...	94·16	6 or 8	,,
70	,,	,, ,,	·231	...	80·07	6 or 8	,,
80	,,	,, ,,	·231	...	66·37	6 or 8	,,
80	,, Combed	,, ,,	·231	...	66·37	6 or 8	,,
90	,, ,,	,, ,,	·208	...	64·16	6 or 8	,,
100	,, ,,	,, ,,	·231	...	63·18	6 or 8	,,
80	,, ,,	Comber 8 hds.	·231	...	66·99	5	,,
90	,, ,,	,, ,,	·208	...	70·8	5·8	,,

100	Egyptian Combed .	Comber 8 hds. .	·231	...	63·72	5·8	Mule
16	Indian or American .	Slubber . . .	½	8·85	...	3·96	Ring and Mule
18	,, ,, .	,, . . .	½	8·85	...	3·6	,, ,,
20	,, ,, .	,, . . .	½	8·85	...	3·6	,, ,,
30	American . . .	,, . . .	⅝	9·41	...	4	,, ,,
40	,, . . .	,, . . .	¾	9·32	...	4·3	,, ,,
40	Egyptian . . .	,, . . .	⅞	9·98	...	4·6	Mule
50	,, . . .	,, . . .	1	9·73	...	4·8	,,
60	,, . . .	,, . . .	1⅛	9·38	...	5·4	,,
70	,, . . .	,, . . .	1¼	8·99	...	5·4	,,
80	,, . . .	,, . . .	1¼	8·99	...	5·4	,,
80	,, Combed .	,, . . .	1¼	8·99	...	5·4	,,
90	,, ,, .	,, . . .	⅞	9·78	...	4·2	,,
100	,, ,, .	,, . . .	1	9·73	...	4·3	,,
16	Indian or American .	Intermediate . .	1	9·9	...	4	Ring and Mule
18	,, ,, .	,, . .	1	10	...	4	,, ,,
20	,, ,, .	,, . .	1⅛	10·08	...	4·5	,, ,,
30	American . . .	,, . .	1½	8·67	...	4·8	,, ,,
40	,, . . .	,, . .	1¾	8·24	...	4·6	Mule
40	Egyptian . . .	,, . .	2¾	9·02	...	6·2	,,
50	,, . . .	,, . .	3	8·79	...	6	,,
60	,, . . .	,, . .	3¼	8·67	...	5·76	,,
70	,, . . .	,, . .	3¾	8·03	...	6	,,
80	,, . . .	,, . .	4	7	...	6·4	,,
80	,, Combed .	,, . .	4	7	...	6·4	,,
90	,, ,, .	,, . .	2	10·17	...	4·4	,,
100	,, ,, .	,, . .	2¼	9·71	...	4·8	,,
16	Indian or American .	Roving . . .	2½	7·96	...	5	Ring and Mule
18	,, ,, .	,, . . .	2¾	7·96	...	5·5	,, ,,
20	,, ,, .	,, . . .	3	7·43	...	5·32	,, ,,
30	American . . .	,, . . .	4	7·61	...	5·2	,, ,,
40	,, . . .	,, . . .	4¾	6·93	...	5·4	Mule
90	Egyptian Combed .	,, . . .	4½	7·03	...	4·5	,,

Nos.	Cotton.	Machine.	Hank Roving.	Hks. per Spindle in 10 hrs.	Lb. per Spindle in 10 hrs.	Draft.	Ring or Mule Yarn.
100	Egyptian Combed .	Jack . . .	5½	6·81	...	4·8	Mule
40	,, . . .	,, . . .	9	7·77	...	6·5	,,
50	,, Combed .	,, . . .	10	7·38	...	6·6	,,
60	,, ,, .	,, . . .	11	6·93	...	6·76	,,
70	,, ,, .	,, . . .	13	6·54	...	6·92	,,
80	,, ,, .	,, . . .	14	6·31	...	7	,,
80	,, ,, .	,, . . .	14	6·31	...	7	,,
90	,, ,, .	,, . . .	15½	5·89	...	6·8	,,
100	,, ,, .	,, . . .	16½	5·78	...	6·0	,,
16	Indian or American .	Mule . . .	16	5·84	...	6·4	
18	,, ,, .	,, . . .	18	5·84	...	6·54	
20	,, ,, .	,, . . .	20	5·84	...	6·66	
30	American . . .	,, . . .	30	5·66	...	7·5	
40	,, . . .	,, . . .	40	4·77	...	8·4	
40	Egyptian . . .	,, . . .	40	4·77	...	8·8	
50	,, . . .	,, . . .	50	4·42	...	10	
60	,, . . .	,, . . .	60	4·07	...	10·9	
70	,, . . .	,, . . .	70	3·80	...	10·76	
80	,, . . .	,, . . .	80	3·54	...	11·4	
80	,, Combed .	,, . . .	80	3·54	...	11·4	
90	,, ,, .	,, . . .	90	3·27	...	11·6	
100	,, ,, .	,, . . .	100	3·00	...	12	
16	Indian or American .	Ring . . .	16	9·51	...	5·8	
20	,, ,, .	,, . . .	20	9·7	...	6·15	
30	American . . .	,, . . .	30	8·32	...	6·6	
40	,, . . .	,, . . .	40	7·21	...	7·27	
14	China . . .	,, . . .	14	7·61	...	5·6	
14	Japan . . .	,, . . .	14	7·61	...	5·6	

CHAPTER VIII

HUMIDITY

Humidity of the air in cotton mills is a subject upon which much has been lately written, and so important as well as interesting is the subject that several writers and able observers have enabled the industry to benefit considerably by the results of their observations, experience and advice. Two names stand out very clearly in this connection—namely, Sir B. A. Dobson and Mr. W. W. Midgeley—and the fruits of their combined labours in books published by Messrs. Dobson and Barlow will also be looked upon for some time as standard works on this subject. Under these circumstances it is not intended to do more than give a mere outline of this feature of mill management.

The essential meaning of humidity is dampness or moisture, and its association with spinning relates to the condition of the atmosphere of the rooms in which spinning operations are in progress. Now this moist condition of the air involves two factors: First, the actual amount of moisture; and, secondly, the relative amount. Strange to say, the actual amount of moisture in a spinning room is not the deciding factor in the case. For instance, yarn may be spun well when the temperature of the room is, say, 70° F., but if that temperature is raised to 90° F., everything else remaining the same, there will be a vast difference in the humidity of the room in spite of the fact that the actual amount of moisture in the air is the same in each case. When one enters a room that is well heated it will be noticed that it is very dry or has a parching effect, but as a matter of fact a cubic foot of air in such a room will have quite as much moisture in it as a cubic foot of outside air. The question may then be asked—Why do we say air is dry or moist? The explanation lies in the fact that these terms dry and moist are not actual but simply relative terms, and that the human body is not capable of deciding from its sensations what the actual humidity of the atmosphere may be. We have an analogous example in the case of temperature. A spinning room may be 90° F. and inconveniently hot, but any one placing a hand on the framing of a machine would feel it very cold, while in reality the iron is at practically the same temperature as the room.

We all know that the air in summer is much drier than in winter, though it is equally well known that there is more moisture in the air during summer than winter. These considerations lead us to the conclusion that the actual amount of moisture in the air is not a deciding factor in our estimate of humidity, so that we must seek for some other element to solve the problem. If water is left to itself in contact with air it will slowly pass into a state of a gas or vapour, the phenomenon being known as evaporation. The water which has thus been transformed into vapour is in an extremely subdivided state, and diffuses very rapidly in the air without increasing the volume of the air with which it has become mixed. It is advisable to point out here that when moisture enters the atmosphere, whether by evaporation or by spraying, no chemical combination takes place: it is purely a mechanical mixture. The vapour of water, therefore, by virtue of its elastic force, which it possesses in common with all other gases, takes up its position between the molecules of the air wherever it is free to do so, and moreover it always remains moist and acts just as it would if it were confined within a vacuum. Now, under a given set of conditions, vapour would continue to rise from the surface of the exposed water until the vapour tension exactly equals the tension which keeps the water in a state of water; after this state has been reached no further evaporation can take place, for the air has now mixed up within itself as much vapour as it can hold, or a much better way to put it is to say that the particles of vapour in the air are putting forth all the pressure they are capable of exerting in keeping each other from changing back into the state of water from which they have arisen. The air is now said to be "saturated," and the particles of vapour are exerting their "maximum pressure" and are also at their "maximum density." If the temperature of this saturated air is now lowered, if it is compressed into a smaller volume, or if an attempt is made to add more moisture to it, the moisture already in it will begin to be deposited in the form of dew, the temperature at which it does this being called the "dew point."

A further characteristic to note is, that water will not evaporate into cold air to the same extent as in warm air. For instance, 2·13 grains of water will evaporate and saturate a cubic foot of air at 32° F., while 19·84 grains will evaporate before it saturates a cubic foot of air that is at a temperature of 100° F. This, of course, leads to the conclusion that the dew point varies according to the temperature of the moisture, or, in other words, the elastic pressure of the particles of vapour increases as their temperature increases. In this connection we may point out that the air itself has nothing whatever to do with humidity, for all the phenomena of saturation, dew point, etc., can be observed in a vacuum, and, as a matter of fact, it is from experiments performed in the absence of air upon which our knowledge of the dew point depends. Those, therefore, who speak of the property of air to retain moisture are wrong in principle; the air happens to be a convenient vehicle for heating the vapour as it arises from the surface of the water, and in so doing increasing its elastic force and enabling still further evaporation to take place; the application of heat to the water itself will cause vapour to be given off, and this, rising in the atmosphere, will heat the air, and so the same result naturally follows.

We can now deal with the relative humidity. If air contains a certain amount of moisture, and this amount is only half of what would cause saturation, the humidity of the air is said to be 50 per cent; so that when we say that air is "dry" we simply imply that the proportion of moisture in the air is small compared with what the air would contain if it was completely saturated; cold air with little moisture in it may be very moist, while warm air with much moisture in it may be very dry.

It is now seen that the point of saturation or dew point is the foundation of our estimate of "humidity," and therefore we must know this before the percentage of humidity in a room can be known. To do this would require skill, but fortunately Mr. Glaisher took advantage of a long series of experiments made in England, America, and India, and from them constructed a series of tables by the use of which the humidity can readily be found. His first set of tables differed considerably from his later ones, published in 1856, but these last ones are now used as a standard by

British observers, though other countries still retain tables based on their own observations, this accounting for the fact that in America, for example, the tables used are different from Glaisher's, and an American would give the humidity of a room slightly differently than we should in this country. It is simply a question of observation and experiment, and the tables are purely empirical.

The instrument used to indicate humidity consists of two thermometers, one of which has its bulb covered with a thin piece of muslin cloth connected by an absorbent strand of material to a small well of water placed at a short distance from the thermometers. One thermometer will register the actual temperature of the air, the other, owing to the moistened covering, indicating a less temperature; this comes about, because water in changing into vapour expends heat and the remaining water becomes so much colder. The water in the muslin evaporates and the heat expended in this action leaves the water slightly colder. As this colder water is in contact with the bulb of the thermometer it causes the instrument to indicate a less temperature, and so we have two readings, one from the wet bulb and the other from the dry bulb thermometer. The difference between the two supplies us with the basis upon which to estimate the humidity, Glaisher's tables giving the amount at a glance.

Since Glaisher's tables are intended to cover extreme conditions of temperature we find that the makers of the instrument just described—which is known as a "Hygrophant"—issue a leaflet containing only the range of temperature likely to exist in the mill. A portion of such a table is given below, and it has some value to the cotton spinner, because it represents what Sir B. A. Dobson and Mr. Midgeley found to be the best relative humidity in the spinning rooms; the complete table will be found in Sir B. A. Dobson's book on *Humidity*.

Dry Bulb.	Wet Bulb.	Relative Humidity.	Dry Bulb.	Wet Bulb.	Relative Humidity.
		per cent			per cent
90	76·7	49	74	63·0	52
89	76·3	50	73	62·3	52
88	75·3	50	72	61·3	52
87	74·6	50	71	61·0	53
86	73·5	51	70	60·0	53
85	72·6	51	69	59·0	53
84	72·0	51	68	58·3	53
83	71·0	51	67	57·3	53
82	70·0	51	66	56·3	53
81	69·3	51	65	55·5	53
80	68·6	52	64	54·5	53
79	67·7	52	63	53·7	54
78	66·7	52	62	53·0	54
77	65·7	52	61	52·0	54
76	65·0	52	60	51·0	54
75	64·0	52			

An instrument is being extensively used now, that avoids the trouble of referring to separate tables. It is an American patent taken out by Huddleston in 1874. It consists of the wet and dry bulb thermometers, but between them is placed a cylinder on which is printed in upright columns a series of figures; each column is headed by a number, which represents the difference in temperature between the two thermometers. Close to the cylinder is a scale similar to the dry bulb thermometer. By turning the cylinder until the column of figures having the number on the top equal to the difference between the thermometers is close to the scale, we read the temperature of the dry bulb on the scale, and opposite to this number is the percentage of humidity in the room. This instrument, not being based upon Glaisher's tables, is not correct for use in England, but a Manchester firm (Casartelli) are now making a copy of this hygrometer having correct readings and specially constructed for mill use; an illustration of it is given in Fig. 218.

When the cotton industry was passing through its initial stages it was soon discovered that two essentials were necessary to obtain good results—namely, a warm atmosphere and a moist atmosphere, and both were obtained in the usual way by heating appliances and spraying the floors of the mill, the moisture arising in the process of evaporation. Improvements were effected and many methods have been adopted, chiefly on sanitary grounds, for obtaining the best and most permanent effects in both directions, for it was found that moisture played a very important part in the production of level and strong yarn. Cotton fibres are hygroscopic in character—that is, they have the property of absorbing moisture, and in doing so they become for the time being less brittle, more pliable, and capable of being incorporated more thoroughly among themselves in the yarn. Electricity in the mill produced by the friction of moving parts—chiefly in the belts—is a disturbing agency among loose fibres, and causes an additional fuzziness in yarn, which is naturally made fuzzy by the spinning operation; in a warm, dry atmosphere this electricity is capable of exerting its full influence on the yarn, but the presence of moisture neutralises its effects considerably, and it is also for this purpose that a reasonable degree of humidity is desirable. Suitable climatic conditions such as exist in Lancashire supply a natural source of moisture, but taking into account the heat of a spinning room, it is found that artificial forms of moistening the air are requisite if the full benefit is to be obtained in the yarn. Mr. Midgeley, by micro-photographs of yarn spun under varying conditions, has been able to demonstrate this

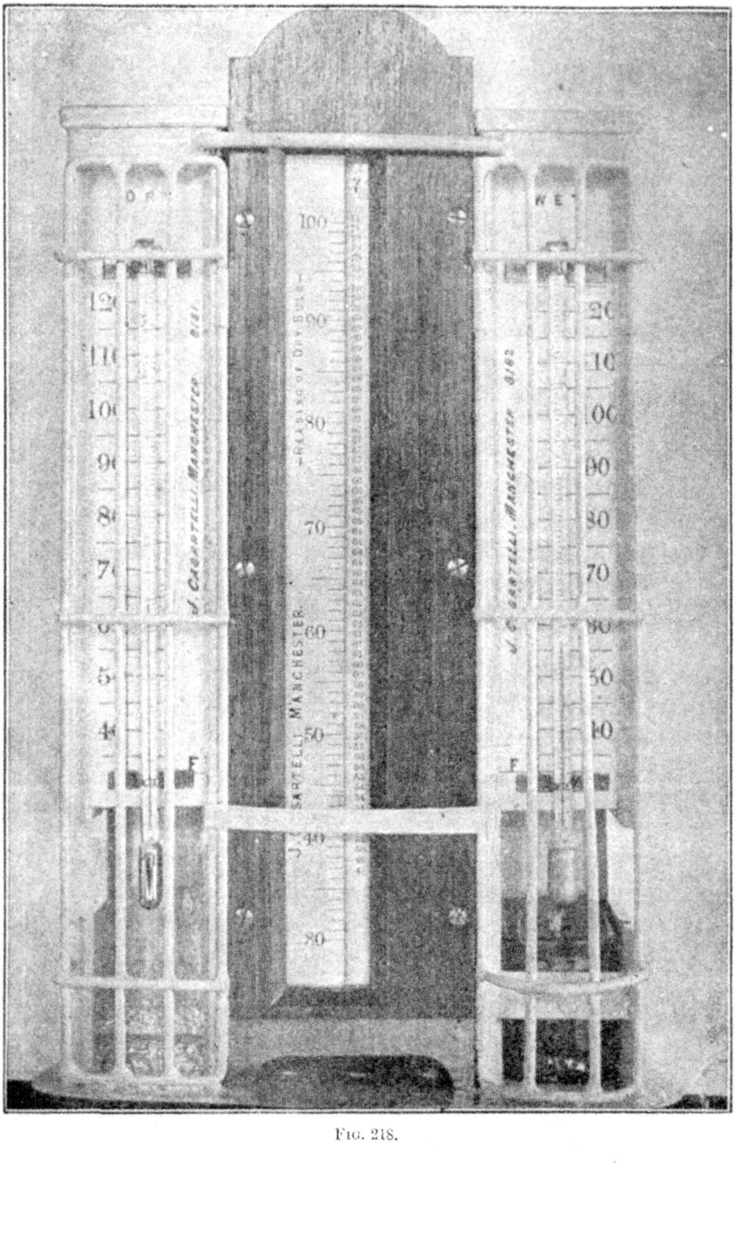

Fig. 218.

fact to a certainty, and his experiments have led to the conclusion that the best results are obtained with the humidity as given in his table just quoted.

Artificial methods of introducing moisture into a room are based upon two properties of water: first, it is capable of being, as it were, pulverised into very fine particles; and, secondly, its evaporation. So far the first method is the one chiefly adopted: the water is forced at a very high pressure in the form of a thin stream through a fine nozzle and made to impinge against a fixed surface; the water is broken up into myriads of fine particles, and in this condition is sent into the room and caused to diffuse either artificially or naturally.

The second method is to place open troughs in suitable positions about the room, fill them with water, and assist evaporation by running small steam pipes through them. Now, although moisture quickly diffuses in the atmosphere, it does not do so to a sufficient extent to give uniform results throughout the room. A recent improvement has been introduced, by means of which currents of air pass over the surface of the water in the trough and disperse the evaporated moisture uniformly in the atmosphere; this is a very important matter, for, in addition to equalising the humidity, there is a constant supply of fresh air admitted to the room.

CHAPTER IX

USEFUL INFORMATION

SOME of the following useful information may be found in other parts of the books, but it is sufficiently important to be gathered together and augmented so as to form a concise and useful reference.

HORSE-POWER OF MACHINES

	I.H.P.
Single Acting Macarthy Gin	1
Double Acting Macarthy Gin	$1\frac{1}{4}$
Bale Breaker	2
Willow	3
Small Porcupine Opener	2
Automatic Hopper Feeder	$1\frac{1}{2}$
Vertical Beater Opener, Single Crighton	4
,, ,, ,, Double Crighton	8
Exhaust Opener	12
Single Opener (without Hopper Feeder)	5
Double Opener (,, ,, ,,)	10
Single Scutcher	4
Double Scutcher	8
Card, Revolving Flat	$\frac{3}{4}$
Sliver Lap Machine	$\frac{1}{2}$
Ribbon Lap Machine (Draw and Lap Machine combined)	1
Comber, single nip, 6 heads	$\frac{5}{8}$
,, ,, ,, 8 ,,	$\frac{3}{4}$
,, double nip, 6 ,,	$\frac{3}{4}$
,, ,, ,, 8 ,,	$\frac{7}{8}$
Draw Frame . . . . per 12 deliveries	1

		I.H.P.
Slubbing Frame	90 spindles per	1
Intermediate Frame	130 ,, ,,	1
Roving Frame	160 ,, ,,	1
Jack Frame	200 ,, ,,	1
Mule, Indian and American cotton	120 ,, ,,	1
Mule, Egyptian and Sea Island cotton	130 ,, ,,	1
Ring Spinning Frame	100 ,, ,,	1
Ring Doubling Frame	60 ,, ,,	1
Twiner, Yorkshire principle	200 ,, ,,	1
Twiner, French principle	140 ,, ,,	1
Quick-Traverse Winding Frame	80 drums ,,	1
Ordinary Winding Frame	300 spindles ,,	1
Gassing Frame	80 drums ,,	1
Reel (Coleby's)	6 reels ,,	1
Improved Reel (for gassed yarn)	8 ,, ,,	1
Single Ordinary Reel	16 ,, ,,	1
Double Ordinary Reel	8 ,, ,,	1
Copping Frame	300 spindles ,,	1
Bundling Press		$\frac{1}{8}$
Banding Machine		$\frac{1}{2}$
Tubular Banding Machine, 3 heads		$\frac{1}{4}$
,, ,, ,, 6 ,,		$\frac{1}{2}$
Balling Machine	per head	$\frac{1}{4}$

The foregoing particulars represent average results, and on testing them on a number of mills through the steam-engine indicator, they were found in some cases to be below, while in others they appeared to be somewhat excessive. They may be taken as fairly accurate, a little judgment being necessary in fixing the spindles per horse-power for the mule and ring frame.

WEIGHTS FOR DRAW FRAME ROLLERS

Cotton.	Front.	2nd.	3rd.	Back.
	lb.	lb.	lb.	lb.
Indian and American Cotton	20	20	20	20
Egyptian cotton	18	18	18	18
Sea Island cotton	16	16	16	16

Some people prefer for American cotton—

1st.	2nd.	3rd.	4th.
20 lb.	18 lb	16 lb.	14 lb.

WEIGHTS FOR FLY FRAME ROLLERS

Kind of Machine.	Kind of Cotton.	Front.	Middle.	Back.
Slubber .	Indian and American	18 lb.	24 lb.	Saddle and Bridle.
Slubber .	Egyptian, etc.	16 lb.	20 lb.	Saddle and Bridle.
Slubber .	do.	14 lb.	12 lb.	Self-Weighted.
Intermediate	Indian and American	16 lb.	20 lb.	Saddle and Bridle.
Intermediate	Egyptian, etc.	14 lb.	18 lb.	Saddle and Bridle.
Intermediate	do.	12 lb.	10 lb.	Self-Weighted.
Roving .	Indian and American	18 lb.	24 lb.	Saddle and Bridle.
Roving and Jack	Egyptian, etc.	16 lb.	20 lb.	Saddle and Bridle.
Roving .	do.	10 lb.	Self-Weighted.	Self-Weighted.
Jack . .	do.	8 lb.	Self-Weighted.	Self-Weighted.

Another firm adopts the following:—

	Front.	Middle.	Back.
Slubbing Frame . . .	18 lb.	14 lb.	10 lb.
Intermediate Frame . .	14 lb.	10 lb.	8 lb.
Roving Frame (double boss) .	18 lb.	14 lb.	12 lb.
Roving Frame (single boss) .	10 lb.	8 lb.	6 lb.

DIAMETERS OF RINGS AND SPACES SUITABLE FOR SPINNING VARIOUS COUNTS OF YARN

For 4's to 20's counts, space $2\frac{3}{4}$ in., dia. of Ring $1\frac{3}{4}$ in.
,, 20's ,, 40's ,, ,, $2\frac{5}{8}$,, ,, ,, $1\frac{5}{8}$,,
,, 40's counts & upwards ,, $2\frac{1}{2}$,, ,, ,, $1\frac{1}{2}$,,

If an anti-ballooning motion is used, then

For 4's to 20's counts, space $2\frac{5}{8}$ in , dia. of Ring $1\frac{3}{4}$ in.
,, 20's ,, 40's ,, ,, $2\frac{1}{2}$,, ,, ,, $1\frac{5}{8}$,,
,, 40's counts & upwards ,, $2\frac{1}{4}$,, ,, ,, $1\frac{1}{2}$,,
,, Weft . . . ,, $2\frac{1}{4}$,, ,, ,, $1\frac{3}{16}$ to $1\frac{1}{4}$

MULTIPLIERS FOR TWIST PER INCH

Fly Frames

INDIAN AND LOW AMERICAN COTTON

Slubber,	sq. root of	hank	roving	multiplied by	1·3
Intermediate,	,,	,,	,,	,,	1·2
Roving,	,,	,,	,,	,,	1·5

AMERICAN AND LOW EGYPTIAN COTTON

Slubber,	sq. root of	hank	roving	multiplied by	1·16
Intermediate,	,,	,,	,,	,,	1·25
Rover,	,,	,,	,,	,,	1·1
Jack, American,	,,	,,	,,	,,	1·1
Jack, Egyptian,	,,	,,	,,	,,	0·9

EGYPTIAN AND SEA ISLAND COTTON

Slubbers,	sq. root of	hank	roving	multiplied by	0·7
Intermediate,	,,	,,	,,	,,	0·78
Rovers,	,,	,,	,,	,,	1·1
Jack, Egyptian,	,,	,,	,,	,,	0·9
Jack, Sea Island,	,,	,,	,,	,,	0·95

In regard to these tables, it may be remarked that some spinners use the multiplier 1·2 throughout the frames.

Mule

Twist, Indian and American cotton, multiply square root of counts by	3·75
Weft, Indian and American cotton, multiply square root of counts by	3·25
Twist, Egyptian cotton, multiply square root of counts by .	3·606
Weft, Egyptian cotton, multiply square root of counts by .	3·183

Ring Frame

Twist, Indian and American cotton, multiply square root of counts by	4·00
Twist, Egyptian cotton, multiply square root of counts by .	3·606

Doubler Frame

Multiply the square root of counts by	4·00

HORSE-POWER FOR COMPLETE MILLS

No. 1 Mill, No. of spls.	(mule, ring and doubling)	53,000 = 48	spindles	i.h.p.
,, 2 ,, ,,	(all mules)	69,300 = 72	,	,,
,, 3 ,, ,,	,,	101,900 = 66	,,	,,
,, 4 ,, ,,	,,	82,000 = 69	,,	,,
,, 5 ,, ,,	,,	80,000 = 66	,,	,,

Preparing machinery is included in all the above mills.

SPEEDS IN THE CARD

Kind of Cotton.	Cylinder. Revs.	Doffer. Revs.	Feed Roller. Revs.	Licker-in.
Indian Cotton	165 to 170	15 to 18	2 to 2·3	About 400 revs.
American ,,	170 to 180	14 to 20	2·3	
Egyptian ,,	160 to 166	9 to 12	2·3 to 2·5	
Sea Islands ,,	150 to 160	5 to 9	2·5 to 2·7	

The flats travel about 3½ inches per minute.

COUNTS OF WIRE IN THE CARD

Kind of Cotton.	Cylinder.	Doffer.	Flats.	Remarks.
Indian—				There are firms who are noted for good work who use the highest counts of wire given in this table.
Lowest	80	90	70 to 90	
Best	90	100 to 110	80 to 110	
American—				
Lowest	100	110	100	
Best	110	120	110	
Egyptian—				
Lowest	110 to 120	120	110 to 130	
Best	120 to 130	130		
Sea Islands	120	130 to 140	130 to 140	

Position of the Wharve on Mule Spindle.—In order to obtain the best results in driving the spindle, the spindle ought to set so that, if a straight edge be placed on the under side of the wharve, it will occupy the following positions :—

For 14 in. spindle straight edge will touch the {under side of the tin roller shaft
,, 15 ,, ,, ,, will be $\frac{3}{16}$ in. below ,, ,,
,, 16 ,, ,, ,, ,, $\frac{3}{8}$,, ,, ,, ,,
,, 17 ,, ,, ,, ,, $\frac{9}{16}$,, ,, ,, ,,

Ends: Piecing-up.—The following table represents the number of ends pieced up per day (caused by breakages only) on the various machines. Three mills are taken, and they are the result of extensive observation for this specific purpose made by the secretary of Mr. Geo. Draper. The table is given by permission of Messrs. Geo. Draper and Sons, U.S.A. :—

Machine.	Breaks No. 1 Mill.	Breaks No. 2 Mill.	Breaks No. 3 Mill.
Card . .	1·90	1·64	13·50
Drawing No. 1	6·18	1·49	5·17
,, ,, 2	1·29	...	...
,, ,, 3	2·57	1·75	3·45
Slubber . .	4·40	7·67	12·57
Intermediate .	13·30	8·50	14·31
Rover . .	46·82	30·60	27·74
Ring Frame, T.	410·00	630·00	1180·00
,, ,, W.	720·00	1120·00	1260·00
Mule . .	...	1670·00	...

The piecing-up on the preparing machine is estimated on the total number of spindles in the mill, while that of the spinning machinery is based on 1000 spindles: for instance, according to the table, a mule of 1000 spindles would have all its ends broken 1·67 times during a day.

ENGLISH WEIGHTS AND MEASURES OF COTTON YARNS

24 grains = 1 pennyweight (dwt. troy).
18 dwts. $5\frac{1}{2}$ grains = 439·5 grains = 1 ounce (oz. avoirdupois).
16 ounces = 7000 grains = 1 pound (lb. avoirdupois).

54 inches = 1 thread or circumference of wrap reel.
4,320 ,, = 80 threads or 1 lea or skein.
30,240 ,, = 560 threads = 7 leas = 1 hank = 840 yards.

The number of hanks in 1 lb. is the count of the yarn.

A bundle of cotton yarn is as many hanks as make 10 lbs.

CONVENIENT MULTIPLIERS

Circles, Areas, and Figures

Diameter of a circle × 3·1416 or $\frac{22}{7}$ = the circumference.
Circumference of a circle × 0·31831 or $\frac{7}{22}$ = the diameter.
Square of diameter × 0·7854 = the area of the circle.
Square of diameter × $\frac{11}{14}$ = the area of the circle.
Square root of area × 1·12837 = the diameter of a circle.
Radius of circle × 6·28318 = the circumference.
Circumference = 3·5449 × $\sqrt{}$ area of circle.
Diameter of a circle × 0·8862 = the side of an equal square.
Side of a square × 1·128 = the diameter of an equal circle.
Area of triangle = the base × $\frac{1}{2}$ the perpendicular height.
Square of the diameter of a sphere × 3·1416 = the convex surface.
Cube of the diameter of a sphere × 0·5236 = the solidity.
Diameter of a sphere × 0·806 = the edge of an equal cube.
Diameter of a sphere × 0·6667 = the length of an equal cylinder.
Surface of a cylinder = area of both ends + length × circumference.
Solidity of a cylinder = area of one end × the length.
Solidity of a cone = area of the base × $\frac{1}{3}$ the perpendicular height.
Area of an ellipse = long axis × short axis × 0·7854.

Conversion of one Denomination to Another

Feet × 0·00019 = miles.
Yards × 0·0006 = miles.
Square inches × 0·00694 = square feet.
Square feet × 144 = square inches.
Cubic feet × 0·037 = cubic yards.
Cubic inches × 0·000579 = cubic feet.
Cubic feet × 6·2355 = gallons.
Gallons × 0·16059 = cubic feet.
Gallons × 10 = lbs. of distilled water.
Cubic feet of water × 62·425 = lbs. avoirdupois.
Cubic inches of water × 0·03612 = lbs. avoirdupois.
Lbs. avoirdupois × 1·2153 = lbs. troy or apothecary.

Lbs. troy or apothecary × 0·8228 = lbs. avoirdupois.
Lbs. avoirdupois × 0·00893 = cwts.
Lbs. avoirdupois × 0·000447 = tons.
Tons of water × 224 = gallons.

ROPE DRIVING

Tables of the Horse-Power of Transmission Rope, by C. W. HUNT. The working strain is 800 lbs. for a 2-inch diameter rope, and is the same at all speeds, due allowance having been made for loss by centrifugal force.

Diam. Rope. Inches.	Speed of the Rope in Feet per Minute.										Smallest Diam. Pulleys. Inches.
	1500	2000	2500	3000	3500	4000	4500	5000	6000	7000	
$\frac{3}{4}$	3·3	4·3	5·2	5·8	6·7	7·2	7·7	7·7	7·1	4·9	30
$\frac{7}{8}$	4·5	5·9	7·0	8·2	9·1	9·8	10·8	10·8	9·3	6·5	36
1	5·8	7·7	9·2	10·7	11·9	12·8	13·6	13·7	12·5	8·8	42
$1\frac{1}{4}$	9·2	12·1	14·3	16·8	18·6	20·0	21·2	21·4	19·5	13·8	54
$1\frac{1}{2}$	13·1	17·4	20·7	23·1	26·8	28·8	30·6	30·8	28·2	19·8	60
$1\frac{3}{4}$	18·0	23·7	28·2	32·8	36·4	39·2	41·5	41·8	37·4	27·6	72
2	23·1	30·8	36·8	42·8	47·6	51·2	54·4	54·8	50·0	35·2	84

LEATHER BELTING

Thickness.—Belts are of various thicknesses, but in a mill they are seldom below $\frac{3}{16}$ in., or above $\frac{1}{4}$ in. The average may be taken as $\frac{7}{32}$ in.

Speed.—It is advisable to keep within the limits of 3500 ft. per minute.

Width.—

$$\text{Width of belt} = \frac{1100 \times \text{Horse-Power of machine}}{\text{Vel. of belt in ft. per min.}}.$$

Power.—

H.P. = Horse-power.
W = Width of belt.
F = driving force in lbs.
T = Tension in belt.
L = Circumference in inches of pulley covered by belt.
V = Velocity of belt in ft. per min.
v = ,, ,, ,, second.
A = Covered area of driven pulley in inches.
l = Circumference in inches of driven pulley covered by belt.

$$T = x + kx.$$

$$F = \frac{W \times T}{2}.$$ T varies from 70 to 150 lbs.

$$W = \frac{33000 \times H.P.}{F \times V}. \qquad W = \cdot 02\ T.$$

$$H.P. = \frac{V \times F.}{33000} \qquad x = \frac{33000\ H.P.}{{}_{\text{z}}V}.$$

$$H.P. = \frac{v \times F}{550}.$$

If a little less than half the pulley, viz. ·4 of it, is covered by the belt, $k = 1 \cdot 1$.

If a little more than half the pulley, viz. ·6 of it, is covered by the belt, $k = \cdot 62$.

$$W = \frac{66000 \times H.P.}{l \times V}$$ for double belting.

$$H.P. = \frac{A \times V}{66000}$$,, ,,

$$A = \frac{66000 \times H.P.}{V}$$,, ,,

$$H = \frac{W \times V}{33000} \times 70 \text{ to } 80.$$

$$W = \frac{36000\ H.P.}{6V \times L}.$$

Diameters.—Pulleys are not working under good conditions if one of the pulleys is more than six times the diameter of the other.

Width.—The pulley ought to be almost $1\frac{1}{4}$ times the width of the belt.

Preservative.—Castor oil applied to the back of the

belt every few weeks, especially if the atmosphere becomes dry.

Splicing.—

Width of belts,	1 in.,	2 in.,	3 in.,	3 to 6 in.,	6 to 8 in.,	over 8 in.
Lap in inches,	2 in.,	4½ in.,	5½ in.,	6 in.,	8 in.,	10 in.

Double Belts.—Double belts transmit 1½ times more power than single belts.

TABLE OF DIVIDENDS

For Ascertaining the Weight of Hank or Decimal Part of a Hank

Rule.—Divide 7000 grains (1 lb. of yarn) by 840 yards = dividend for 1 yard.

Yards.	Dividends.	Yards.	Dividends.
1	8·333	10	83·333
2	16·666	15	125·000
3	25·000	20	166·000
4	33·333	30	250·000
5	41·666	40	333·333
6	50·000	60	500·000
7	58·333	80	666·666
8	66·666	100	833·333
9	75·000	120	1000·000

Examples.

If 2 yards of card sliver weigh 80 grains, what hank is it? Divide the dividend for 2 yards by 80 = 0·208 hank.

If 30 yards of roving frame roving weigh 62½ grains, what hank is it? Divide the dividend for 30 yards by 62½ = 4 hank roving.

What ought 60 yards of a 4½ hank roving to weigh? Divide the dividend for 60 yards by 4½ = 111 grains.

SQUARE ROOTS

No.	Square Root.	No.	Square Root.	No.	Square Root.	No.	Square Root.
0·0625	0·250	0·4375	0·661	0·65	0·806	0·86	0·927
0·125	0·353	0·44	0·663	0·66	0·812	0·87	0·933
0·1875	0·433	0·45	0·671	0·67	0·819	0·875	0·935
0·25	0·500	0·46	0·678	0·68	0·825	0·88	0·938
0·26	0·510	0·47	0·686	0·6875	0·829	0·89	0·943
0·27	0·520	0·48	0·693	0·69	0·831	0·90	0·949
0·28	0·529	0·49	0·700	0·70	0·837	0·91	0·954
0·29	0·539	0·50	0·707	0·71	0·843	0·92	0·959
0·30	0·548	0·51	0·714	0·72	0·849	0·93	0·964
0·31	0·557	0·52	0·721	0·73	0·854	0·9375	0·968
0·3125	0·559	0·53	0·728	0·74	0·860	0·94	0·970
0·32	0·566	0·54	0·735	0·75	0·866	0·95	0·975
0·33	0·574	0·55	0·742	0·76	0·872	0·96	0·980
0·34	0·583	0·56	0·748	0·77	0·878	0·97	0·985
0·35	0·592	0·5625	0·750	0·78	0·883	0·98	0·990
0·36	0·600	0·57	0·755	0·79	0·889	0·99	0·995
0·37	0·608	0·58	0·762	0·80	0·894	1·00	1·0
0·375	0·612	0·59	0·768	0·81	0·900	7·5	2·739
0·38	0·616	0·60	0·775	0·8125	0·901	8·0	2·828
0·39	0·624	0·61	0·781	0·82	0·906	8·5	2·915
0·40	0·632	0·62	0·787	0·83	0·911	9·0	3·0
0·41	0·640	0·625	0·790	0·84	0·917	9·5	3·082
0·42	0·648	0·63	0·794	0·85	0·922	10·0	3·162
0·43	0·656	0·64	0·800				

Note:—For the square roots of higher numbers refer to the Yarn Table opposite.

YARN TABLE OF TWISTS PER INCH AND SQUARE ROOT OF COUNTS

Counts.	Square Root of Counts.	Indian and American Cotton.			Egyptian Cotton.		
		Mule Twist.	Mule Weft.	Ring Frame Twist.	Mule Twist.	Mule Weft.	Ring Frame Twist.
1	1·000	3·75	3·25	4·00	...	...	...
2	1·414	5·30	4·60	5·65	...	...	...
3	1·732	6·49	5·62	6·92	...	...	...
4	2·000	7·50	6·50	8·00	...	...	...
5	2·236	8·38	7·26	8·94	...	...	...
6	2·449	9·18	7·96	9·79	...	...	...
7	2·645	9·92	8·59	10·58	...	...	...
8	2·828	10·60	9·19	11·31	...	...	...
9	3·000	11·25	9·75	12·00	...	...	...
10	3·162	11·85	10·27	12·64	11·44	10·10	11·44
11	3·316	12·43	10·77	13·26	11·95	10·55	11·95
12	3·464	12·99	11·25	13·85	12·47	11·01	12·47
13	3·605	13·52	11·71	14·42	13·00	11·57	13·00
14	3·741	14·03	12·16	14·96	13·46	11·89	13·46
15	3·872	14·52	12·48	15·49	13·96	12·32	13·96
16	4·000	15·00	13·00	16·00	14·40	12·72	14·40
17	4·123	15·46	13·40	16·49	14·86	13·12	14·86
18	4·242	15·90	13·78	16·97	15·27	13·48	15·27
19	4·358	16·34	14·16	17·43	15·71	13·87	15·71
20	4·472	16·77	14·53	17·88	16·09	14·21	16·09
22	4·690	17·58	15·24	18·76	16·88	14·91	16·88
24	4·898	18·37	15·92	19·59	17·63	15·57	17·63
26	5·099	19·11	16·57	20·39	18·35	16·21	18·35
28	5·291	19·84	17·19	21·16	19·04	16·83	19·04
30	5·477	20·54	17·80	21·90	19·75	17·42	19·75
32	5·656	21·21	18·38	22·62	20·40	18·00	20·40
34	5·830	21·86	18·95	23·32	21·02	18·55	21·02
36	6·000	22·50	19·50	24·00	21·64	19·09	21·64
38	6·164	23·11	20·03	24·65	22·23	19·61	22·23
40	6·324	23·71	20·55	25·29	22·81	20·13	22·81
42	6·480	24·30	21·06	25·92	23·37	20·62	23·37
44	6·633	24·87	21·55	26·53	23·92	21·10	23·92
46	6·782	25·43	22·04	27·12	24·45	21·58	24·45
48	6·928	25·98	22·51	27·71	24·98	22·04	24·98
50	7·071	26·51	22·98	28·28	25·50	22·50	25·50
52	7·211	...	...	...	26·00	22·94	26·00
54	7·348	...	...	...	26·50	23·38	26·50
56	7·483	...	...	...	26·98	23·81	26·98
58	7·615	...	...	...	27·46	24·23	27·46

YARN TABLE OF TWISTS—*Continued*

Counts.	Square Root of Counts.	Indian and American Cotton.			Egyptian Cotton.		
		Mule Twist.	Mule Weft.	Ring Frame Twist.	Mule Twist.	Mule Weft.	Ring Frame Twist.
60	7·745	...	...	...	27·93	24·54	27·93
62	7·874	...	...	...	28·39	25·05	28·39
64	8·000	...	...	...	28·85	25·45	28·85
66	8·124	...	...	...	29·29	25·87	29·29
68	8·246	...	...	...	29·73	26·23	29·73
70	8·366	...	...	...	30·17	26·62	30·17
72	8·485	...	...	...	30·60	27·00	30·60
74	8·602	...	...	...	31·02	27·37	31·02
76	8·717	...	...	...	31·44	27·74	31·44
78	8·831	...	...	...	31·85	28·10	31·85
80	8·944	...	...	...	32·25	28·47	32·25
82	9·055	...	...	...	32·65	28·81	32·65
84	9·165	...	...	...	33·05	29·16	33·05
86	9·273	...	...	...	33·44	29·50	33·44
88	9·380	...	...	...	33·83	29·84	33·83
90	9·486	...	...	...	34·21	30·18	34·21
92	9·591	...	...	...	34·59	30·52	34·59
94	9·695	...	...	...	34·96	30·85	34·96
96	9·797	...	...	...	35·33	31·17	35·33
98	9·899	...	...	...	35·70	31·50	35·70
100	10·000	...	...	...	36·06	31·83	36·06
102	10·099	...	...	...	36·41	32·14	36·41
104	10·198	...	...	...	36·77	32·46	36·77
106	10·295	...	...	...	37·12	32·76	37·12
108	10·392	...	...	...	37·47	33·07	37·47
110	10·488	...	...	...	37·81	33·32	37·81
112	10·583	...	...	...	38·16	33·68	38·16
114	10·677	...	...	...	38·50	33·98	38·50
116	10·770	...	...	...	38·83	34·28	38·83
118	10·862	...	...	...	39·17	34·57	39·17
120	10·954	...	...	...	39·50	34·86	39·50

APPENDIX

IMPROVEMENTS IN THE LONG LEVER MULE

IT has been thought necessary to give a few words of explanation of further improvements that have been effected upon the arrangement illustrated in Figs. 108, 109, 110, and 114. These improvements are quite recent, but they have proved so valuable in enabling changes to be rapidly and certainly made that the machine-makers who make this type of mule are adopting them on all their newest mules. The young reader is advised to read up thoroughly all that has been said on the specific actions of the mule in the previous pages; if this is done, the following brief summary of the actions now to be described will become comprehensive to him.

Referring to Fig. 219 it will be noted that the long lever is retained, being fulcrumed on the side of the framing on the stud 2; it carries several studs or stops, as at 3, 4, 5, 6, and 7, the purpose of which will be subsequently explained. The drawing shows the positions of the various parts, as when the carriage is running out and spinning is in progress, under these circumstances:—

The strap is on the fast pulley on the rim shaft.

The backing-off lever D is kept from permitting the backing-off cone wheel to go into gear with the fast

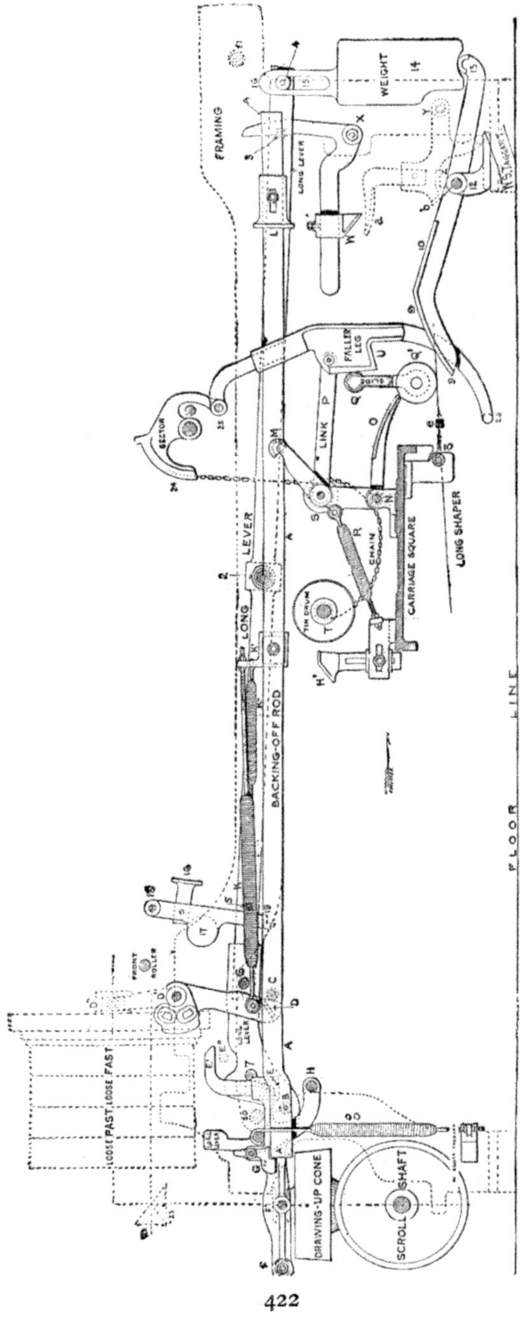
FRAMING
WEIGHT
LONG LEVER
FALLER LEG
LINK
SECTOR
CHAIN
TIN DRUM
CARRIAGE SQUARE
LONG SHAPER
BACKING-OFF ROD
FRONT ROLLER
LOOSE FAST LOOSE FAST
DRIVING-UP CONE
SCROLL SHAFT
FLOOR LINE

Fig. 219.

pulley on the rim shaft by the stud C on the backing-off rod A.

The drawing-up cone is kept out of gear with the scroll shaft by stud 7 on the long lever, and by the catch G which is carried by the drawing-up lever centred at F. The catch G, it will be noticed, rests upon the end of the backing-off rod, and in this position it prevents the drawing-up lever, which is fulcrumed at F, from falling into gear with the scroll shaft.

The long lever is kept from changing by the stud 3 being hooked under the recess in the lever which is fulcrumed at X, also by weight 14 resting directly upon its end.

It will also be observed that the spring K is in tension, and is tending to pull forward the backing-off lever D, but is prevented from doing so by the stud C on the backing-off rod. The spring "g" is also in tension, and is tending to pull the drawing-up lever, centred at F, into gear with the scroll shaft, but is prevented from doing so by the catch G and the pin 7 on the long lever.

Since spinning is in progress, the faller leg is not connected to the shaper bowl Q'.

As the carriage nears the completion of the run-out, a bowl 8 on the carriage square comes into contact with the inclined end 9 of a lever centred at 12; this has the effect of depressing it, and lifting up the end 13, upon which the weight 14 rests, and which is, as a consequence, raised out of contact with the end of the long lever.

The end M of a lever fulcrumed at N now comes against a projection L on the backing-off rod and moves it forward, thus freeing the backing-off lever D and causing the spring K to pull it forward and so putting the backing-off cone wheel into gear with the rim shaft. Backing-off now takes place; the scroll on the tin roller shaft T winds on the backing-off chain and causes the faller leg to rise until the recess U is pulled over the upper part Q of the slide which carries the shaper bowl Q′. This action of course causes the lever M to be drawn backwards as well, and the backing-off rod, being now free from M, instantly shoots backwards under the influence of the spring K″, and the backing-off cone is taken out of gear. When the backing-off rod is moved forward by M a stud B is brought under a part of the drawing-up lever at E, so that during "backing-off" the drawing-up lever is locked; the catch G is also disconnected by the same movement from the end of the backing-off rod.

On the release of the backing-off rod and a simultaneous release of the lever centred at X the long lever is free to change, so that the drawing-up lever at once puts the drawing-up cone into gear with the scroll shaft and the carriage is drawn in. When the long lever has changed, a stud 5 carried by it falls under a recess 19 on a pendent lever carried on a stud at 18, so that the long lever therefore becomes locked in this position. At the same time as the carriage runs in, the lever centred at 12 is free from contact with the stud 8, and consequently the weight is now only supported by hanging from the end of the long lever.

The carriage now approaches the roller beam, and as it does so an incline H′ comes against a stud bowl H on the drawing-up lever and lifts the drawing-up cone out of gear with the scroll shaft, thus stopping the carriage; at the same time the fallers come against the projection 16 on the lever centred at 18 and release the recess 19 from the stud 5 on the long lever. There is now no resistance to the movement of the long lever, so that the weight 14 hanging on one end of it at once falls, and in doing so causes the catch box on the back shaft to be put into gear, thus connecting the front roller with the back shaft ready for the run-out, which immediately commences.

It will be noticed that it has not been considered necessary to go into detail as to the precise action of the various changes, these already having been thoroughly described and explained in the previous pages; to those who understand the actions of the mule the drawing given will be practically self-explanatory.

SHORT SHAPER

From page 135 to page 165 will be found a very complete description of the mule shaper, together with a full explanation of its principle. The short shaper, however, has not been mentioned, though it may be remarked that the greater part of the explanation is equally as applicable to the short as to the long shaper. An illustration is here given of the short shaper, and the following remarks will be sufficient to enable its working to be clearly understood.

A section of the carriage square is shown in Fig. 220; to the under side of it is bolted a strong framing in the form of a slide cover R. Into the grooves of R there is fitted a slide Q, as shown in the section. To the slide Q is connected a short rack S, and into this rack the small pinion T gears; on the boss of the wheel T is a larger wheel U, the two wheels T and U thus forming a compound carrier which runs on a stud carried by a bracket from the carriage square. The large wheel U now gears with the teeth of a long rack which is fastened to the floor, so that when the carriage moves backwards and forwards the wheel U will revolve by virtue of its being in gear with the long rack V. As U revolves so will the pinion T; but T being smaller than U, it will only cause the rack S to move a proportionate distance to the carriage that the pinion T is smaller than U. If T has 13 teeth and U has 43 teeth and the carriage travels 60 inches, then the rack S, and consequently the slide Q to which it is attached, will move $\frac{13 \times 60}{43} = 18\frac{1}{7}$ inches, and, moreover, this movement of the slide will be in the opposite direction to the carriage. On a projection to the lower part of the slide Q rests the shaper, composed of the back, middle, and front plates; these plates are connected to the slide Q by the nut M working on the screw which is carried from the slide by the brackets N and O.

Upon the shaper plates rest the shaper rail and shell through the pins A and B, the shell CGH being loose from the rail K for the purpose of adjustment. The pin B, in addition to resting upon the back incline, also projects into a vertical cut into the slide Q. From this description we can now see that any movement of the carriage will cause the slide Q, the shaper plates, and the shaper rails to be moved in the opposite direction, but to a less degree; in other words, the shaper rail moves forward under the shaper bowl as the carriage runs in, and whilst the carriage travels 60 inches inwards the shaper travels 18 inches outwards, both movements occupying exactly the same time.

As the carriage moves in, the end X of the slide Q comes against the lever Y carrying the catch Z, and this, gearing with the ratchet wheel P on the end of the shaper screw, moves the shaper plate so that the shaper rail is lowered and the cop is lengthened.

For various purposes there are several points of adjust-

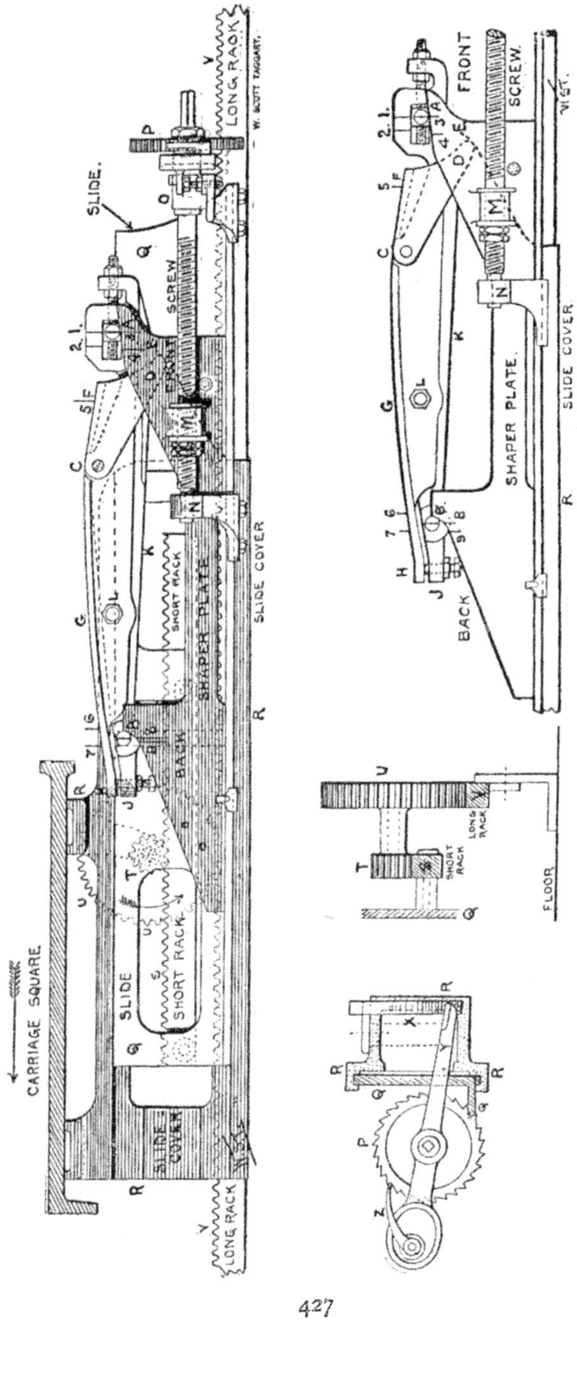
CARRIAGE SQUARE
SLIDE.
LONG RACK
SHORT RACK
SHAPER PLATE
SLIDE COVER
SCREW
FRONT
BACK
SLIDE
FLOOR
W. SCOTT TAGGART.

Fig 229.

ment: for instance the ratchet wheel P can be changed; the number of teeth taken can be regulated; the position of the pins A and B can be adjusted; and by means of the wheels T and U the distance moved by the shaper rail can be made to suit any diameter of cop required.

When the pin A is at 1 and is set over 3 on the front plate, and the pin B is at 7 and set over 8 on the back plate, twist cops can be made. When the pin A is at 2 and is set over 4 on the front plate, and the pin B is at 6 and is set over 9 on the back-plate pin, weft cops can be made. The starting-point in all cases for the shaper bowl is at 5F, the finishing-point for weft cops being at 6 and for twist cops at 7; the wheels T and U of course requiring to be altered to suit the stretch. The usual wheels used for changing are, for T 11 to 16 and for U 42 and 43. Short shapers are used now only for very fine spinning mules, and their advantage lies in the fact that the stretch can be altered without changing the shaper: in the long shaper only one stretch can be made; any variation from this would mean a new shaper.

A further example of the **Short Shaper** is given in Fig. 221. Its connection to the copping-faller is clearly shown as well as other details connected with the locking and unlocking of the faller leg.

Fine Spinning Mule.—The following descriptions and drawings are given to amplify the notes given on pages 244 to 253.

Backing-off Motion, etc.—A general view of the Single-speed Mule, Low Headstock, is given in Fig. 222. The backing-off motion is the chief feature illustrated. As the carriage completes the outward run the regulating bracket X is moved and the long backing-off lever, centred at Q, is unlocked. The end P of this lever falls and a

COPPING MOTION.

THRELFALL SELF-ACTING MULE.

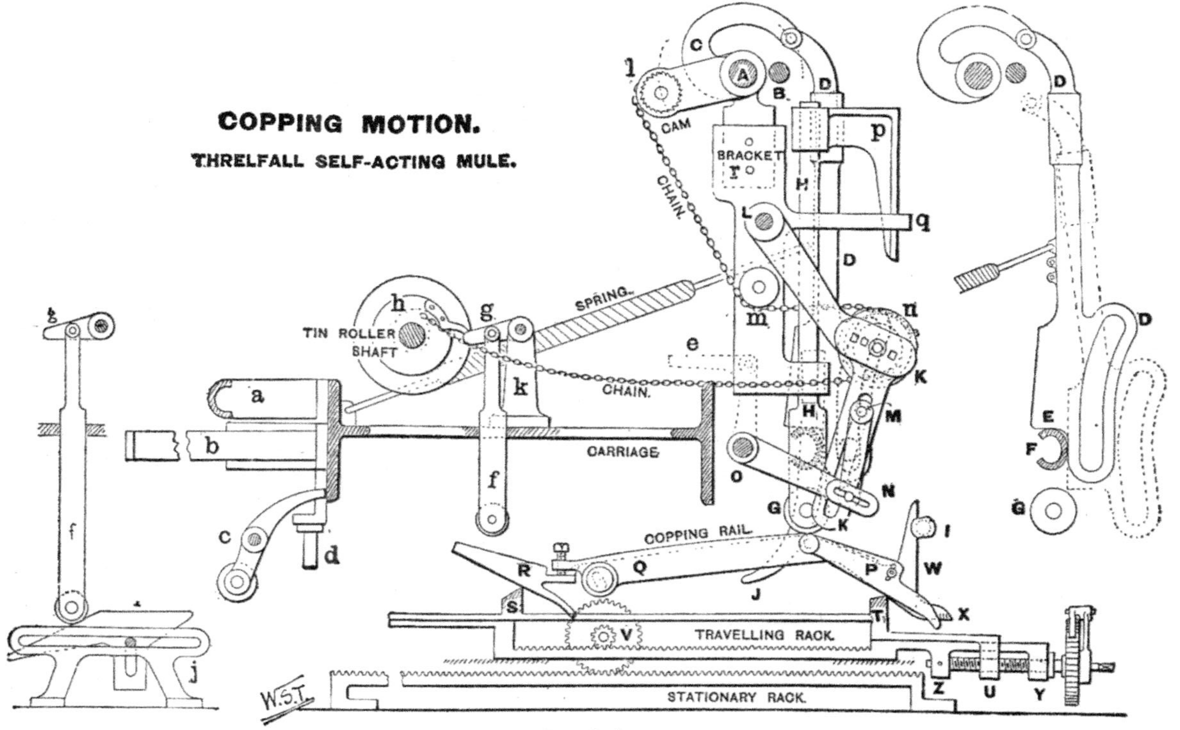

FIG. 221.

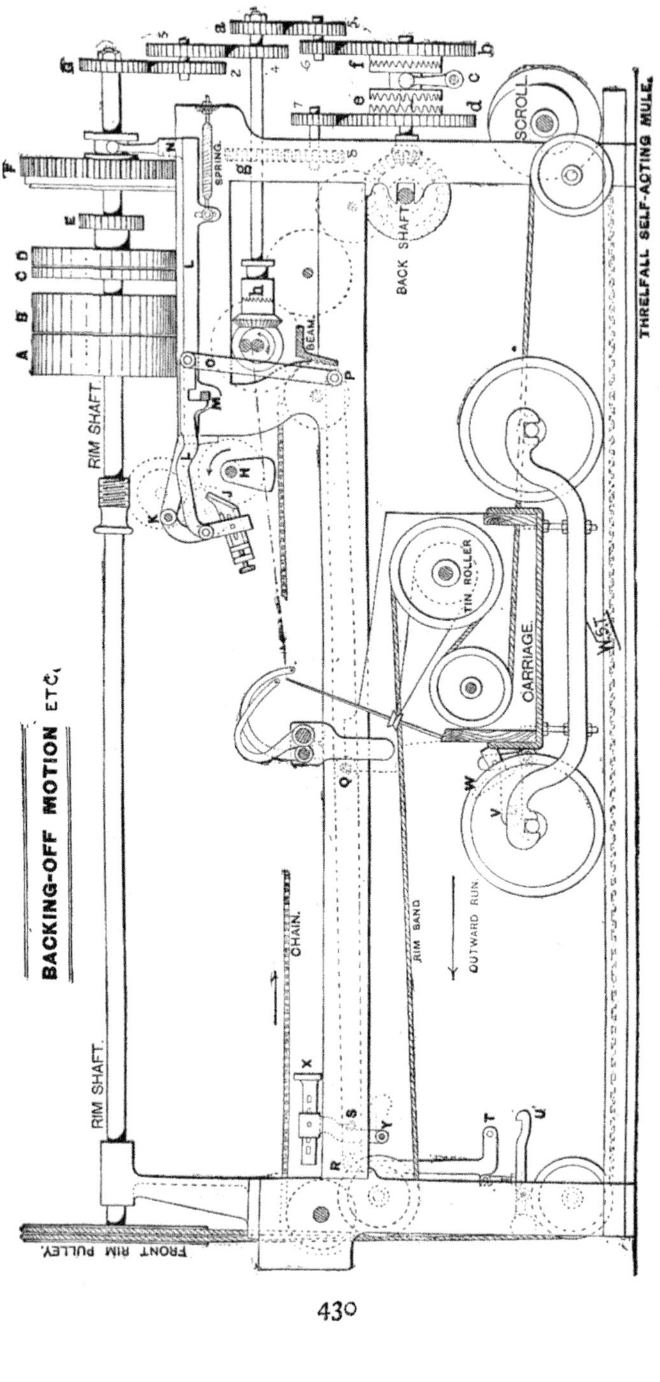

BACKING-OFF MOTION ETC.
RIM SHAFT.
RIM SHAFT.
FRONT RIM PULLEY.
SPRING.
BEAM.
BACK SHAFT.
SCROLL
TIN. ROLLER
CARRIAGE.
CHAIN.
RIM BAND
OUTWARD RUN.
THRELFALL SELF-ACTING MULE.

Fig. 222.

recess in the rod L falls over the square stud M on the long backing-off rod L. The backing-off cam H, driven from the rim shaft through the worm, comes into contact with the swing J centred at K, and moves it forward. In so doing the slide L is dragged forward, and, through the lever at N, puts the backing-off wheel into gear with the cone clutch. In the meantime the carriage has been locked by means of the lever U locking on the square stud V on the square, and the down lever stud T is in contact with the strike finger W. During the locking of the faller leg, the strike finger W swivels downwards and depresses the down lever stud T. This action pulls down the long backing-off lever and locks it at S; at the same time the slide L is released from the stud M, and, a spring pulling L backward, takes the backing-off wheel out of action. The depression of T also lowers U and unlocks the carriage, so the carriage is free to commence its inward run. A general idea of the other changes for moving the straps, etc., can be obtained by reference to the other detail drawings which follow.

Setting-on and Drawing-up Motions.—The drawing, Fig. 223, gives a general view of the mechanism for these motions. Some portions are shown displaced from their correct position in order to bring them into view.

A single rim shaft with one rim pulley is used, the double speed being obtained through gearing. C is the single-speed driving pulley, whilst A through $\frac{U \times X}{W \times V}$ gives the increased speed. A bent lever centred at U carries a stud that fits into the three notches S, R, T, on the setting-on rod. The other arm of this bell-crank lever has an incline *a* which is, at the correct moment, moved aside by the cam driven from the worm *t*. The end of the

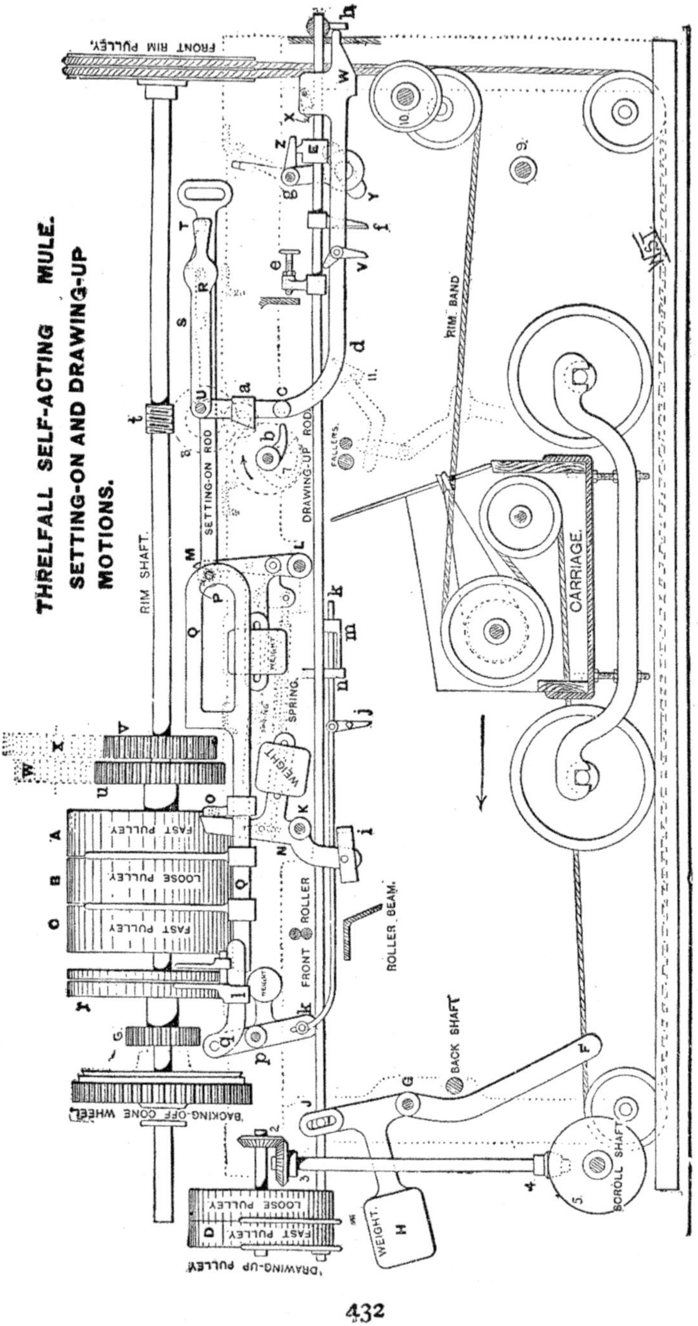
THRELFALL SELF-ACTING MULE.
SETTING-ON AND DRAWING-UP MOTIONS.
FRONT RIM PULLEY
RIM SHAFT
FAST PULLEY
LOOSE PULLEY
FAST PULLEY
BACKING-OFF CONE WHEEL
LOOSE PULLEY
FAST PULLEY
DRAWING-UP PULLEY
WEIGHT
SPRING
SETTING-ON ROD
DRAWING-UP ROD
FALLERS
FRONT ROLLER
ROLLER BEAM
RIM BAND
CARRIAGE
BACK SHAFT
SCROLL SHAFT
WEIGHT H

FIG. 223.

short arm has centred on it at *c* a rod *d* carrying the tumbler V and safety catch X. The drawing-up rod is released by the swivel arm 11 on the carriage, coming into contact with Y when backing off, and so releasing the finger Z from the stop E, thus permitting the weight H on the gun lever to pull over the strap fork on to the fast

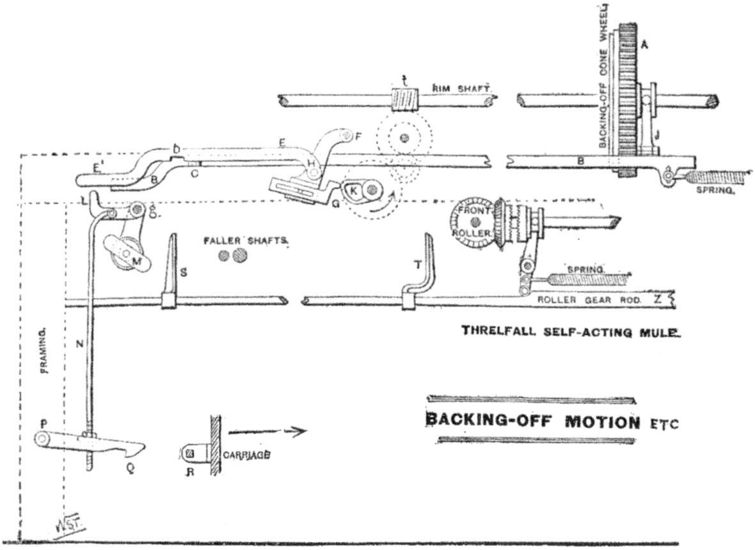

Fig. 224.

drawing-up pulley D. The carriage, as the inward run is being completed, moves the gun lever at F and puts the drawing-up belt on the loose pulley.

For opposite side of tall headstock see Fig. 224.

Backing-off Motion.—As the carriage runs out, the notch D, Fig. 224, is occupied by the projection C on the rod B. The backing-off cam K forces G forward round the centre F, and so forces the rod B outwards and puts the backing-off wheel A into contact with the backing-off cone. When the faller leg is changed, a lever moves into contact with the incline M and lifts the lever L, thus unlocking D from C. The spring now pulls the backing-off wheel out of gear. At the same time, the carriage is released from the catch Q by virtue of the link rod which connects the two levers L and P.

Part of the roller gear rod is shown, but the back mechanism of the headstock is so similar to that shown in

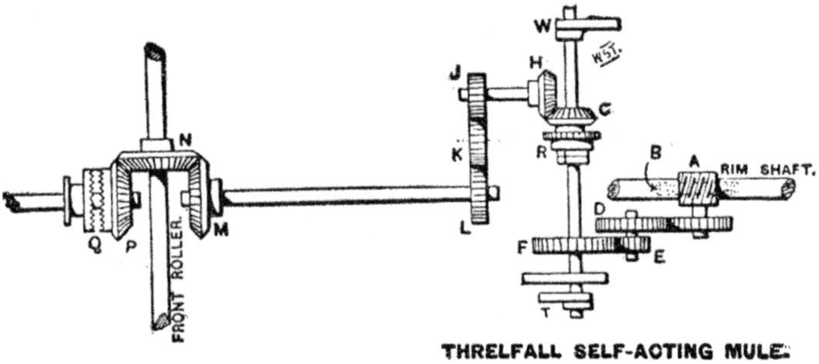

FIG. 225.

Fig. 222 that it has not been considered necessary to repeat it here.

Twist Motion.—This twist motion, Fig. 225, is a detail of Fig. 223; it is very simply arranged, and is actuated from the worm A on the rim shaft which drives the worm wheel B on cross shaft. On the end of this shaft is the change twist wheel C; this gears into D part of compound carrier, whilst E part of same drives a 72's wheel on twist shaft, and thereby gives a motion to this shaft of one revolution per draw.

The twist shaft carries three cams—W, S, and T. The cam W actuates the backing-off motion (see Fig. 224).

Fig. 226.

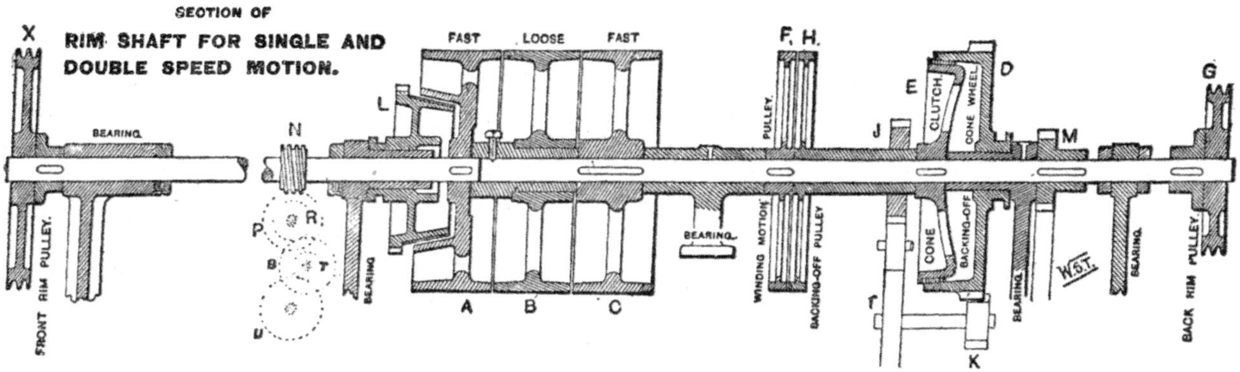

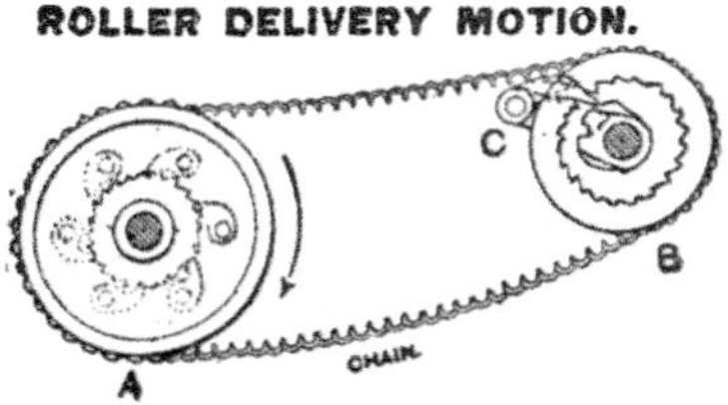

Fig. 228.

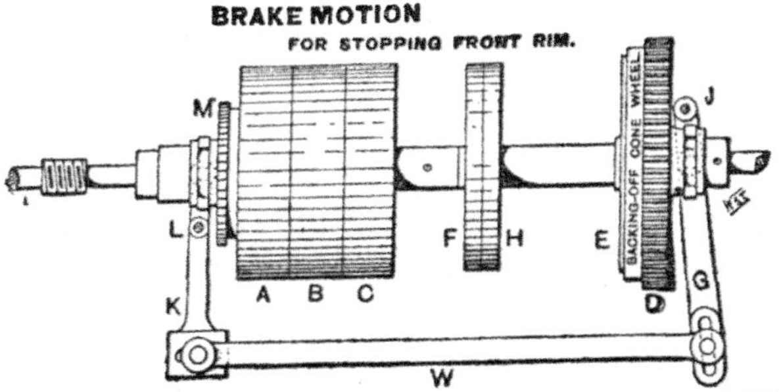

Fig. 227.

THRELFALL SELF-ACTING MULE.

Fig. 229.

The cam S is for setting on (see Fig. 223), whilst the cam T is for twist (see Fig. 223).

Roller-delivery Motion.—This motion, Fig. 225, is driven from the twist shaft by means of wheel G and ratchet wheel R, which are in one piece, and can only revolve when the catch or pawl is allowed to fall into gear by the action of the two cams on which one end of the catch rests. These cams can be adjusted to give motion to the rollers, so as to deliver the necessary amount of yarn which is required.

This motion is mostly used when spinning twist.

Special Mule. Section of Rim Shaft (see Fig. 226). —The drawing is practically self-explanatory. The *single* and *double speeds* are obtained by making the rim shaft in two parts and using two rim pulleys as G and X. Each length of rim shaft is driven by a separate pulley A and C. The pulley C is the one through which the single-speed rim G is driven, as well as the one through which the general gearing receives its motion. Pulley A drives the double-speed rim pulley X and also drives, through the worm N, the roller-delivery motion whilst twisting at the head. The backing off is actuated separately from the pulley H on the boss of which is a wheel J. This wheel drives K almost continuously, but the backing-off cone wheel D is only effective in driving the rim shaft when D is moved into contact with the cone clutch E.

The wheel M is the point through which the general gearing receives its motion (see plan of gearing, Fig. 234).

Brake Motion (see Fig. 227).—When backing off is about to take place, the belt is moved from pulley A to the loose pulley B. The backing-off cone wheel D is now forced into contact with the cone clutch E, and at the same moment the front part of the rim shaft is stopped dead through the cone clutch M being forced into contact with the fast pulley A.

Setting-on and Drawing-up Motions, Figs. 230 and 231.—A general view is given of the above motions in Fig. 230, whilst an enlarged view of the out end of the headstock is shown in Fig. 231. The sketch illustrates the disposition of the mechanism as drawing up takes place, and the main driving belt in on the loose pulley B. The belt on the drawing-up pulley D drives the scroll shaft and through it the carriage. The setting-on rod is locked in position by a square stud at S. This stud will be raised when the carriage comes into contact with the end F of the gun lever centred at G, so that the setting-on rod will be at liberty to move the belt on to the fast pulley C; the movement of F will also move the drawing-up belt on to the loose pulley. After this happens the lifting lever at A^1 is depressed by lever and bowl on carriage square, and the balance lever M taken away from the pin P.

As the carriage completes the run out, the carriage moves V forward and releases the stud S from the notch T, and puts in tension the spring between the two balance levers M and N. The tension of the spring moves the setting-on rod forward, and transfers the belt from C to A, and so puts the double speed in action.

The twist latch O is now released, and the setting-on rod moves backward and puts the belt on pulley C. The stud S is now in the notch T where it remains during the drawing out.

In the event of the carriage overrunning the catch, the faller would come in contact with finger Y, thus moving forward the safety rod; this would then move the L lever *u* and cause lever *r* to raise the twist latch *o*, thereby moving the strap on to the loose pulley B.

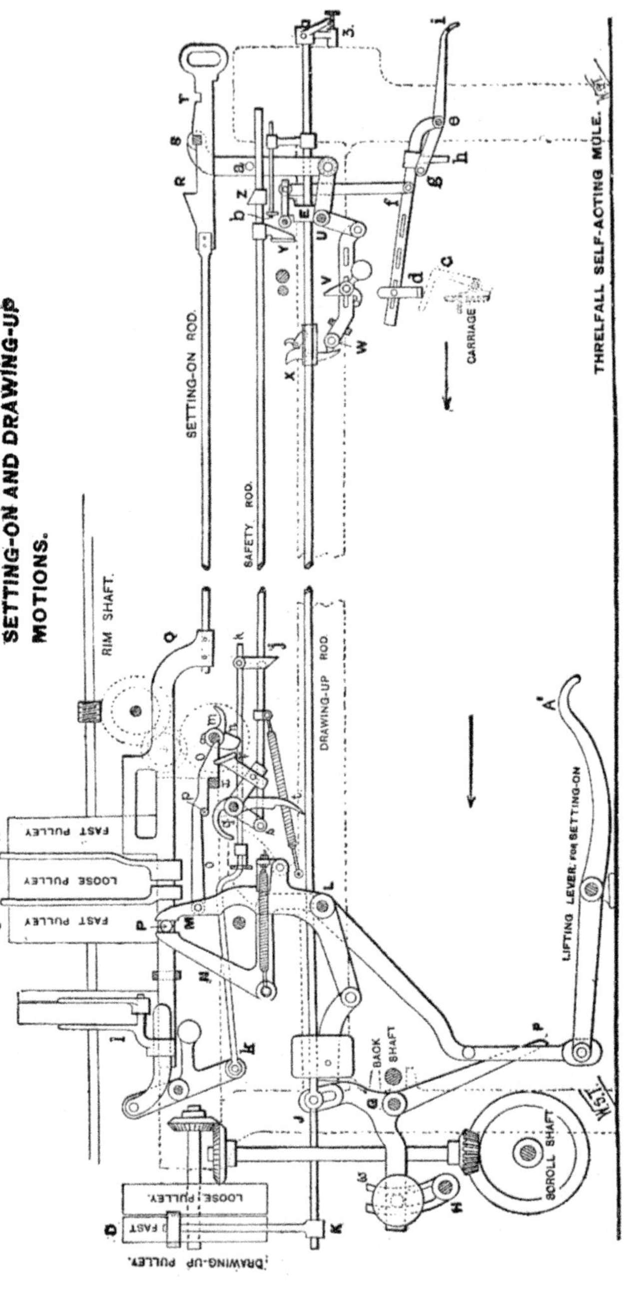
SETTING-ON AND DRAWING-UP MOTIONS.
RIM SHAFT.
FAST PULLEY
LOOSE PULLEY
FAST PULLEY
SETTING-ON ROD.
SAFETY ROD.
DRAWING-UP ROD.
CARRIAGE
LIFTING LEVER FOR SETTING-ON
BACK SHAFT
SCROLL SHAFT
LOOSE PULLEY.
FAST
DRAWING-UP PULLEY.
THRELFALL SELF-ACTING MULE.

Fig. 230.

The stop Z is used for lifting square stud at S through lever *a* when running single speed.

During the drawing out, the drawing-up rod has been locked in position through the finger K resting on the

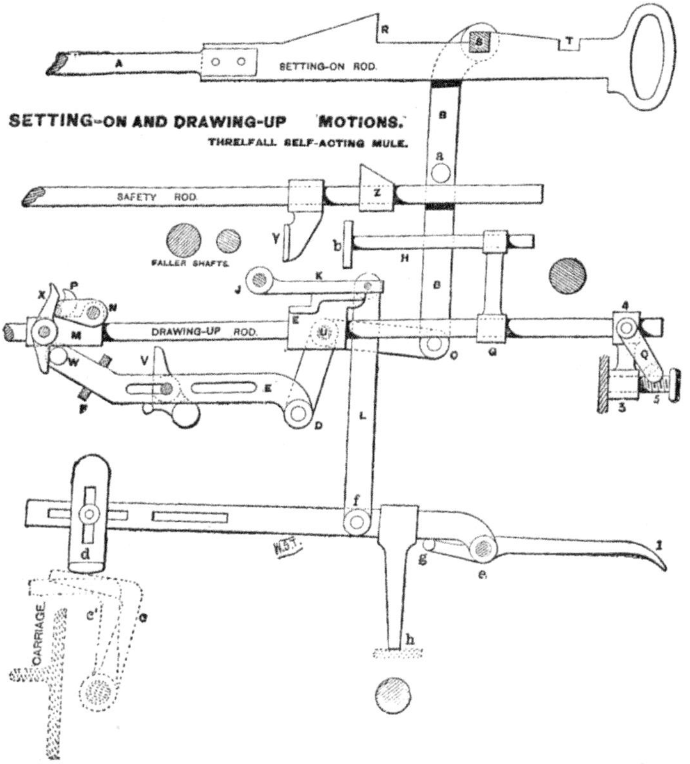

FIG. 231.

recessed boss E. At backing off the lever *c* is raised (see Fig. 231) and the finger K released, thus allowing the drawing-up rod to place the belt on the fast pulley D.

The stud *b* comes in contact with faller, and can be regulated so as to allow a small portion of the strap to go on fast drawing-up pulley D, thereby preventing the carriage from starting up too quickly.

The drawing-up rod can be released if necessary by the knee lever *i*. Its release can be prevented when necessary by turning over the lever Q, Fig. 231.

Fig. 231 also shows at N and P a small lever which locks X to the drawing-up rod. If P is turned over, the tumbler X is free to move aside without moving the incline W, and consequently the carriage will come to a standstill because the stud S is not raised, and so keeps the setting-on rod locked in position.

Drawing-out Motion, Fig. 232.—The driving strap is on the fast pulley C and the back rim pulley is driving the spindles.

The rim shaft pinion (see A in the plan of gearing, Fig. 234) is driving the front roller, the catch box Y being closed.

The rim shaft pinion is also driving the back shaft through the closed catch box *y*. This is also clearly shown in the gearing plan, Fig. 234.

We thus have the spindles, carriage, and front roller driven from the rim shaft through the pulley C during the outward run.

Soon after the carriage has started on the outward run the bowl *a* on the back of the square depresses the end V of the lifting lever centred at S, and raises the other end T; a link connects the end T to a lever I, centred at H, so that the lever I is raised together with the weight carried by I. The upper end of this weight carries one end of a spring, which end is attached to an arm U of a T lever centred at H. The T lever is locked in position at L, so the spring (by the lifting of the weight) is put in tension ready to pull the end U upwards when the T lever is unlocked.

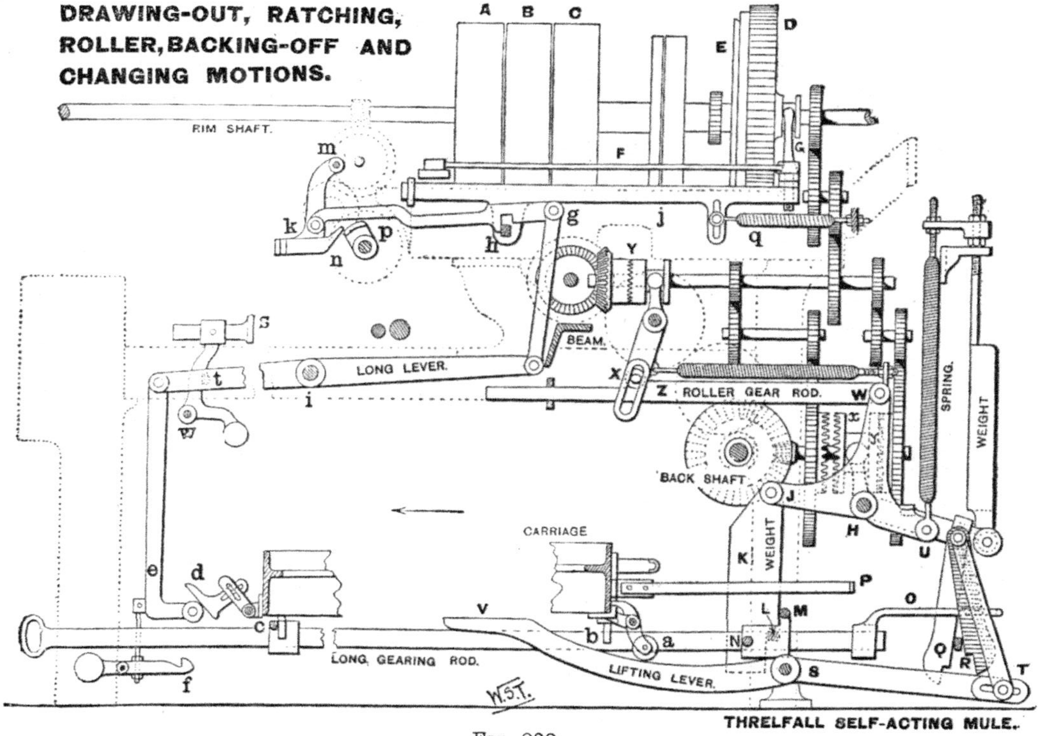

DRAWING-OUT, RATCHING, ROLLER, BACKING-OFF AND CHANGING MOTIONS.
RIM SHAFT.
LONG LEVER.
BEAM.
ROLLER GEAR ROD.
BACK SHAFT
SPRING.
WEIGHT
WEIGHT
CARRIAGE
LONG GEARING ROD.
LIFTING LEVER.
W.S.T.
THRELFALL SELF-ACTING MULE.

FIG. 232.

The lifting of the lever I also raises the two pendent links or gearing legs Q and R, the longer one Q locking itself on a square stud carried by a bracket fixed on the floor. This stud is shown in the drawing between Q and R, but the floor bracket carrying it is not shown.

As shown in the drawing, the end W of the T lever is coupled to the roller gear rod Z, on a reduced portion of which rests a bowl X carried by the lever which actuates the catch box Y. The catch box Y is therefore locked so long as the T lever is locked in its present position.

The end J of the T lever carries a weight K which rests on a bracket L fixed on the headstock back.

Ratching, Jacking or After-stretch Motion.—This motion is one that stops the front rollers before the stretch is completed, but enables the carriage to complete its outward run. As the carriage nears the termination of the outward run, a projection or finger on the front of the square comes in contact with a finger *c* on the long gearing rod and moves the rod forward. A projection M near the back end of the rod bears against the weight K and moves it off the stud L. This action causes the weight K to fall, as well as allowing the spring to pull up the end U of the T lever. The effect of the change is to cause the end W to move the roller gear rod forward and so put the catch box Y out of gear, thus stopping the rollers, and the same change takes the catch box *y* out of gear and puts the catch box *x* into gear. At the same time the strap moves from pulley C to pulley A, thus putting the spindles on to increased speed, the rollers are stopped, and the carriage is now driven through the jacking wheels and the catch box *x* for the remainder of the outward run.

Assistant Winding Motion, Fig. 233.—This motion is arranged to prevent snarls forming in the loose yarn at the moment the fallers change on the completion of the inward run. The amount of yarn set free varies throughout the set, and as it is beyond the control of the quadrant, a pulley A keyed to the rim shaft is utilised so that at the precise moment required the belt on B is moved on to the

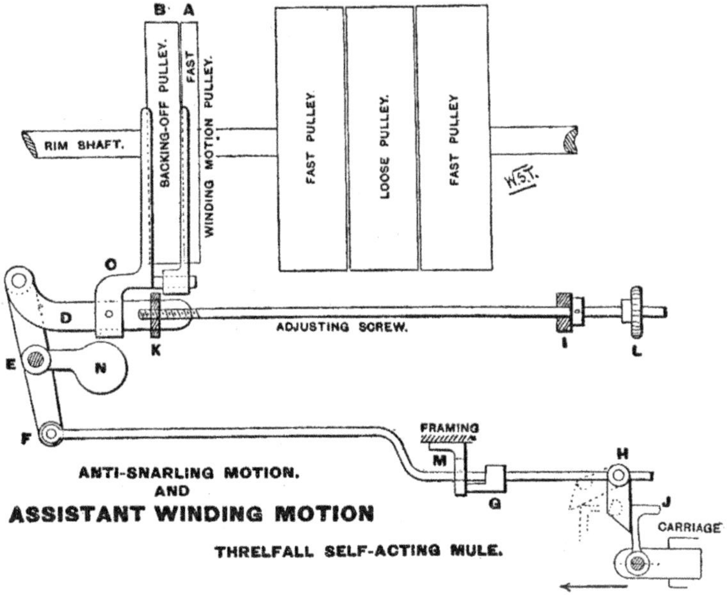

Fig. 233.

pulley A and so drives the rim shaft and, consequently, the spindles for a fraction of a minute, thus winding on the yarn that would otherwise be free to form snarls. The motion, it will be observed, is not one that allows snarls to form and then stretches such snarls out: its advantage lies in the fact that it prevents snarls forming on spindle.

The motion is extremely simple. The bracket J is fixed to the carriage, and near the end of the run-in J

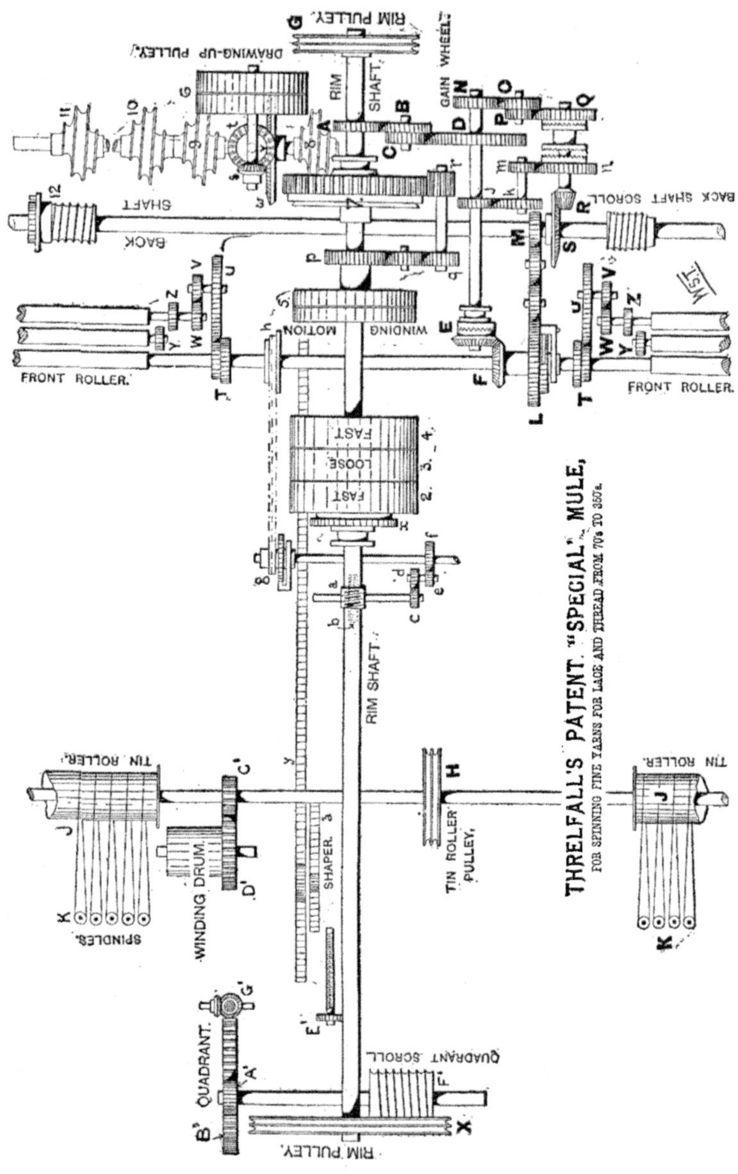

THRELFALL'S PATENT "SPECIAL" MULE,

FOR SPINNING FINE YARNS FOR LACE AND THREAD FROM 70's TO 350's.

FIG. 234.

comes into contact with a pendent lever centred on a rod at H. The lever is raised (as shown in dotted lines) and a projection on it comes into contact with the rod and raises it, thus allowing a stop finger G to pass through a slot in a fixed bracket M. A weight N on a lever centred at E pulls over the end F and forces forward a slide D on the other end. The slide D carries the strap fork so the strap is moved from pulley B to pulley A. The amount of strap allowed to move on to A can be carefully adjusted by the stop rod or adjusting screw passing through a fixed bracket K, the adjustment being effected by a handle or nut L conveniently reached by the minder. The pendent lever at H can be set so that the carriage can actuate it at any required distance before finishing the inward run.

Jacking Motion, Fig. 235.—This motion is for the purpose of driving the carriage during the last few inches of the outward run, after the rollers are stopped to stretch the yarn, as is customary in spinning fine counts. The speed wheel carrier, bevel A, drives the front roller through the bevel C and the catch box B and an internal disc which is keyed on the front roller shaft. When the catch box is put out of gear the front roller is stopped. The bevel C is keyed on the boss of the jacking box D and E, and runs loosely on the front roller shaft. Inside this jacking box two pinions F and G are mounted, which are keyed together, but run loosely inside the box. These pinions gear with two wheels H and J, H being keyed on front roller shaft and J on the long boss K. By reason of the wheels F and G being of different sizes, and being carried round the outside of the wheels H and J by the jack box, motion is transmitted to the wheel J, which, being keyed on the boss K, on the other end of which is secured the wheel L, drives the back shaft in the ordinary way but at a reduced speed.

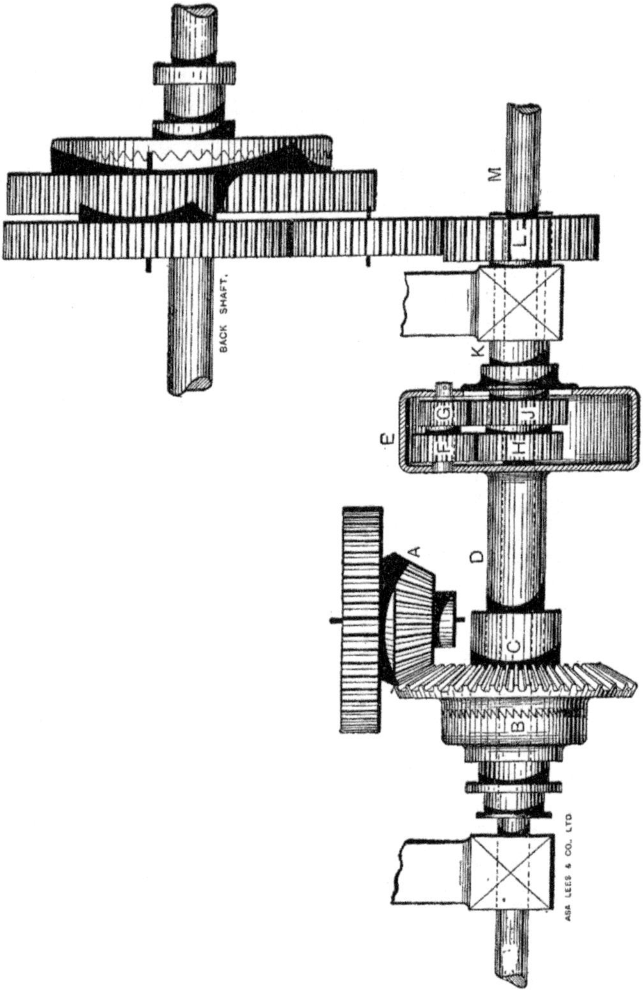

Fig. 235.

This motion can be operated to cause any desired amount of stretching of the yarn.

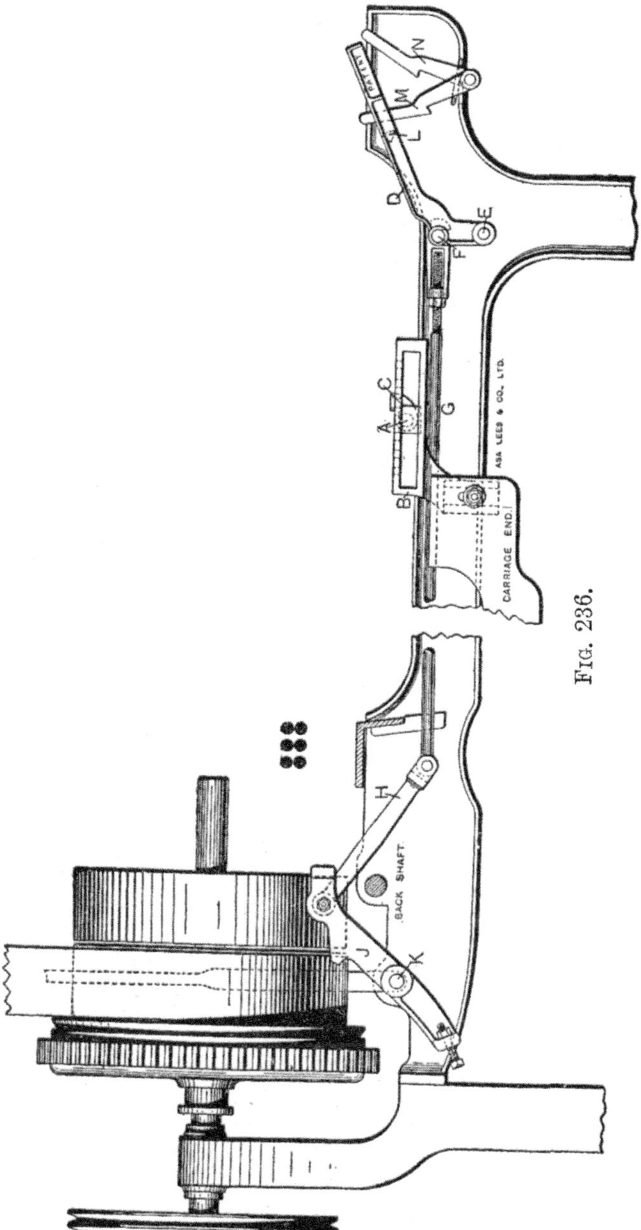

PATENT
N
M
L
D
E
F
A
C
G
B
ASA LEES & CO. LTD.
CARRIAGE END.
H
BACK SHAFT
J
K

FIG. 236.

Strap Relieving and Locking Motions, Fig. 236.—The object of this motion is to move the strap from the fast to the loose pulley when the carriage is within 2 inches to 12 inches of the completion of the outward run, the momentum of the carriage at this stage being sufficient to complete the outward run.

Fixed on the carriage end is a frame B carrying a stud A. As the carriage moves outwards this stud comes in contact with the inclined surface of lever D, which it depresses, and, being centred at E, the swivel F is taken with it. To this swivel F the end of the rod G is secured, the other end of which is connected to the lever H, which in turn is secured to the lever J. This lever is centred at K and the strap lever is secured to it.

It will be seen that if lever D is now pressed downwards the rod G will be pulled outwards, which, through the various levers, will operate on the strap fork lever and so bring about the desired change.

The moment at which this motion comes into operation may be regulated by sliding the stud A in its frame B, the stud being held in any desired position by turning down the catch C into one of the holes drilled in the upper surface of the frame B.

The locking device is for stopping the carriage at any part of the outward run, and operates as follows:—When the lever D is pressed down, the lug L, which is cast upon it, is caught by the catch M, thus preventing the levers resuming their original position, and keeping the carriage stationary until released. When it is not required to stop the carriage the catch is turned back to the position shown at N.

Twist Motion. Driven from Tin Roller, Fig. 237.—This motion is designed to put the required twist in every draw alike. The motion, being driven by a worm on the tin roller shaft, ensures this shaft, and therefore the

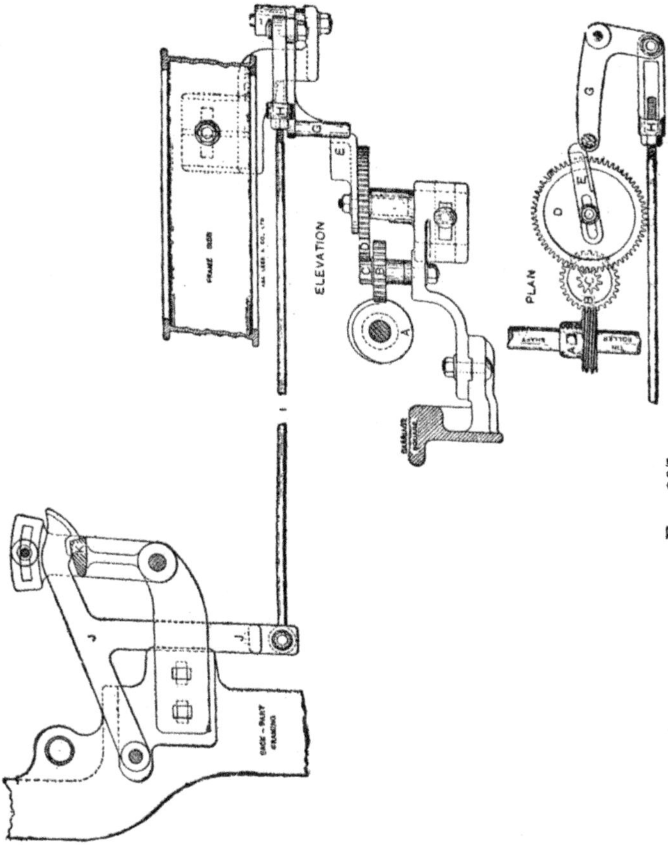

Fig. 237.

spindles, making the same number of revolutions each draw.

The twist catch J is hinged at the back of the headstock as usual, and is connected by the rod I to the bell-crank lever G, which is pivoted on a bracket secured to the head-stock side. A bracket is fixed on the square which carries the twist motion wheels C and B, and to this bracket a slide is fitted carrying the twist wheel D. This slide can be adjusted to take any size of wheel from 50 to 100 teeth. It will thus be seen that the worm A on the tin roller shaft transmits motion to the wheels B and C, which in turn drive the change wheel D, whilst on the same stud, and secured to wheel D, is a finger E revolving with wheel D.

When the carriage has completed the outward run, the tin roller shaft continues to revolve until the finger E comes in contact with the bell-crank lever G, which, being turned on its centre, exerts a pull on the rod I, and thus lifts the catch J, allowing the strap lever K to move the strap on to the loose pulley on the rim shaft previous to the mule backing off.

Backing-off Motion, Fig. 238.—The object of the above motion is to unwind the coils of yarn formed on the spindle blade during spinning, in order that the spun yarn may be wound on the cop, and this is done by turning the spindles in the opposite direction, the slack yarn thus formed being taken up temporarily by the counter faller.

The backing-off wheel A is mounted loosely on the rim shaft and is driven constantly in an opposite direction to the rim shaft during spinning. When the carriage completes its outward run the lever K, which is mounted in the square, depresses the lever J, which moves the rod F and allows the spring G to turn lever D on its centre and so force the backing-off wheel, the inside of which is turned conical, on to the friction cone connected with the fast pulley and thus driving through to the spindles in the usual manner but in the opposite direction.

The winding faller is pulled down by the backing-off

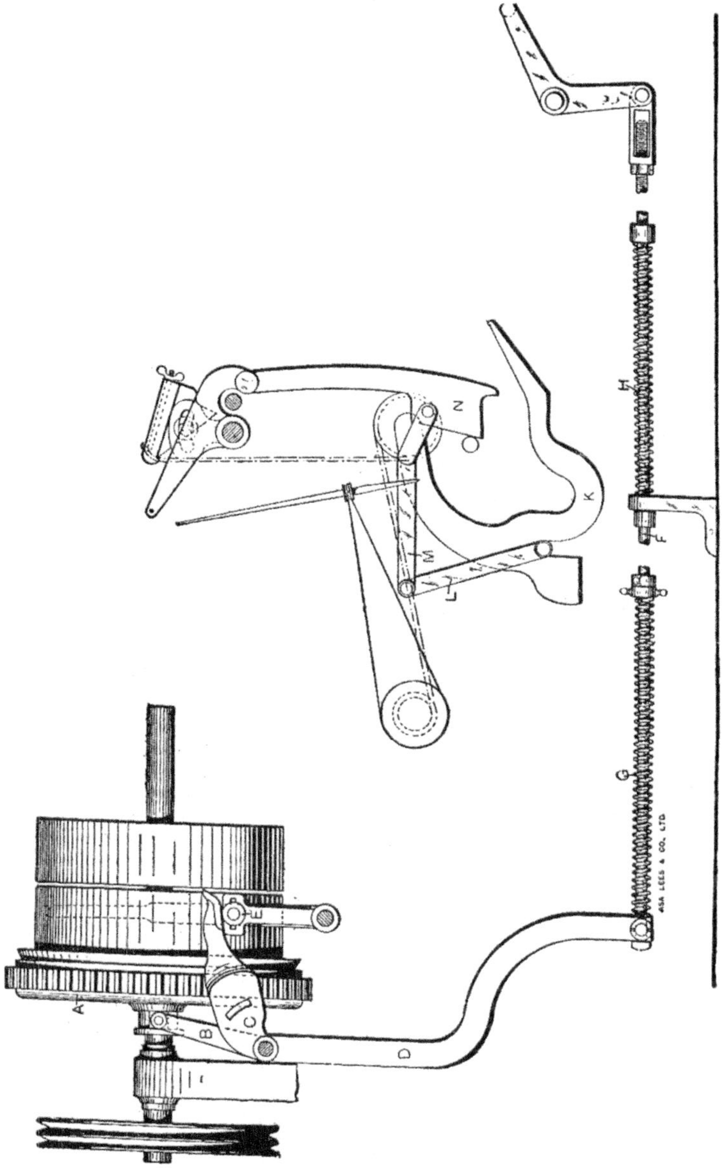

FIG. 238.

chains in the usual manner—through a click wheel on the tin roller shaft—until the bottom end of the boot-leg N rests on the locking bowl connected with the slide on the shaper rail; in which position the fallers are said to be "locked." During this operation the lever K is raised, through the levers L and M, and the lever J is released, allowing the spring H to draw the backing-off wheel out of gear.

To prevent this motion coming into operation too soon, a finger C rests on a bowl E connected to the strap lever, and prevents it dropping until the cam is changed and the strap is moved from the fast to the loose pulley, thus preventing the backing-off friction going into gear before the strap is moved entirely on to the loose pulley.

Whilst backing off, the button-head on rod F must be $\frac{1}{4}$ inch clear of the long lever D.

Gearing Plan of S.-A. Mule.—In Fig. 239 is given a plan view of the gearing of a S.-A. Mule that is becoming more widely known, and so will be interesting to students. It may be remarked that the copping faller ought to have been shown thicker than the counter faller.

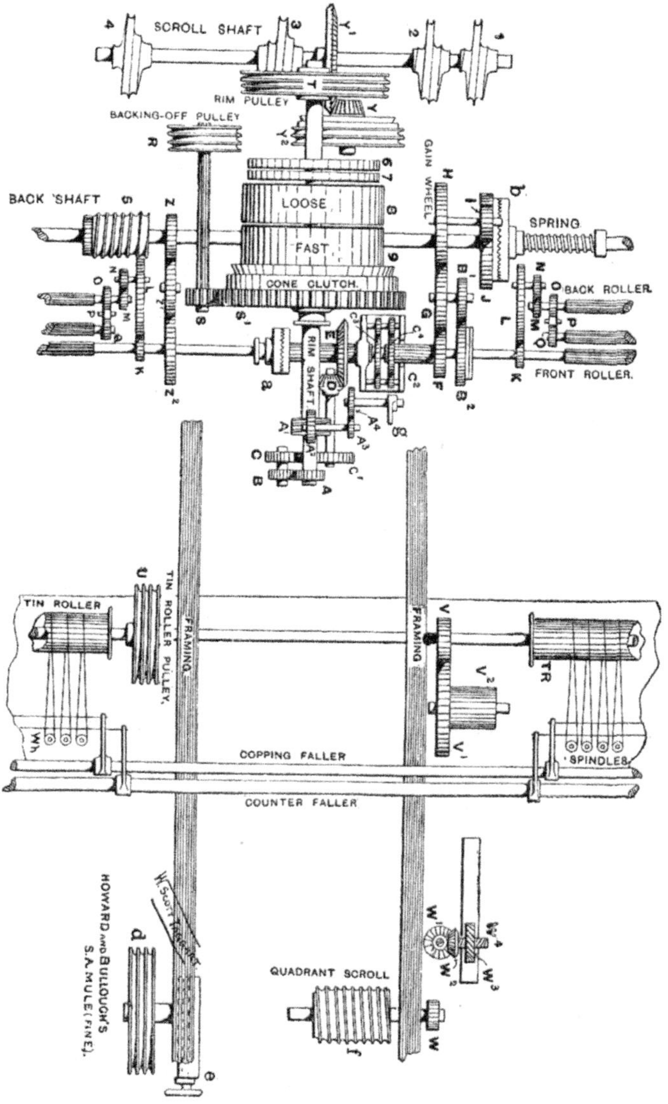

GEARING PLAN OF SELF-ACTING MULE

FIG. 239.

INDEX

Printed by R. & R. Clark, Limited, *Edinburgh*.

Printed in Dunstable, United Kingdom

88387943R00278